Density Matrix Theories in Quantum Physics

Authored by

Boris V. Bondarev

Moscow Aviation Institute (National Research University),
Moscow,
Russia

Density Matrix Theories in Quantum Physics

Author: Boris V. Bondarev

ISBN (Online): 978-981-14-7541-2

ISBN (Print): 978-981-14-7539-9

ISBN (Paperback): 978-981-14-7540-5

First published in 2020.

Bentham Science Publishers Pte. Ltd.
80 Robinson Road #02-00
Singapore 068898
Singapore
Email: subscriptions@benthamscience.net

BENTHAM
SCIENCE

CONTENTS

PREFACE

In this book, the author opens up new possibilities for the main quantities in quantum physics – the statistical operator $\hat{\varrho}$ and the density matrix_ϱ_{nm}. The meaning of the density matrix is that its diagonal elements ϱ_{nn} are equal to the probability w_n that the system in the quantum state n. The point in this book is the Lindblad equation for the statistical operator $\hat{\varrho}$, where the main element of influence on the system of its environment is the dissipative operator $\widehat{D}$:

$$\mathrm{i}\,\hbar\,\partial\hat{\varrho}/\partial t = [\,\widehat{H}\;\hat{\varrho}\,] + \mathrm{i}\,\hbar\,\widehat{D}\,.$$

This operator is written in the most General form. In order for the Lindblad equation to be solved, the $\widehat{D}$ operator must be specified. The author wrote down the dissipative diffusion and attenuation operators that will allow us to find the $\widehat{D}$ operator. Now, this operator depends on the temperature T and describes the effect of the thermostat R on the quantum system S. This new equation is not difficult to write for the particle density matrix in coordinate representation as compared to the Wigner equation, which coincides with the Fokker – Planck equation. This proved the equivalence of quantum physics and classical statistical physics.

The author wrote the Lindblad equation for a harmonic oscillator and inserted a dissipative attenuation operator into it. And without any approximation, he derived the equation of damped oscillations for the average value of the $\bar{x}$(t) coordinate with absolute accuracy.

Bondarev based on the Lindblad equation with another operator $\widehat{D}$ developed the theory of the harmonic oscillator, in which he found the density matrix and proved the Heisenberg relation.

He further developed the theories of the light diode and ball lightning. In light diode theory, he used the diffusion and attenuation operators and derived the Fokker – Planck equations for electrons and holes. These equations present the terms that are responsible for radiation.

The theory of ball lightning is based on the assumption that the gas inside the ball is completely ionized and electrons, due to their lightness in comparison with nuclei, evenly fill this ball. The equation for the statistical operator $\hat{\varrho}$ nuclei contains operators of diffusion and damping. This equation is a second-degree equation with respect to the coordinate and momentum operators. The probability of distribution of nuclei over the volume of a ball lightning is found.

Bondarev derived von Neumann equation from the Liouville, which is valid for a non-equilibrium system S and an equilibrium thermostat R, the equations for the density matrix S of a single particle and a system of identical particles. These equations have a remarkable property. When the density matrix has a diagonal form, they get turned into quantum kinetic equations for probabilities, which are obtained in the wave graphical representation.

The book presents new theories of such experimentally discovered phenomena as step kinetics of bimolecular reactions in solids, superconductivity, superfluidity, energy spectrum of an arbitrary atom, laser, spaser and graphene.

Kinetics is called as a stepwise process, in which the reaction suddenly stops at a constant temperature even in the presence of a lot of reagents. But as soon as the temperature is raised, the reaction starts again. The reason for this reaction is the tunnel effect, which is observed only in solids, when there are molecules in the bodies that hold the reagents near them. In liquids, these molecules can move along with the reagents and enter into a reaction that goes all the way while there are reagents. The reaction in liquids always obeys the Arrhenius law. To describe stepwise kinetics, the author came up with a correlation theory.

So, when processing the results of the step kinetics experiment using correlation theory, it was found that the Arrhenius law is also fulfilled here. And there was also an increase in localization volume with increasing temperature, as predicted by the tunnel effect.

Superconductivity can be described by the law of changing the probability w_k of filling the state of electrons with the wave vector k as a function of temperature. This law has long been known. It depends on the energy $\varepsilon_{kk'}$ of the interaction of electrons with the wave vectors k and $\boldsymbol{k}'$. When $\varepsilon_{kk'} = 0$, the probability w_k obeys the Fermi – Dirac law. Our goal was to find the energy $\varepsilon_{kk'}$ of the interaction of electrons.

We denote the matrix elements of the interaction Hamiltonian of two particles as $H_{12,1'2'}$, where 1 is the spin quantum number of the particles. If the particles are bosons, then the matrix elements must be antisymmetric, i.e. then the matrix elements must change the sign when replacing variables 1 and 2, or $1'$ and $2'$. This is possible if the matrix elements represent the sum of two terms of different characters. In the wave representation, the energy ε_(kk^') will also represent two terms of different signs. But in this case, it is very difficult to solve the equation. Therefore, we roughly denote these terms as

$$\varepsilon_{kk'} = \mathrm{I}\ \delta_{k+k'} - \mathrm{J}\ \delta_{k-k'} ,$$

where I and J are positive constants, δ_k is the Kronecker symbol. Now we can substitute this function into the equation and get

$$\ln\ [\ (1 - w_k)/w_k\ ] = \beta\ (\ \varepsilon_k + I\ w_{-k} - J\ w_k - \mu\ ) .$$

This equation has a remarkable property. For some areas of ***k*** will this inequality be true

$$w_k \neq w_{-k} .$$

The property that is expressed by this inequality is called anisotropy. The appearance of this property here is superconductivity.

Solving this equation, we obtain for T = 0 functions that have five values for one argument value. Since this function describes stationary states, the lowest energy is the value of the function where the electrons remain indefinite. This will be a superconducting state.

In theory, the parameter represents $f = (J - I)/(J + I)$. This parameter divides superconductivity into two kinds. If $0 \leq f \leq 1$, then it is a I-type superconductor, and if $-1 \leq f < 0$, then it is a II-type superconductor. Critical temperature is defined as $T_c = (I + J)/(4\ k_B)$. All the main effects and properties of superconductors are covered by this theory.

In the theory of superfluidity for liquid helium, He^3 and He^4, all values that express the properties of this mixture are described by functions having multiple values in a certain temperature range. As a consequence, the heat capacity tends to infinity when the temperature approaches the temperature T_λ of the lambda transition.

The theory of the energy spectrum of an arbitrary atom begins with determining the energy using statistical operators:

$$\mathrm{E} = \int \hat{H}^{(1)}\ \hat{\varrho}^{(1)}\ \mathrm{dq} + 1/2 \int \hat{H}^{(2)}\ \hat{\varrho}^{(2)}\ \mathrm{d}q_1\ \mathrm{d}q_2 ,$$

where $\hat{H}^{(1)}$ is the Hamiltonian of one electron, $\hat{H}^{(2)}$ is the Hamiltonian of two interacting electrons, $\hat{\varrho}^{(1)}$ and $\hat{\varrho}^{(2)}$ are the statistical operators of one and two electrons. The matrix $H_{\alpha_1\alpha_2,\,\alpha_1'\alpha_2'}$ of the Hamiltonian $\hat{H}^{(2)}$ must be antisymmetric. To do this, it is taken equal to

$$H_{\alpha_1\alpha_2,\,\alpha_1'\alpha_2'} = \int \Phi^*_{\alpha_1\alpha_2}(q_1, q_2)\ \hat{H}^{(2)}(q_1, q_2)\ \Phi_{\alpha_1'\alpha_2'}(q_1, q_2)\ \mathrm{d}q_1\ \mathrm{d}q_2 ,$$

where

$$\Phi_{\alpha_1\alpha_2}(q_1, q_2) = 1/\sqrt{2}\,[\,\varphi_{\alpha_1}(q_1)\,\varphi_{\alpha_2}(q_2) - \varphi_{\alpha_1}(q_2)\,\varphi_{\alpha_2}(q_1)\,]$$

there is an antisymmetric Slater function. The eigenfunctions of electrons in the hydrogen atom are taken as functions $\varphi_\alpha(q)$. After a series of calculations, an equation is obtained from which one can obtain the eigenfunction and energy $\varepsilon_{n\,m\,l\,\sigma}$ of electrons of an arbitrary atom.

In the following chapters, new theories of the laser and spaser are constructed, which are similar to each other in the content of the main quantum approaches to describing the phenomena occurring in them. The basis of these theories is the Lindblad equation. The equation for the density matrix will be written in a coordinate form with a known Hamiltonian and an unknown dissipative matrix. To find this matrix, we need to remember that we know the kinetic equation for active atoms, which follows from the equation for the density matrix in the representation where it has a diagonal form. So, a representation needs to be find out where the density matrix has a diagonal form. The closest to this representation is the representation in which the Hamiltonian of active atoms will also have a diagonal form. Thus, the Hamiltonian has two representations. One α-representation is a coordinate representation. The other is the κ-representation, in which the Hamiltonian has a diagonal form. These two Hamiltonians are connected by the unitary matrix $U_{\alpha\kappa}$. The density matrices $\varrho_{\alpha\alpha'}$ and $\tilde{\varrho}_{\kappa\kappa'}$ will also be connected by the same unitary transformation:

$$\varrho_{\alpha\alpha'} = \sum_{\kappa\kappa'} U_{\alpha\kappa}\, \tilde{\varrho}_{\kappa\kappa'} U^*_{\alpha'\kappa'}\ .$$

We find the dissipative matrix in the κ-representation.

Now we need to create another equation in the α-representation. This is the most important equation in laser theory. This is the equation for the spectral energy density of radiation. To solve it, we will use the density matrix $\varrho_{\alpha\alpha'}$. As a result, we will have the spectral energy density of the radiation from the laser.

Almost free electrons wander along the surface of graphene. For every carbon atom, there is one such electron. To obtain the kinctic equation of these electrons, their Hamiltonian must be reduced to a diagonal form. After these transformations, we will have a system of two equations that are equivalent to the equation obtained in the theory of superconductivity.

CONSENT FOR PUBLICATION

Not applicable.

CONFLICT OF INTEREST

The authors confirm that there is no conflict of interest.

ACKNOWLEDGEMENT

Declared none.

Boris V. Bondarev
Moscow Aviation Institute (National Research University)
Moscow
Russia

INTRODUCTION

1. Fundamentals of Quantum Mechanics. Schrödinger Equation

In quantum mechanics, the Schrödinger equation was obtained:

$$\mathrm{i}\,\hbar\,\partial\psi/\partial t = \hat{H}\psi, \qquad \textbf{(1)}$$

where the unknown

$$\psi = \psi(t, q) \qquad \textbf{(2)}$$

is called the wave function, $\hat{H}$ is the Hamilton operator энергии. Here q is a quantum variable on which the Hamiltonian acts. The physical meaning of the ψ function is that the product

$$w = \psi^{*}\,\psi \qquad \textbf{(3)}$$

is the probability of finding the system in state q. The probability satisfies the normalization condition

$$\int w\,(t, q)\,\mathrm{d}q\ = 1 \qquad \textbf{(4)}$$

The energy operator $\hat{H}$ is by definition, such an action on the wave function $\psi(t, q)$ to obtain the known energy $E(t, q)$ of the system:

$$\hat{H}\,\psi(t, q) = E(t, q)\,\psi(t, q)$$

2. Liouville – von Neumann Equation

Soon another function was invented, it is called the density matrix:

$$\varrho = \varrho(\,t, q, q') \qquad \textbf{(5)}$$

At first, this function was equal to the product

$$\varrho(\,t, q, q') = \psi^{*}(t, q)\,(t, q') \qquad \textbf{(6)}$$

In this case, the state of the quantum system is called **pure**. This function satisfies the equation that can be easily deduced from the Schrödinger equation:

$$\mathrm{i}\,\hbar\,\partial\varrho/\partial t = [\,\hat{H}\ \varrho\,] \tag{7}$$

then there was invented the density matrix

$$\varrho_{nn'}(t) = \int \psi_n^*(t, q)\ \hat{\varrho}\ \psi_{n'}(t, q)\ \mathrm{d}q \tag{8}$$

Here $\psi_n(t, q)$ is a set of wave functions, and $\hat{\varrho}$ is called the statistical operator.

The state described by the density matrix (8) is called **mixed**.

The physical meaning of the density matrix is that its elements

$$\varrho_{nn} = w_n \tag{9}$$

are probabilities to find a quantum system in states n. The probability is such that

$$\sum_n w = 1 \tag{10}$$

The equation for the mixed state density matrix:

$$\mathrm{i}\,\hbar\,\partial\varrho_{nn'}/\partial t = \sum_m (\,H_{nm}\,\varrho_{mn'} - \varrho_{nm}\,H_{mn'}\,) \tag{11}$$

It can be shown that the statistical operator $\hat{\varrho}$ satisfies the equation

$$\mathrm{i}\,\hbar\,\partial\hat{\varrho}/\partial t = [\,\hat{H}\ \hat{\varrho}\,] \tag{12}$$

The equations (11) and (12) are called Liouville – von Neumann equations. Both of these equations follow from the Schrödinger equation.

3. Lindblad Equation

The Schrödinger equation includes only the operator $\hat{H}$ of the quantum system energy. The equation for the statistical operator was first supplemented by Lindblad [1]:

$$\mathrm{i}\,\hbar\,\partial\hat{\varrho}/\partial t = [\,\hat{H}\ \hat{\varrho}\,] + \mathrm{i}\,\hbar\,\hat{D}\,, \tag{13}$$

where the operator $\widehat{D}$ can be called a dissipative operator. According to Lindblad, this operator is

$$\widehat{D} = \sum_{jk} C_{jk} \{ 2\, \hat{a}_j\, \hat{\varrho}\, \hat{a}_k^+ - \hat{a}_k^+ \hat{a}_j\, \hat{\varrho} - \hat{\varrho}\, \hat{a}_k^+ \hat{a}_j \}, \tag{14}$$

C_{jk} are some numbers, $\hat{a}_j$ is an arbitrary operator. The operator $\widehat{D}$ can be written as follows:

$$\widehat{D} = \sum_{jk} C_{jk} \{ [\, \hat{a}_j\, \hat{\varrho}\, ,\, \hat{a}_k^+] + [\hat{a}_j\, ,\, \hat{\varrho}\, \hat{a}_k^+] \}$$

Operators $\hat{a}_j$ are still to be found.

4. Equation for the Density Matrix

The Liouville – von Neumann equation (11) is applied to the composite system *R* + *S*, where *R* is a thermostat and *S* is an arbitrary system that is much smaller than the thermostat. The author of this work has derived from equation (11) the equation for the density matrix of the system *S* [2]:

$$\mathrm{i}\,\hbar\; \partial\varrho_{nn'}/\partial t = \sum_m \left(H_{nm}\varrho_{mn'} - \varrho_{nm}H_{mn'} \right) + \mathrm{i}\,\hbar\, D_{nn'}, \tag{15}$$

where $D_{nn'}$ is a **dissipative** matrix that equals to

$$D_{nn'} = \sum_{m\,m'} \gamma_{nm,m'n'}\, \varrho_{mm'} - 1/2 \sum_m \left(\gamma_{nm}\, \varrho_{mn'} + \varrho_{nm}\, \gamma_{mn'} \right), \tag{16}$$

$\gamma_{nm,m'n'}$ – a matrix,

$$\gamma_{nn'} = \sum_m \gamma_{mn',nm} \tag{17}$$

5. Quantum Kinetic Equation

At the moment of time when the density matrix ϱ_{nm} is diagonal:

$$\varrho_{nm} = w_n\, \delta_{nm}\, , \tag{18}$$

where δ_{nm} is the Kronecker symbol. Then from equation (16) follows the kinetic equation

$$\partial w_n/\partial t = \sum_m (p_{nm} w_m - p_{mn} w_n), \quad \textbf{(19)}$$

where

$$p_{nm} = \gamma_{nm,mn} = (2\pi/\hbar) \sum_N \sum_M |v_{nN,mM}|^2 W_M \delta(\varepsilon_n - \varepsilon_m + E_N - E_M) \quad \textbf{(20)}$$

is the probability of transition of the system S from the state m to the state n per unit of time,

$$W_N = \nu \exp(-\beta E_N)$$

is a probability that the system R is in a state N with energy E_N, ν is the normalization factor,

$$\beta = 1/(k_B T)$$

is the inverse temperature of the thermostat; $v_{nN,mM}$ is the matrix element of the energy of interaction of the system S with thermostat R. Formula (20) is the **Fermi Golden rule**. Equations (13) and (15) are used in the articles included in this book.

6. Connection Dissipative Matrix and Dissipative Operator

Equations (14) and (16) are connected by the ratio

$$\gamma_{nm,m'n'} = 2 \sum_{jk} C_{jk}\, a_{nm,j}\, a^{+}_{m'n',k}\,, \quad \textbf{(21)}$$

where $a_{nm,j}$ – matrix elements of operators $\hat{a}_j$.

7. Probability of Transition and Relaxation

Rule (20), together with the Boltzmann principle, allows to record the probability of transition in the form of

$$p_{nm} = p^{(o)}_{nm} \exp[-\beta(\varepsilon_n - \varepsilon_m)/2],$$

where

$$p^{(o)}_{nm} = p^{(o)}_{mn}\,.$$

Let's use formula (20):

$$p_{nm} = \gamma_{nm,mn} \cdot$$

We express the matrices $\gamma_{nm,m'n'}$ and $\gamma_{nn'}$ by the transition probability p_{nm}. We will have

$$\gamma_{nm,m'n'} = \sqrt{p_{nm}\, p_{n'm'}}\,, \qquad \gamma_{nn'} = \sum_m \sqrt{p_{mn'}\, p_{mn}}$$

Substituting these matrices into formula (6.2), we obtain the dissipative matrix in the form

$$D_{nn'} = \sum_{mm'} \{ \sqrt{p_{nm}\, p_{n'm'}}\, \varrho_{mm'} -$$
$$- 1/2 \,(\sqrt{p_{m'm}\, p_{m'n}}\, \varrho_{mn'} + \varrho_{nm} \sqrt{p_{m'n'}\, p_{m'm}} \,) \}$$

If we put the density matrix $\varrho_{nn'} = w_n\, \delta_{nn'}$ and $n = n'$ in this formula, we get the dissipative matrix

$$D_{nn} = \sum_m \,(p_{nm}\, w_m - p_{mn}\, w_n \,)$$

If the matrix p_{nm} is such that

$$\pi_{nm} = \gamma_n\, \delta_{nm},$$

then

$$\gamma_{nm,m'n'} = \sqrt{\gamma_n\, \gamma_{n'}}\, \delta_{nm}\, \delta_{n'm'},$$

Here with

$$\gamma_{nn'} = \gamma_n\, \delta_{nn'},$$

the dissipative matrix will be equal to

$$D^{(r)}_{nn'} = - \Gamma_{nn'}\, \varrho_{nn'},$$

where

$$\Gamma_{nn'} = 1/2\,(\,\gamma_n + \gamma_{n'}) - \sqrt{\gamma_n\,\gamma_{n'}} \geq 0$$

The value

$$\Gamma_{nn} = 0.$$

The matrix $D^{(r)}_{nn'}$ describes the relaxation of the system *S*, which is the pursuit of zero non-diagonal elements of the density matrix.

8. Heisenberg Uncertainty Relation

The seventh chapter discusses two dissipative matrices that are used for the quantum oscillator. First, in a dissipative operator, we put the operator

$$\hat{a} = (\,\mathrm{i}\;\hat{p}/\sqrt{m} + \sqrt{\kappa}\,\hat{x}\,)/\sqrt{2\,\hbar\,\omega}\,,$$

which is used to describe the harmonic oscillator. Equations for density matrix in different representations are written. An equilibrium density matrix is found from these equations. The equation for the Wigner function is written. The equilibrium solution of these equations is found and with the help of this function, the Heisenberg uncertainty relation for the quantum harmonic oscillator is found:

$$\overline{x^2}\cdot\overline{p^2} = \hbar^2/4\left[\,(\,e^{\beta\,\hbar\,\omega}+1)/(\,e^{\beta\,\hbar\,\omega}-1)\right]^2$$

In the future articles, the dissipative operator using the operator

$$\hat{a} = \hat{x} + \mathrm{i}\,\hbar\,\beta\,\hat{p}/(4\,m)\,,$$

is applied to describe the motion of the damped oscillator. It has been proved that from the Lindblad equation with such a dissipative operator, the Newton equation for the mean value of $\bar{x}(t)$ follows exactly, which describes the damped oscillations of the pendulum.

REFERENCES

[1] G. Lindblad, “On the generators of quantum dynamical semigroups,” Commun. Math. Phys., vol. 48, pp. 119-130, 1976.

[2] B.V. Bondarev, “Derivation of the quantum kinetic equation from the equation Liouville – von Neumann”, Theor. Math. Phys., vol. 100, pp. 33-43, 1994.

CHAPTER 1

New Theory of Step Kinetics

1. STEP KINETICS OF REACTIONS IN SOLIDS CORRELATION THEORY

1.1. Introduction

One of the most characteristic features of the kinetics of low-temperature reactions occurring in the condensed phase is the kinetic stop of the reaction. The phenomenon of kinetic stopping is observed in the recombination of radicals (radical R is a particle with an unpaired electron), in the interaction of radicals with oxygen, *etc*. The particles entering these reactions appear due to the irradiation of a solid at a low temperature by ionizing radiation. After irradiation of the solid body, the reaction is not observed. If the temperature is increased to a certain level and kept constant, the reaction begins to proceed and then stops. If the temperature is increased once again by some value, the reaction resumes. The kinetics of this reaction is called stepwise.

In this paper, the theory of solid-phase reactions is presented, the essence of which is that the reactivity of particles is characterized not only by one constant of velocity, but during the reaction the correlation function is determined by the mutual arrangement of particles changes [1]. Consider two particles A and B, which appear in a solid under the action of ionizing radiation, when the temperature of the solid body is not very high. These particles might react, but for some reason, they do not react. If the temperature is not much increased, then the reaction will proceed. But after a while, it will stop again. This is called step kinetics of a bimolecular reaction.

In solids, there are various kinds of heterogeneities. When new particles appear in a solid, they are localized in the vicinity of these inhomogeneities. Fig. (**1**) shows particles A and B, arising at a constant temperature in the body under irradiation. Black dots represent the heterogeneity of the solid body. Particles A and B are localized in some volumes in the vicinity of these inhomogeneities. The particles are moving all the time inside these volumes of localization, making the "tunnel" transitions at the node of the crystal lattice. In Fig. (**1**), it can be observed that particles A and B can not meet and react, since the volumes of their localization are

Boris V. Bondarev

small. In the next article, it will be shown that, according to the laws of quantum mechanics, the volume of localization increases with the increase in temperature.

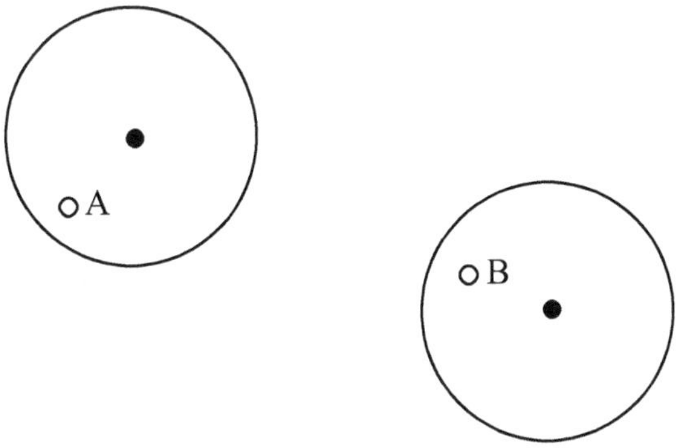

Fig. (1). The volumes of localization of particles **A** and **B** at constant temperature are small. Therefore, the particles cannot meet and react.

If the temperature is increased to a certain level, the volume of localization will increase, particles A and B will meet and react (see Fig. **2**).

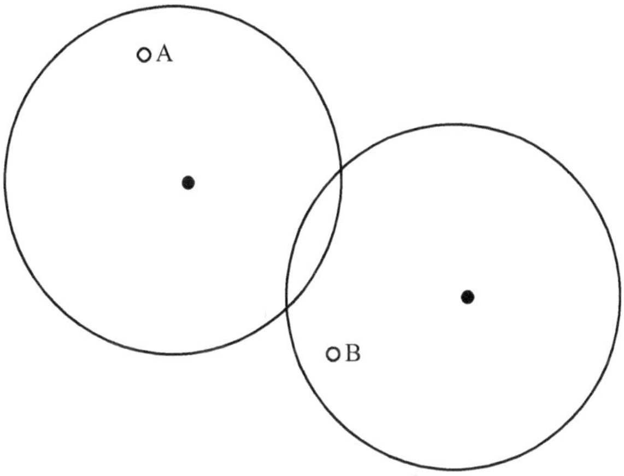

Fig. (2). The volume of localization of particles with increasing temperature expanding and after a while particles **A** and **B** meet and react.

This paper presents the theory of kinetics of solid-phase reactions, taking into account the correlation of the distributions of reacting particles. Kinetic equations are derived. Their solution for the homogeneous case is given. The obtained solution explains the basic laws of "step" kinetics.

1.2. Kinetic Theory of Solid-Phase Reactions

Let the particles A and B stabilize in some matrix, which can react as follows:

$$A + B \longrightarrow AB\,.$$

Each of these particles is localized around some center, which will be called the stabilization center and the particles move through diffusion in the microregions surrounding the center. The area in which the microdiffusion of the particle occurs increases with temperature. At a certain temperature, the size of this area becomes so large that the particle can be considered almost "free". In this case, the particle movement in the matrix volume is determined by the macrodiffusion process.

$$n_A = n_A(\boldsymbol{r}_A, t) \qquad \text{and} \qquad n_B = n_B(\boldsymbol{r}_B, t)$$

denote concentrations of the centers of particle stabilization A and B, respectively.

By definition

$$\int n_A\, dV_A = N_A\,, \quad \int n_B\, dV_B = N_B \tag{1.1.1}$$

where N_A and N_B are the numbers of particles A and B stabilized in volume V of the matrix. Let the probability be given as:

$$F_{AB} = F_{AB}(\boldsymbol{r}_A, \boldsymbol{r}_B, t)$$

where F_{AB} is formation per unit time of a composite particle AB from two arbitrarily selected particles A and B, whose stabilization centers are at points $\boldsymbol{r}_A$ and $\boldsymbol{r}_B$. The function F_{AB} satisfies the following relations:

$$\int F_{AB}\, dV_A = \int F_{AB}\, dV_B = k\,, \tag{1.1.2}$$

where k is the reaction rate constant given as:

$$k = 4\,\pi\, r_o\, D\, p\, e^{-E/(k_B T)}\,(\,1 + r_o/\sqrt{\pi\, D\, t}\,) \tag{1.1.3}$$

where r_o is the distance between the particles at which their active interaction begins; D is the microdiffusion coefficient, p is the steric factor, and E is the activation energy of the reaction.

The number of pairs of particles A and B, the stabilization centers which are located in volumes dV_A and dV_B, by definition of the correlation function

$$n_{AB} = n_{AB}(\boldsymbol{r}_A, \boldsymbol{r}_B, t)$$

is equal to $n_{AB}\, dV_A\, dV_B$. At the same time, there is equality

$$\int n_{AB}\, dV_A = N_A\, n_B\,, \qquad \int n_{AB}\, dV_B = N_B\, n_A\,. \qquad \textbf{(1.1.4)}$$

Using the values of F_{AB} and n_{AB}, we can write the equations for the rates of change of concentrations of n_A and n_B of the centers of the stabilization of the particles as follows:

$$\begin{cases} dn_A/dt = -\int F_{AB}\, n_{AB}\, dV_B\,, \\ dn_B/dt = -\int F_{AB}\, n_{AB}\, dV_A\,. \end{cases} \qquad \textbf{(1.1.5)}$$

If there is no correlation between the particles, then

$$n_{AB} = n_A\, n_B\,. \qquad \textbf{(1.1.6)}$$

If, in addition, the particles are evenly distributed over the volume, the equations (1.1.5) are reduced to the following equation:

$$dn_A/dt = dn_B/dt = -k\; n_A\, n_B\,, \qquad \textbf{(1.1.7)}$$

which describes the kinetics of bimolecular reactions controlled by diffusion.

The binary function satisfies the equation given as:

$$\partial n_{AB}/\partial t = -F_{AB}\, n_{AB} - \int F_{AB'}\, n_{ABB'}\, dV_{B'} - \int F_{A'B}\, n_{AA'B}\, dV_{A'}, \qquad \textbf{(1.1.8)}$$

where $n_{ABB'}(\boldsymbol{r}_A, \boldsymbol{r}_B, \boldsymbol{r}_{B'}, t)$ – three-particle function.

We introduce a correlation function as follows:

$$g_{AB} = g_{AB}(\boldsymbol{r}_A, \boldsymbol{r}_B, t)$$

Ratio

$$n_{\mathrm{AB}} = n_{\mathrm{A}}\, n_{\mathrm{B}} - g_{\mathrm{AB}}\,. \tag{1.1.9}$$

In the pair collision approximation, the three-particle function is given as:

$$n_{\mathrm{ABB'}} = n_{\mathrm{A}}\, n_{\mathrm{B}}\, n_{\mathrm{B'}} - n_{\mathrm{A}}\, g_{\mathrm{BB'}} - n_{\mathrm{B}}\, g_{\mathrm{AB'}} - n_{\mathrm{B'}}\, g_{\mathrm{AB}}\,. \tag{1.1.10}$$

The pair collision approximation assumes that of the three particles A, B and B′ one, for example, *B*, is far from the other two. Therefore, in (1.1.10) we should put $g_{\mathrm{AB}} = 0$ and $g_{\mathrm{BB'}} = 0$. The equation (1.1.10) takes the form

$$n_{\mathrm{ABB'}} \cong n_{\mathrm{A}}\, n_{\mathrm{B}}\, n_{\mathrm{B'}} - n_{\mathrm{B}}\, g_{\mathrm{AB'}} = n_{\mathrm{B}}\, n_{\mathrm{AB'}}\,. \tag{1.1.11}$$

Similarly, it can be written as:

$$n_{\mathrm{AA'B}} \cong n_{\mathrm{A}}\, n_{\mathrm{A'B}}\,, \tag{1.1.12}$$

Substituting (1.1.11) and (1.1.12) in (8), we get the equation, which, taking into account the system (1.1.5) is converted into the following form:

$$\partial(n_{\mathrm{AB}} - n_{\mathrm{A}}\, n_{\mathrm{B}})/\partial t = -\, F_{\mathrm{AB}}\, n_{\mathrm{AB}}\,. \tag{1.1.13}$$

By excluding (1.1.9) the function n_{AB} from (1.1.5) and (1.1.13), we obtain a system of equations for the functions n_{A}, n_{B} and g_{AB} as follows:

$$\begin{cases} \partial n_{\mathrm{A}}/\partial t = \int F_{\mathrm{AB}}\,(n_{\mathrm{A}}\, n_{\mathrm{B}} - g_{\mathrm{AB}})\, \mathrm{d}V_{\mathrm{B}}\,, \\ \partial n_{\mathrm{B}}/\partial t = \int F_{\mathrm{AB}}\,(n_{\mathrm{A}}\, n_{\mathrm{B}} - g_{\mathrm{AB}})\, \mathrm{d}V_{\mathrm{A}}\,, \\ \partial g_{\mathrm{AB}}/\partial t = F_{\mathrm{AB}}\,(n_{\mathrm{A}}\, n_{\mathrm{B}} - g_{\mathrm{AB}})\,. \end{cases} \tag{1.1.14}$$

Using the third equation of the system (1.1.14), we transform the first two to the form:

$$\partial n_{\mathrm{A}}/\partial t = \int \partial g_{\mathrm{AB}}/\partial t\;\, \mathrm{d}V_{\mathrm{B}}\,, \qquad \partial n_{\mathrm{B}}/\partial t = \int \partial g_{\mathrm{AB}}/\partial t\;\, \mathrm{d}V_{\mathrm{A}}\,.$$

The last equations are reduced to the relations:

$$\begin{cases} \partial(\, n_{\mathrm{A}} + \int g_{\mathrm{AB}}\, \mathrm{d}V_{B}\,)/\partial t = 0\,, \\ \partial(\, n_{\mathrm{B}} + \int g_{\mathrm{AB}}\, \mathrm{d}V_{A}\,)/\partial t = 0\,, \end{cases} \tag{1.1.15}$$

which are convenient to use in solving kinetic equations (1.1.14).

1.3. Uniform Distribution of Particles in the Matrix

Let the particles A and B be distributed uniformly over the volume of the matrix. In this case, the concentrations n_A and n_B do not depend on coordinates and the function F_{AB} and the correlation function g_{AB} depend only on $\boldsymbol{R} = \boldsymbol{r}_A - \boldsymbol{r}_B$ and on time t. The system (1.1.14) takes the form:

$$\begin{cases} dn_A/dt = dn_B/dt = -\,k\,n_A\,n_B + \int F\,g\,dV\,, \\ \qquad dg/dt = F\,(n_A\,n_B - g)\,, \end{cases} \qquad \textbf{(1.1.16)}$$

where

$$F \equiv F_{AB} \text{ and } g = g_{AB}.$$

It is obvious that the probability of F formation of a composite particle decreases with increasing distance R between the centers of particle stabilization A and B. We describe approximately the dependence of F on $\boldsymbol{R}$ by a simplified formula

$$F(\boldsymbol{R}, t) = \begin{cases} F(t) & \text{at } \boldsymbol{R} \in G\,, \\ 0 & \text{at } \boldsymbol{R} \ni G\,, \end{cases} \qquad \textbf{(1.1.17)}$$

where G is some area in the space of $\boldsymbol{R}$ variable. The G region determines the relative position of the stabilization centers of two particles A and B, at which the probability F of their reaction is different from zero. From equality (1.1.2) follows the ratio

$$k = F\,v, \qquad \textbf{(1.1.18)}$$

where v is the volume of the region G, which is a monotonically increasing function of temperature if the particles are in thermal equilibrium with the matrix.

Consider the kinetics of a bimolecular reaction under isothermal conditions. Let the temperature of the matrix be constant and equal to T_0 until the time $t = 0$, and at the time $t = 0$ it increases "abruptly" to the value $T > T_0$ and remains unchanged in the future. The concentrations of n_A and n_B of the particle stabilization centers at $T \geq 0$ do not change with time. Let's denote their values n_{Ao} and n_{Bo}. The correlation function must be equal to:

$$g\,(\boldsymbol{R},0) = \begin{cases} n_{\mathrm{Ao}}\, n_{\mathrm{Bo}} & \text{at} \quad \boldsymbol{R} \in G_{\mathrm{o}} \equiv G(T_{\mathrm{o}})\,, \\ 0 & \text{at} \quad \boldsymbol{R} \ni G_{\mathrm{o}}\,, \end{cases} \qquad \textbf{(1.1.19)}$$

where G_{o} denotes the region G corresponding to the temperature T_{o}. Substituting this function in the right-hand side of equations (1.1.16), it is easy to make sure that they vanish. The form of the function (1.1.19) means that the particles whose distance between the stabilization centers is small ($\boldsymbol{R} \in G_{\mathrm{o}}$) have already reacted with the particles for which $\boldsymbol{R}$ is large ($\boldsymbol{R} \ni G_{\mathrm{o}}$) and cannot further converge and react.

The solution of the system (1.1.15) in the homogeneous case has the form:

$$\begin{cases} n_{\mathrm{A}} + \int g\, \mathrm{d}V = \ n_{\mathrm{Ao}} + \ n_{\mathrm{Ao}}\, n_{\mathrm{Bo}}\, v_{\mathrm{o}}\,, \\ n_{\mathrm{B}} + \int g\ \mathrm{d}V = n_{\mathrm{Bo}} + \ n_{\mathrm{Ao}}\, n_{\mathrm{Bo}}\, v_{\mathrm{o}}\,. \end{cases} \qquad \textbf{(1.1.20)}$$

where the initial condition (1.1.19) was used when calculating the right parts; $v_{\mathrm{o}} = v(T_{\mathrm{o}})$ – the volume of the domain G_{o}.

There is another integral of the system (1.1.16) which follows the equation

$$\mathrm{d}n_{\mathrm{A}}/\mathrm{d}t = \mathrm{d}n_{\mathrm{B}}/\mathrm{d}t\,.$$

and is expressed by the equality

$$n_{\mathrm{A}} - n_{\mathrm{Ao}} = n_{\mathrm{B}} - n_{\mathrm{Bo}}\,. \qquad \textbf{(1.1.21)}$$

Excluding with the help of (1.1.20) and (1.1.21) the functions n_{B} and g from the system (16), and taking into account (1.1.17) and (18), we come to the equation for the concentration n_{A}:

$$\mathrm{d}n_{\mathrm{A}}/\mathrm{d}t = -\,k\,(\,n_{\mathrm{A}}^2 + p\,n_{\mathrm{A}} - q\,)\,, \qquad \textbf{(1.1.22)}$$

where

$$p = n_{\mathrm{Bo}} - n_{\mathrm{Ao}} + 1/v\,, \qquad q = n_{\mathrm{Ao}}\,(\,1 + \ n_{\mathrm{Bo}}\, v_{\mathrm{o}})/v\,.$$

The solution of the equation (1.1.22) has the form:

$$n_{\mathrm{A}}(t) = [\,n_2 - n_1\,\mu\,\mathrm{e}^{-\lambda(t)}\,]/[\,1 - \ \mu\,\mathrm{e}^{-\lambda(t)}\,]\,, \qquad \textbf{(1.1.23)}$$

where

$$\lambda(t) = \int_0^t 1/\tau(t)\,\mathrm{d}t\,, \qquad \tau(t) = 1/[\,k\,(n_2 - n_1)\,]\,, \tag{1.1.24}$$

$$\mu = (n_{\mathrm{Ao}} - n_2)/(n_{\mathrm{Ao}} - n_1)\,,$$

n_2 and n_1 are the roots of the equation

$$n_{\mathrm{A}}^2 + p\,n_{\mathrm{A}} - q = 0\,. \tag{1.1.25}$$

The solution to this equation, as is known, has the form:

$$n_{12} = \;[-p \pm \sqrt{p^2 - 4q}\,]/2\,.$$

According to the formula (23), the concentration of particles A decreases monotonically from n_{Ao} at $t = 0$ to n_2 at $t \to \infty$. The rate of particle death at the initial site, as follows from the formula (1.1.24), decreases, since the reaction rate constant k decreases with time (see formula (1.1.3) and the volume of the region G is relaxed from the value $v_o = v\,(T_o)$ at $t = 0$ to the value $v\,(T)$. Outside the initial interval, the time constant τ takes a steady-state value:

$$\tau = \mu/[\,k(T)\,n_{\mathrm{Ao}}\,(1 - \gamma)(1 - \mu\,)\,]\,, \tag{1.1.26}$$

where

$$\gamma = n_2/n_{\mathrm{Ao}}\,, \tag{1.1.27}$$

is the fraction of particles A, which can not react at the temperature T.

The reaction kinetics is convenient to describe

$$\alpha(t) = [\,n_A(t) - n_2\,]/n_{Ao}\,, \tag{1.1.28}$$

which is the fraction of particles A, which will die in the process of bimolecular reaction. Initial value

$$\alpha_o = [\,n_{Ao} - n_2\,]/n_{Ao} = 1 - \gamma \tag{1.1.29}$$

is the fraction of all particles A killed in the course of the reaction due to the temperature increase from T_0 to T. Dependence of α on time at T = const has the form

$$\alpha(t) = \alpha_o\,[\,(1 - \mu)\,e^{-\lambda(t)}\,]/[\,1 - \mu\,e^{-\lambda(t)}\,], \quad \textbf{(1.1.30)}$$

where

$$\mu = \nu_A\,(1 - \gamma)\,(1 - \gamma + \nu_B)/[\,\nu_B\,(1 + \gamma\,\nu_A + \nu_B\,)\,], \quad \textbf{(1.1.31)}$$

$$\nu_A = n_{Ao}\,v_o\,, \qquad \nu_B = n_{Bo}\,v_o\,.$$

Solving equation (1.1.25), we find that

$$\gamma = 1/(2\,\nu_A)\,\{\,-\,v_o/v + \nu_A - \nu_B + \sqrt{(\nu_B - \nu_A + v_o/v\,)^2 + 4\,v_o\,\nu_A\,(1 + \nu_B)/v}\,\} \quad \textbf{(1.1.32)}$$

As can be seen from the formula (1.1.32), the reaction depth γ is determined by the values of the initial and final temperatures and initial concentrations of particles. It is easy to show that the value of γ does not depend on the heating conditions.

If the depth of the reaction is small ($\gamma \approx 1$), or the concentration of particles *B* is much greater than the concentration of particles A ($\nu_B \gg \nu_A$), then, according to (31), the parameter μ is much less than unity then the function (1.1.30) is reduced

$$\alpha(t) = \alpha_o\,e^{-\lambda(t)}\,. \quad \textbf{(1.1.33)}$$

According to this formula, beyond the accelerated initial the reaction proceeds according to the law of kinetics of the first order, and the kinetic curves for sufficiently large values of time should be straightened in the coordinates ln α and t. Generally,

$$\ln\,\alpha(t) = \alpha_0 - \lambda(t) - \ln\,\{\,[\,1 - \mu\,e^{-\lambda(t)}\,]/(\,1 - \mu\,)\,\} \quad \textbf{(1.1.34)}$$

Finally, we find the dependence of n_A on time t. Dependence (1.1.23) takes the form

$$n_A(t) = n_{Ao}\,[\,\gamma + (\,1 - \gamma - \mu\,)\,e^{-\lambda(t)}\,]/[\,1 - \mu\,e^{-\lambda(t)}\,], \quad \textbf{(1.1.35)}$$

When the parameter μ is small ($0 \ll 1$), which is the case for $n_A \ll n_B$ or $\gamma \approx 1$, the formula (35) leads to the function

$$n_A(t) = n_{Ao}\,[\,\gamma + (\,1 - \gamma\,)\,e^{-\lambda(t)}\,]\,. \tag{1.1.36}$$

The graphs of the function (1.1.36) are shown in Fig. (**3**) for different γ values. The graphs of the function (1.1.35) for different values of μ do not differ much from these graphs.

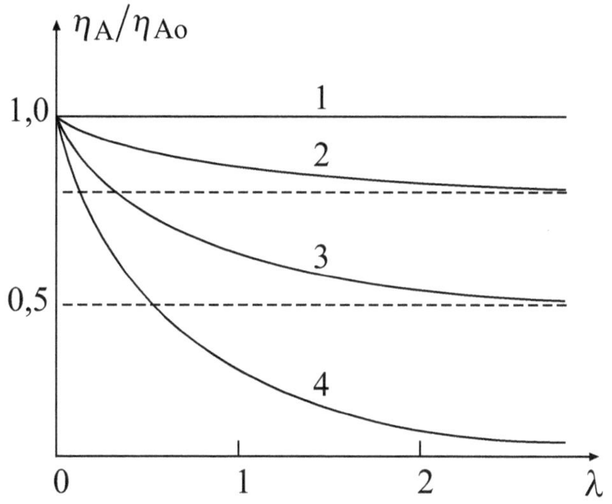

Fig. (3). Dependence of the relative concentrations of particles A of the parameter λ when the parameter value $\mu = 0$ and different values of γ: $\gamma = 1$ (1; 0,8 (2); 0,5 (3); 0 (4).

1.4. Conclusion

The so-called defrosting curve is removed by quasi-static heating of the test sample so that at each time its state is close to equilibrium. The rate of change in concentration with slow heating can be made arbitrarily small. Therefore, to find the theoretical dependence of the concentration on temperature, we can put in equation (22) the right part equal to zero. Then, we come to equation (25), whose solution (32) describes the defrosting curve.

2. INFLUENCE OF HYDROGEN ATOM TUNNEL JUNCTION ON THE "STEP" KINETICS OF SOLID-PHASE REACTIONS WITH PARTICIPATION OF RADICALS IN ORGANIC SUBSTANCES

2.1. Introduction

Solid-phase reactions of recombination and radical transformation are possible provided that the radicals can move in the matrix at distances > 100 Å. There are two possibilities for the implementation of the movement of free valence. Free valence can move with the molecule or part of the molecule to which it belongs, or free valence can migrate through the substance. In the case of organic substances, the migration of free valence can be imagined as a "jump" of the hydrogen atom to an unpaired electron from the nearest molecule. This movement of the hydrogen atom in quantum mechanics is called "tunnel" transitions.

A significant feature of the isothermal solid-phase reactions with participation of radicals is "stepwise". In a study [1], the kinetic theory of solid-phase reactions was developed, which was based on two hypotheses: 1) the motion of each of the reacting particles is limited to an area whose size increases with temperature; 2) the spatial distributions of the particles are not independent, which leads to the need to take into account their correlations.

In another study [2], it was shown that the free valence localized in the organic region near the irregularity of the matrix structure migrates along it by means of tunnel transitions along the lattice nodes. It is shown that the size of this region increases with temperature. Migration is carried out as a result of tunnel transitions of the hydrogen atom from the RH molecule to the radical R.

2.2. Tunnel Junction

It is possible to explain the microdiffusion of a radical in a bounded region whose volume increases with temperature by the tunneling transitions of a hydrogen atom in a bound state localized in the vicinity of any violation of the translational symmetry of the crystal lattice or polymer chain. In solids, there are always various kinds of irregularities (point defects, dislocations, fractures of polymer chains, *etc.*), and under the action of ionizing radiation, they are formed in addition to the existing ones. As known from the quantum theory of a solid body, the so-called bound States of particles arise in the vicinity of irregularities. Let the binding energy E^*_{sv} of the hydrogen atom in the molecule (RH)*, located in close proximity to the irregularity, be less than its binding energy E_{sv} in other molecules RH. If the distance between the radical and the molecule (RH)* is not very large, a "capture" of the free valence of this molecule may occur. Molecule (RH)* is a kind of donor of hydrogen atom. However, due to the tunneling effect, the radical can be detected with some probability at a sufficient distance from the molecule (RH)*.

The probability of finding an unpaired electron near the link of the polymer chain is [3].

$$W_j = \mathrm{th}\ (\alpha\, d)\, \mathrm{e}^{-2\alpha d |j|}\ , \tag{1.2.1}$$

where $j = 0,\ \pm 1, \pm 2, \dots; j = 0$ corresponds to the molecule located near the defect; d is the distance between the RH molecules in the polymer chain, the parameter α satisfies the equality

$$\mathrm{e}^{\alpha d} - \mathrm{e}^{-\alpha d} = \varepsilon/\bar{V}\ , \tag{1.2.2}$$

$$\varepsilon = E_{\mathrm{SV}} - E^*_{\mathrm{SV}},$$

$\bar{V}$ – the average value of the matrix element of interaction between the radical R and the hydrogen atom of the RH molecule. As follows from formula (1.2.1), the probability of finding an unpaired electron at a distance of $l = 1/\alpha$ from the defect decreases by e^2 times. Therefore, we can assume that the free valence migration occurs on the section of the chain with a length of 2 l. The temperature dependence of $\bar{V}$ is due to the thermal vibrations of the molecules of the polymer chain, since at temperatures

$$T < \hbar/(\pi\, k_{\mathrm{B}}\, d)\, \sqrt{2\, E_{\mathrm{SV}}/m_{\mathrm{H}}}\ , \tag{1.2.3}$$

when the tunnel junction prevails over the above-barrier one, the thermal excitation of the vibrational states of the R-H bond can be neglected. In formula (1.2.3), k_{B} – Boltzmann constant, m_{H} – mass of hydrogen atom.

The wave function of the hydrogen atom at a distance of $r \sim d$ from the RH molecule has the form

$$\varphi(r) \sim \mathrm{e}^{-\beta r}, \tag{1.2.4}$$

where

$$\beta = \sqrt{2\, m_H\, E_{sv}}\, / \hbar\,. \tag{1.2.5}$$

Formula (1.2.4) leads to the dependence of the matrix element V on the relative displacement ξ of the RH molecule and the radical R:

$$V = V_0\, e^{-\beta\, \xi}\,. \tag{1.2.6}$$

We assume that the value ξ obeys the Boltzmann distribution law:

$$w(\xi) = \sqrt{c/(2\,\pi\, k_B\, T)}\, \exp[-c\,\xi^2/(2\, k_B\, T)]\,, \tag{1.2.7}$$

where c is the coefficient of "stiffness" of the radical molecule bond. In this case, the average value of (1.2.6) is

$$\bar{V} = V_0 \exp[\beta^2 k_B\, T/(2\,c)]\,. \tag{1.2.8}$$

The force constant c can be conveniently expressed in terms of the reduced mass M and the frequency of ω relative oscillations of the radical-molecule system:

$$c = M\,\omega^2. \tag{1.2.9}$$

Substituting (1.2.8) in (1.2.2), we find approximately that

$$l = 1/\alpha \approx 2\,\bar{V}\, d/\varepsilon = [2\, V_0\, d/\varepsilon]\exp[\beta^2 k_B\, T/(2\,c)]. \tag{1.2.10}$$

If we neglect the jump of the hydrogen atom from one polymer chain to another, the volume v [3], available for free valence, can be considered proportional to the length l. The logarithm ln v will be a linear function of temperature:

$$\ln v = \ln v_0 + T/T^*, \tag{1.2.11}$$

where

$$T^* = 2\,c/(\beta^2 k_B) = M(\hbar\,\omega)^2/(m_H\, E_{sv}\, k_B). \tag{1.2.12}$$

We determine the temperature interval in which the mechanism under consideration has a decisive influence on the kinetics of solid-phase reactions involving radicals in organic substances. To do this, substitute in the expression (1.2.3) the values $d = 1$ Å and $E_{sv} = 40$-170 kJ/mol. Calculations lead to inequality

$$T \leq 220 - 440\ \text{K}. \qquad \textbf{(1.2.13)}$$

Put $\omega = (2 - 3) \cdot 10^{13}$ sec^{-1} and $M/m_{\text{H}} = 6$. Substituting these values in the formula (12), for the same values E_{sv} we obtain

$$T^* = 7 - 64\ \text{K}. \qquad \textbf{(1.2.14)}$$

According to the theory, the volume v is related to the depth

$$\gamma = [\text{R}]/[\text{R}]_{\text{o}} = \gamma(T)$$

reaction R + A →, occurring at a constant temperature T by relations:

$$[\text{R}]_{\text{o}}\, v(T) = \{\, 1 - \gamma + [\text{R}]_{\text{o}}\, v(T_{\text{o}})\,\}/\gamma^2\,, \qquad \textbf{(1.2.15)}$$

if the particle A with which the radical reacts is the same radical (A = R); or

$$[\text{A}]_{\text{o}}\, v(T) = \{\, 1 - \gamma + [\text{A}]_{\text{o}}\, v(T_{\text{o}})\,\}/\gamma^2\,, \qquad \textbf{(1.2.16)}$$

if the concentration of particles A is much greater than the concentration of radicals. In formulas (1.2.15) and (1.2.16) of $[\text{R}]_{\text{o}}$ and $[\text{A}]_{\text{o}}$ there are concentrations corresponding to temperatures at which the reaction R + A → in this matrix is impossible. Knowing the dependence $\gamma(T)$, we can find the function $v(T)$ by these formulas.

2.3. Experimental Data Processing

Processing by formulas (1.2.15) and (1.2.16) of limit curves of thawing reactions involving radicals shows that ln v is a linear function of temperature. The results of the calculation of the values of the characteristic temperature T^* are given in the table. From the comparison of the temperature regions of the studied reactions with the inequality (1.2.13) and the temperature values T^* from the table with the values (14), it follows that the proposed mechanism of free valence displacement and the theory [1] does not contradict the results of experiments.

For low molecular weight compounds, the volume v must be proportional to l^3. The characteristic temperature will be equal to

$$T^* = 2\,c/(3\,\beta^2 k_{\text{B}}) = M(\hbar\,\omega)^2/(3\,m_{\text{H}}\,E_{\text{sv}}\,k_{\text{B}}). \qquad \textbf{(1.2.17)}$$

and for the same values of the quantities included in (1.2.17), in this case, we will have

$$T^* = 2{,}4 - 21\ \mathrm{K}\,.$$

This probably explains the lower temperatures of T^* for the low molecular weight compounds shown in the table. For the same substance, the value of T^* the higher the degree of crystallinity (*S*) (see table, polypropylene). This may be due to an increase in the stiffness coefficient c or a decrease in the binding energy E_{sv}.

Mathematical processing of experimental kinetic curves of solid-phase isothermal reactions shows that these curves, as predicted by the theory [1], are straightened in coordinates

$$\ln\{\ ([\mathrm{R}]_t - [\mathrm{R}]_\infty)/[\mathrm{R}]_\mathrm{o}\ \} - t\,.$$

In this case, the temperature dependence and the rate constants satisfy the Arrhenius law. The values of activation energy E and pre-exponential factor k_o are given in the Table **1**.

The value of V_o in the expression (1.2.6) can be estimated by the formula

$$V_\mathrm{o} = 2\,E_{\mathrm{sv}}\,\mathrm{e}^{-\beta d}\,.$$

For values $d = 1$ Å and $E_{\mathrm{sv}} = 40$-170 kJ/mol time $\tau = \pi\hbar/V_\mathrm{o}$, for which the hydrogen atom makes a tunnel transition from molecule to radical, ranges from 10^{-8} to $1{,}8 \cdot 10^{-2}$ c. As expected, the tunneling time τ is much less than the time constant characterizing the reaction rate.

2.4. Conclusion

For diffusion-controlled reactions, the velocity constant is proportional to the product of the diffusion coefficient D at a distance r_o, at which the active interaction of reacting particles begins:

$k = 4\pi D r_\mathrm{o}$. Putting $D \sim d^2/\tau$ and $r_\mathrm{o} \sim d$, estimate the value of the pre-exponential factor k_o. The calculations result in values of $k_\mathrm{o} \approx 10^{-15} - 10^{-21}$ cm^3/sec, which are consistent with those in the table. Low activation energy values (see table) also support the tunnel mechanism of free valence migration.

3. DEATH KINETICS OF STABILIZED ELECTRONS IN POLYETHYLENE

Experimental kinetic curves of isothermal death of stabilized electrons in PE obtained by R. M. Keyser [4] are investigated from the point of view of the correlation theory of the kinetics of bimolecular reactions in condensed media proposed by the author [1, 5]. The conclusions of the theory are consistent with the results of kinetic measurements, assuming that the death of stabilized electrons occurs during the bimolecular reaction $R^{\cdot} + e_t^- \rightarrow R^-$. Processing of experimental kinetic curves according to the formulas of correlation theory leads to the Arrhenius law for the bimolecular velocity constant and to the exponential dependence on the volume temperature, which characterizes the mobility of particles in the solid matrix at low temperatures [2].

3.1. Introduction

Irradiation of a substance with high-energy particles leads to the ionization of molecules and the formation of free electrons, which, before their thermalization occurs, can be removed at considerable distances from their positive ion. Free electrons whose energy is the result of collisions with molecules decreased to the level of thermal energy captured by various molecular formations. Unlike ion-radicals the so-called stabilized electrons e_t^- interact with several molecules of the matrix [4] when the electron is captured by a single molecule. In other words, it is assumed that the stabilized electrons are localized in the "physical traps", which have an intermolecular origin.

Stabilized electrons can be directly observed by optical and EPR-methods spectroscopy [5]. The type and properties of the EPR spectrum of stabilized electrons allow us to distinguish these electrons from electrons captured by individual molecules quite clearly. Absorption bands of stabilized electrons lie in the near-infrared region of the spectrum.

Among the models explaining the nature of physical traps, the most common model is presented in the study [2, 6], in which it is assumed that electrons are localized in intermolecular cavities caused by structural inhomogeneities of the matrix.

Another explanation for the existence of stabilized electrons in some solid-phase matrices is the theory, according to which the disordered crystal structure of the matrix is the cause of localized States (Anderson localization). This explanation is consistent with the fact that the ability of the polymer to stabilize electrons increases with increasing crystallinity [4]. In amorphous polymers, electrons either do not stabilize at all or stabilize at a very low yield. This means that the centers around which the electrons are stabilized are located in the crystalline regions of the polymer characterized by a significant disordering of the structure. In low-molecular substances, on the contrary, the ability of the matrix to stabilize electrons decreases with increasing crystallinity. This can be explained by the fact that low-molecular crystals are characterized by a greater ordering of the structure than the crystal regions in polymers. Therefore, in ideal low-molecular crystals, according to the theory, electrons do not stabilize at all, and the appearance of the structure disorder determines the possibility of their stabilization.

3.2. Study of Loss of Electrons

The results of kinetic studies of the death of stabilized electrons in PE by R. M. Keyser are presented in the temperature range 77 – 127 K. Samples of linear PE Marlex-6050 were irradiated with γ-quanta in the dark at 77 K. The number of stabilized in the sample electrons was determined by the intensity of a narrow singlet line in the EPR of polyethylene. The isothermal kinetics of the death of stabilized electrons was investigated at temperatures

$$T_i = 77 + 10\,(\,i - 1)\ \mathrm{K}\,,$$

where $i = 1, 2, \dots, 6$. The observed kinetic curves have a "step" character and are satisfactorily described by the dependence

$$n(t) = n_\infty + (\,n_0 - n_\infty)\,\mathrm{e}^{-Kt}, \tag{1.3.1}$$

where $n(t)$ – concentration of stabilized electrons at time t; $n_0 = n(0)$ – initial concentration; n_∞ – the limit electron concentration, which is set for long-term storage of the sample at a constant temperature; K – the reaction rate constant of the first order. After each rapid increase in temperature of 10 K, time is susceptible again.

We introduce the value

$$\alpha(t) = [\,n(t)\ -\ n_\infty]/n_0\,.$$

According to equation (1) , this value satisfies the equality

$$\ln \alpha = -Kt + \ln(1 - \gamma), \quad \textbf{(1.3.2)}$$

where

$$\gamma = n_\infty / n_0$$

is reaction depth.

From equality (2), it follows that the experimental kinetic curves are straightened in coordinates

$$\ln \alpha \quad - \quad t.$$

Calculated from the angles of inclination of these lines, the *K* constant values for different temperatures are given below.

Table 1. *K* constant values for different temperatures.

T, K	77	87	97	107	117	127
$K \cdot 10^4$, sec^{-1}	4,0	4,5	5,5	5,8	5,7	6,5

The absolute error of determining the values of *K* is ~ $0{,}8 \cdot 10^{-4}\ s^{-1}$. Fig. (**4**) shows the defrosting curve of the death of stabilized electrons, *i.e.* the temperature dependence of the limit concentration n_∞, divided by the initial electron concentration e_t^- at 77 K

$$\gamma^*(T) = n_\infty(T)/n_0(T = 77 \text{ K}).$$

The graph of this dependence is shown in Fig. (**4**).

In some works, it was suggested that free radicals are effective electron traps. This assumption was confirmed by experimental studies [4]. Following these works, we assume that the stabilized electrons die during the bimolecular reaction

$$R^{\cdot} + e_t^- \rightarrow R^-. \quad \textbf{(1.3.3)}$$

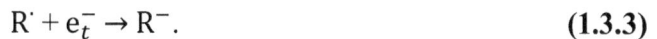

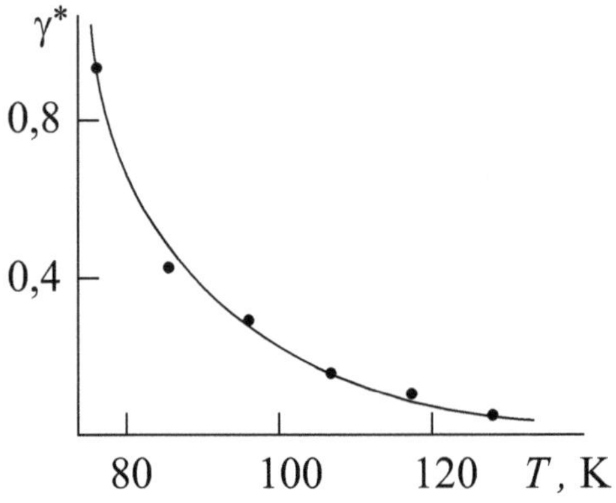

Fig. (4). Defrosting curve of death stabilized electrons.

The most common explanation of the "stepwise" character of the kinetics of many solid-phase reactions is the polychromatic model. In early work, the kinetic theory of bimolecular reactions in condensed media is offered, explaining various deviations of the kinetics of these reactions from the laws of the first or second order. It is assumed that a particle stabilized in the matrix performs microdiffusion movements in a limited region of space. The size of this area increases with temperature. It is the spatial limitation of particle movements in the matrix that appears to be the main reason for the stepwise kinetics of some bimolecular reactions in solids.

3.3. Explanation Experiment Theory

According to the theory [5], reaction kinetics (1.3.3) is described by a system of equations

$$\begin{aligned} \partial n_e/\partial t &= \int F\,(n_e\, n_R - g)\, dV_R\,, \\ \partial n_e/\partial t &= \int F\,(n_e\, n_R - g)\, dV_R\,, \\ \partial g/\partial t &= F\,(n_e\, n_R - g)\, dV_R\,, \end{aligned} \tag{1.3.4}$$

where $n_e = n_e(\boldsymbol{r}_e, t)$ and $n_R = n_R(\boldsymbol{r}_R, t)$ – concentrations of centers around which electrons e_t^- and free radicals are localized R$^{\cdot}$, respectively; $g = g(\boldsymbol{r}_e, \boldsymbol{r}_R, t)$ – correlation function, characterizing the distribution centers stabilization according to the distances between them; $F = F(\boldsymbol{r}_e, \boldsymbol{r}_R, t)$ – the probability of reaction per unit time of the electron e_t^- and the radical R$^{\cdot}$, stabilization centers which are located at the points $\boldsymbol{r}_e$ and $\boldsymbol{r}_R$; V – volume of the matrix.

Suppose that the stabilized electrons and free radicals are uniformly distributed over the volume of the matrix, we assume that

$$n_e = n_e(t)\,, \qquad n_R = n_R(t), \qquad g = g(\boldsymbol{r}, t), \qquad F = F(\boldsymbol{r}, t),$$

where $\boldsymbol{r} = \boldsymbol{r}_e - \boldsymbol{r}_R$.

We introduce a simplified dependence of the reaction probability on the vector $\boldsymbol{r}$:

$$F(r, t) = \begin{cases} k/v & \text{at} \quad r \in G\,, \\ 0 & \text{at} \quad r \ni G\,. \end{cases}$$

where v is the volume of the region G, k is the bimolecular constant of the reaction rate.

In this case, the system (1.3.4) is reduced to one equation

$$dn/dt = -\,k\,(\,n^2 + b\,n - a\,)\,, \tag{1.3.4}$$

where

$$n = n_e(t)\,, \qquad a = n_o\,(\,1\, + [R^{\cdot}]\,v_0)/v\,, \qquad b = [R^{\cdot}] - n_o + 1/v\,;$$

where $n_o = n_e(0)$, $[R^{\cdot}]_o = n_R)$, v_0 is values corresponding to the initial time $t = 0$.

The values of k and v depend on time and temperature. The dependence of k and v on time at a constant temperature determines the unsteadiness of the isothermal reaction at the initial site. Outside the non-stationary (accelerated) initial phase, when k and v take steady-state and correspond to given temperature values, the solution of equation (1.3.5) is of the form (1) provided that the electron concentration e_t^- is significantly less than the concentration of radicals:

$$n \ll [R^{\cdot}]\,. \tag{1.3.5}$$

The constant K is defined by the formula

$$K = k\,[R^{\cdot}]\;(\,1 + \nu)/(\,1 - \,\gamma + \nu\,)\,. \tag{1.3.6}$$

where $\nu = v_o\,[R^{\cdot}]$.

The relationship between the v value of the G volume at a given temperature T and the stationary n_∞ concentration can be found if the right-hand side of equation (1.3.5) is equal to zero Given the inequality (1.3.6), we come to the formula

$$v\,[\mathrm{R}^{\cdot}] = (1 - \gamma + \nu)/\gamma\,. \qquad \textbf{(1.3.7)}$$

Let i = 1, 2, ... , 6 – the number of the isothermal quotient curve for the death of stabilized electrons in PE [4]. Values are defined for each curve K_i, $\gamma_i^* = n_{\infty\,i}/n_0$, $\gamma_i = n_{\infty\,i}/n_{0\,i}$, (see Fig. **4**). It is obvious that

$$\gamma_i^* = \gamma_i\,\gamma_{i-1}\,, \qquad \textbf{(1.3.8)}$$

because $n_i = n_{\infty\,(i-1)}$. We transform the formula (7) to be written as

$$k_i\,[\mathrm{R}^{\cdot}] = (1 - \gamma_i + \nu_i)/(1 + \nu_i)\,K_i\,, \qquad \textbf{(1.3.9)}$$

where $\nu_i = v_{0\,i}\,[\mathrm{R}^{\cdot}]$.

Initial volume $v_{0\,i}$ for i-th curve is equal to volume $v_{i-1} = v(T_i)$. Given this, the parameter γ_i can be calculated by the formula (8)

$$\nu_i = \nu_{i-1}\,[\mathrm{R}^{\cdot}] = (1 - \gamma_{i-1} + \nu_i)/\gamma_{i-1}\,, \qquad \textbf{(1.3.10)}$$

where $i \geq 2$. The recurrence relation (11) together with the equality (1.3.9) leads to the expression

$$\nu_i = (1 - \gamma_{i-1}^* + \nu_{i-1})/\gamma_{i-1}^*\,. \qquad \textbf{(1.3.11)}$$

Formulas (1.3.10) and (1.3.12) allow us to establish the temperature dependencies of the velocity constant $k = k(T)$ and the volume $v = v(T)$ on the results of kinetic measurements. These dependencies are shown in Figs. (**5**) and (**6**).

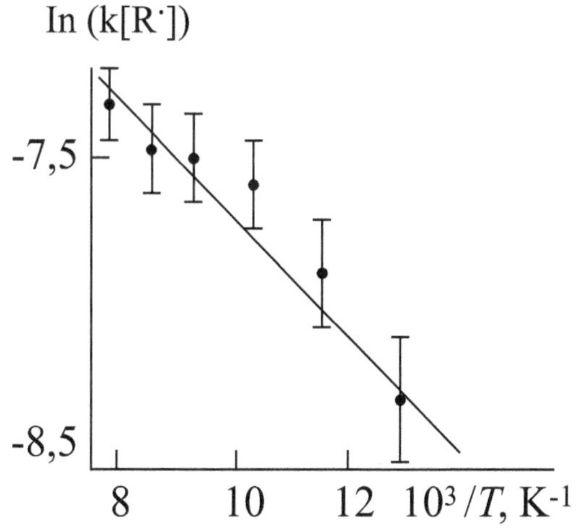

Fig. (5). Arrhenius dependence of the death rate of stabilized electrons upon capture by free radicals.

Parameter v_i put equal to one. Under this condition, all points in Fig. (**2**) within the measurement errors lie satisfactorily on the line in the Arrhenius coordinates [1].

$\ln(k\,[R^{\cdot}])$ – $1/T$, and in Fig. (**6**) – in coordinates $\ln(v\,[R^{\cdot}])$ – T.

At $v_i \ll 1$, two points correspond to temperatures 77 and 87 K, Fig. (**5**) and 3 "go" deep down", while other points practically do not change their position. Straight to Fig. (**5**), evidence of the fairness of the Arrhenius law

$$k = k_0\, e^{-E/(k_B T)}.$$

Determined from the slope of the straight line the activation energy $E = 1,5 \pm 0,3$ kJ/mol ($0,36 \pm 0,07$ kcal/mol). Samples PE were irradiated with quanta to a dose $1,1 \cdot 10^{19}$ eV /g [5], at which the concentrations of stabilized electrons are $3 \cdot 10^{16}$ cm $^{-3}$, and the concentration of radicals $[R^{\cdot}] = 3 \cdot 10^{17}$ cm $^{-3}$. These values satisfy inequality (6). Arrhenius line (Fig. **5**) cuts off on the ordinate axis segment $\sim 4 \cdot 10^{-3}$ sec $^{-1}$. Dividing this value by the concentration of radicals, we find that the pre-exponential factor $k_0 \sim 10^{-20}$ cm ^{3} sec $^{-1}$.

Direct in Fig. (**6**) leads to dependence $v(T) \sim e^{T/T^*}$, where the characteristic temperature $T^* = 16$ K. At this temperature, 70 K corresponds to the volume $\sim 3 \cdot 10^{-18}$ cm ^{3}, and at 120 K the volume $\sim 8 \cdot 10^{-17}$ cm ^{3}. If we assume that the region G is spherical, then its radius at 70 K will be 90 Å, and at 120 K – 270 Å.

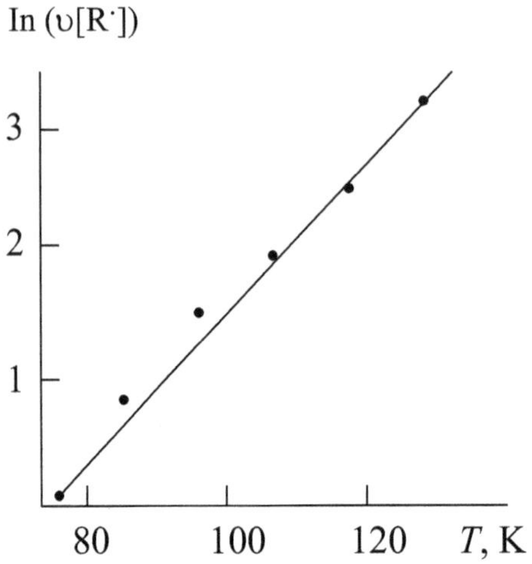

Fig. (6). Dependence of ln (v [R˙]) on temperature.

Calculation formulas (1.3.10) and (1.3.12) are derived from kinetic equations (1.3.4) under the assumption that the matrix is homogeneous on average. From this, it follows that the probability of the reaction depends only on the mutual location of the stabilization centers of the electron-radical pair under consideration and does not depend on where this pair is located [2]. Amorphous-crystalline PE is not such a homogeneous system.

Following work [4], we assume that the electrons stabilize only in the crystalline regions of the polymer. In this work, the technique can be used to calculate the kinetic parameters of the death of stabilized electrons in the amorphous and crystalline PE (crystallinity of 82%), if the size of the crystalline regions is large compared to the size of the domain of G stabilizing electron: $V_{\text{cr}} \gg v$, here V_{cr} – the volume of a crystallite. This inequality is obviously satisfied in this case, since the crystallite sizes are equal to hundreds and even thousands of Å, and the radius of the region G, according to the above estimates, is 90 – 270 Å.

The relative error of the calculated formulas (1.3.10) and (1.3.12) due to the finite size of crystallites can be estimated by v/V_{cr}.

3.4. Conclusion

The literature kinetics curves of isothermal death of stabilized electrons in polyethylene are discussed from the view-point of the correlation theory of kinetics of bimolecular reactions in condensed media proposed by the author. The conclusions of the theory coincide with results of kinetic measurements with an assumption about the disorder of stabilized electrons following the bimolecular reaction $R^{\cdot} + e_t^- \rightarrow R^-$. The treatment of experimental kinetic curves with equations of the correlation theory results in the Arrhenius law for the bimolecular rate constant and exponential dependence on the temperature of the volume characterizing mobility of particles in the solid matrix at low temperature.

REFERENCES

[1] B.V. Bondarev, "Correlation theory of solid-phase reactions kinetics," *Kinet. Catal.,,* vol. 23, no. 2, pp. 334-339, 1982.

[2] B.V. Bondarev, "Influence of hydrogen atom tunnel junction the "step" kinetics of solid-phase reactions with participation of radicals in organic substances," *Kinet. Catal.*, vol. 24, no. 3, pp. 568-571, 1983.

[3] A.M. Stoneham, "Theory of Defects in Solids: The Electronic Structure of Defects in Insulators and Semiconductors", *Mir: Moscow*, 1978, vol. 2, pp. 102.

[4] R. M. Keiser, K. Tzui, F. W. Williams, "ESR and Optical Studies of Trapped Electrons in Glasses and Polymers", In: *Radiation chemistry of macromolecules*. M. Dole (Ed.), Atomizdat: Moscow, pp. 135, 1972.

[5] B.V. Bondarev, "Death kinetics of stabilized electrons in polyethylene," *Polym. Sci.,* vol. A 27, no. 12, pp. 2589-2593, 1985.

CHAPTER 2

Density Matrix

2.1. EQUATION FOR DENSITY MATRIX DERIVATION OF QUANTUM MARKOV KINETIC EQUATION FROM THE LIOUVILLE – VON NEUMANN EQUATION

In the framework of the second-order kinetic perturbation theory, the quantum Markov kinetic equation is derived from the Liouville – von Neumann equation [1]. This equation holds when the density matrix is diagonal. An arbitrary representation of the resulting equation is called the equation for the density matrix. The last equation written in the operator form is called the Lindblad equation [2].

2.1.1. Introduction

From a practical point of view of the problems of statistical physics, the kinetics of a stochastic system interacting with its environment is crucial. One of the main goals of the experimental or theoretical study of an open system, which was initially in some arbitrary nonequilibrium state, is the study of relaxation processes that lead this system to an equilibrium state. The basis of kinetic studies of the open system is the equation that controls the evolution of its States in time. The derivation of this equation, which is called a generalized kinetic equation, is generally a very complex problem.

A mathematically rigorous theory can not raise doubts about its validity only if it is built on the first principles. In the quantum theory of nonequilibrium processes, the first principle is expressed by the Liouville – von Neumann equation for the density matrix. Therefore, various methods for obtaining a generalized kinetic equation from the Liouville – von Neumann equation are of interest. The simplest form of the generalized kinetic equation takes when it describes a random Markov process.

The kinetic theory of an open system is usually formulated on the basis of the "system – reservoir" model, when the "small" system S interacts with the "large" system R, which is considered as an infinitely capacious heat reservoir. It is assumed that the composite system $S + R$ is closed and the evolution of its States is described in the framework of quantum theory by the Liouville – von Neumann equation for the statistical operator $\hat{\rho}(t) = \hat{\rho}_{S+R}(t)$:

Boris V. Bondarev

$$\mathrm{i}\hbar\,\dot{\hat{\rho}} = [\,\hat{\mathcal{H}}\hat{\rho}\,]\,, \tag{2.1.1}$$

where the complete Hamiltonian

$$\hat{\mathcal{H}} \equiv \hat{\mathcal{H}}_{S+R} = \hat{H}_S + \hat{H}_R + \hat{V} \tag{2.1.2}$$

consists of Hamiltonian $\hat{H}_S$ and $\hat{H}_R$ of system S and reservoir R, respectively, and Hamiltonian of their interaction $\hat{V}$.

Statistical operator

$$\hat{\varrho}_S = \hat{\varrho}(t)$$

of system S is determined by the ratio

$$\hat{\varrho}(t) = \mathrm{Tr}_R\,\hat{\rho}(t)\,. \tag{2.1.3}$$

The main task is to obtain an equation for the operator (3). There are two ways to solve this problem: 1) the generalized kinetic equation can be derived from the Liouville – von Neumann equation and 2) the quantum Markov kinetic equation can be obtained phenomenologically taking into account the basic properties of the statistical operator, such as Hermitian, positive certainty, and normalization. The first way attracts the rigor and sequence of actions that lead to the desired equation. But this path is very difficult. Mathematically correct and practically useful results can be obtained in this way only in extreme cases. It is usually assumed that the interaction of the system with its environment is "weak", and at a certain stage of calculations make a thermodynamic limit transition. In addition, you have to make others strong and well-educated guesses. As for the second way, it should be noted that the phenomenologically obtained quantum Markov kinetic equation makes it possible to explain successfully many qualitative and quantitative regularities of kinetic phenomena in quantum open systems. It is therefore interesting to compare these two approaches and derive the quantum Markov kinetic equation directly from the Liouville – von Neumann equation.

Formally, the general solution of equation (2.1.1) can be represented as

$$\hat{\rho}(t) = \exp\,[\,-\,\mathrm{i}/\hbar\int_0^t \hat{\mathcal{H}}(t')\,\mathrm{d}t'\,]\;\hat{\rho}(0)\;\exp\,[\,\mathrm{i}/\hbar\int_0^t \hat{\mathcal{H}}(t')\,\mathrm{d}t'\,]. \tag{2.1.4}$$

It is usually assumed that at the initial time $t = 0$, the states of the system and the reservoir are statistically independent, and the latter is in a state of statistical equilibrium. This assumption corresponds to the equality

$$\hat{\rho}(0) = \hat{\varrho}_S(0)\, \hat{\varrho}_R^{(\mathrm{eq})} , \qquad \textbf{(2.1.5)}$$

where $\hat{\varrho}_S(0)$ is an arbitrary statistical operator describing the initial state of the system;

$$\hat{\varrho}_R^{(\mathrm{eq})} = \nu \exp(-\beta\, \hat{H}_R) \qquad \textbf{(2.1.6)}$$

is the statistical operator of the equilibrium state of the reservoir, ν is the normalizing multiplier,

$$\beta = 1/(k_{\mathrm{B}}\, T)$$

is return temperature. Using formulas (2.1.3) – (2.1.6), it is not difficult to obtain a relation linking the statistical operators $\hat{\varrho}_S(0)$ and $\hat{\varrho}_S(t)$. Unfortunately, this formal solution to the problem is almost useless because of its complexity.

As it is known, the density matrix $\varrho_{nn'}$ of the system S must be self-adjoint, normalized, and positive definite at any time t. The quantum Markov kinetic equation, which guarantees the preservation of these properties of the density matrix, in the most general case has the form

$$\mathrm{i}\, \hbar\, \dot{\varrho}_{nn'} = \sum_m (h_{nm}\, \varrho_{mn'} - \varrho_{nm}\, h_{mn'}) + \mathrm{i}\, \hbar\, [\sum_{mm'} \gamma_{nm,\, m'n'}\, \varrho_{mm'} - 1/2 \sum_m (\gamma_{nm}\, \varrho_{mn'} + \varrho_{nm}\, \gamma_{mn'})] , \qquad \textbf{(2.1.7)}$$

where h_{nm} are matrix elements of the effective Hamiltonian of the system, variables *n, m,...* the essence of the quantum numbers that characterize its state,

$$\gamma_{nm,\, m'n'} = \sum_j a_{nm,j}\, a^*_{n'm',j} , \qquad \textbf{(2.1.8)}$$

$a_{nm,j}$ are a system of arbitrary linearly independent matrices, $\gamma_{nm} = \sum_l \gamma_{lm,nl}$.

The purpose of this paper is to derive the quantum Markov kinetic equation (7) from the Liouville – von Neumann equation and to obtain calculation formulas for its coefficients. This will be done within the framework of the unsteady perturbation theory.

2.1.2. Method of Density Matrix and Non-stationary Perturbation Theory

Consider a system whose evolution in time is governed by the Liouville – von Neumann equation (2.1.1). Suppose that the Hamiltonian of a system can be represented as a sum

$$\hat{\mathcal{H}} = \hat{H} + \hat{v} \,,$$

where $\hat{H}$ is the" unperturbed "Hamiltonian, $\hat{v}$ is the" small " perturbation. The decomposition of the statistical operator $\hat{\rho}$ in the perturbation powers $\hat{v}$ is well known [2], can be found by using formula (4), and is written in the form

$$\hat{\rho}(t) = \mathrm{e}^{-\mathrm{i}\hat{\varphi}(t)} \{ \hat{\rho}(0) - \mathrm{i}/\hbar \, [\, \hat{u}(t) \,, \hat{\rho}(0) \,] - 1/\hbar^2 \int_0^t \, [\, \dot{\hat{u}}(t') \,, [\, \hat{u}\,(t'), \hat{\rho}(0) \,] \,] \, \mathrm{d}t' + \ldots \} \, \mathrm{e}^{\mathrm{i}\hat{\varphi}(t)}, \qquad \textbf{(2.1.9)}$$

where

$$\hat{\varphi}(t) = 1/\hbar \int_0^t \hat{H}(t') \, \mathrm{d}t',$$
$$\hat{u}(t) = \int_0^t \mathrm{e}^{\mathrm{i}\hat{\varphi}(t')} \, \hat{v}(t') \, \mathrm{e}^{-\mathrm{i}\hat{\varphi}(t')} \, \mathrm{d}t'.$$

Function (2.1.9) is an exact solution of equation (2.1.1) satisfying the given initial conditions.

Let the unperturbed Hamiltonian $\hat{H}$ not depend explicitly on time. There is a representation in which this Hamiltonian is diagonal, *i.e.* its matrix elements in this representation have the form

$$H_{\nu\nu'} = E_\nu \, \delta_{\nu\nu'} \,, \qquad \textbf{(2.1.10)}$$

where ν and ν' are quantum numbers characterizing the state of the system; E_ν are the eigenvalues of the Hamiltonian $\hat{H}$, $\delta_{\nu\nu'}$ are Kronecker symbols. In the energy representation, equation (2.1.9) will look like this:

$$\rho_{\nu\nu'}(t) = \mathrm{e}^{-\mathrm{i}\,\omega_{\nu\nu'}\,t} \; \{ \, \rho_{\nu\nu'}(0) - \mathrm{i}/\hbar \sum_\mu \; [\, u_{\nu\mu}(t) \, \rho_{\mu\nu'}(0) - \rho_{\nu\mu}(0) \, u_{\mu\nu'}(t) \,] -$$

$$- 1/\hbar^2 \sum_{\mu\mu'} \int_0^t [\, \dot{u}_{\nu\mu}(t') \,(\, u_{\mu\mu'}(t')\, \rho_{\mu'\nu'}(0) - \rho_{\mu\mu'}(0)\, u_{\mu'\nu'}(t')\,) -$$
$$-(\, u_{\nu\mu}(t')\, \rho_{\mu\mu'}(0) - \rho_{\nu\mu}(0)\, u_{\mu\mu'}(t')\,)\, \dot{u}_{\mu'\nu'}(t')\,]\, \mathrm{d}t' + \ldots \} , \qquad \textbf{(2.1.11)}$$

where

$\omega_{\nu\nu'} = 1/\hbar\, (\, E_\nu - E_{\nu'}\,)$, $\quad u_{\nu\mu}(t) = \int_0^t v_{\nu\mu}(t')\, \mathrm{e}^{\mathrm{i}\,\omega_{\nu\mu}\, t'}\, \mathrm{d}t'$, $\rho_{\nu\mu}$, $u_{\nu\mu}$, and $v_{\nu\mu}$ are matrix elements of operators $\hat{\rho}$, $\hat{u}$ and $\hat{v}$, respectively.

If the V operator also does not depend on time, then after integration by t', equality (2.1.11) can be converted to

$$\rho_{\nu\nu'}(t) = \mathrm{e}^{-\,\mathrm{i}\,\omega_{\nu\nu'}\, t}\, \{\, \rho_{\nu\nu'}(0) -$$
$$- \mathrm{i}/\hbar \sum_\mu \,[\, v_{\nu\mu}\, \rho_{\mu\nu'}(0)\, f(t, \omega_{\nu\mu}) - \rho_{\nu\mu}(0)\, v_{\mu\nu'}\, f(t, \omega_{\mu\nu'})\,] -$$
$$- 1/\hbar^2 \sum_{\mu\mu'} \int_0^t [\, v_{\nu\mu}\, v_{\mu\mu'}\, \rho_{\mu'\nu'}(0)\, F(t', \omega_{\nu\mu}, \omega_{\mu\mu'}) -$$
$$- v_{\nu\mu}\, \rho_{\mu\mu'}(0)\, v_{\mu'\nu'}\, (\, \mathrm{F}(t', \omega_{\nu\mu}, \omega_{\mu'\nu'}) + F(t, \omega_{\mu'\nu'}, \omega_{\nu\mu})\,) +$$
$$+ \rho_{\nu\mu}(0)\, v_{\mu\mu'}\, v_{\mu'\nu'}\, F(t', \omega_{\mu'\nu'}, \omega_{\mu\mu'})\,) + \ldots \} , \qquad \textbf{(2.1.12)}$$

where

$$f(t, \omega) = \int_0^t \mathrm{e}^{\mathrm{i}\,\omega\, t'}\, \mathrm{d}t', \qquad \textbf{(2.1.13)}$$

$$F(t, \omega_1, \omega_2) = \int_0^t \mathrm{e}^{\mathrm{i}\,\omega_1\, t'}\, f(t', \omega_2)\, \mathrm{d}t'. \qquad \textbf{(2.1.14)}$$

2.1.3. Quantum Markov Kinetic Equation

We present a Hamiltonian $\hat{V}$ describing the interaction of the system S with the heat reservoir R as a sum

$$\hat{V} = \mathrm{Tr}_R\, (\, \hat{V}\, \rho_R^{(\mathrm{eq})}) + \hat{v}, \qquad \textbf{(2.1.15)}$$

where the operator $\hat{v}$ satisfies the condition

$$\mathrm{Tr}_R\, (\, \hat{v}\, \rho_R^{(\mathrm{eq})}) \equiv 0\, . \qquad \textbf{(2.1.16)}$$

The Hamiltonian (2.1.2) of the composite $S + R$ system can now be be written as follows:

$$\widehat{\mathcal{H}} = \widehat{H} + \widehat{v} , \qquad \textbf{(2.1.17)}$$

where the operator

$$\widehat{H} = \widehat{h} + \widehat{H}_R \qquad \textbf{(2.1.18)}$$

we consider as an unperturbed Hamiltonian including the averaged Hamiltonian of the system S

$$\widehat{h} = \widehat{H}_S + \mathrm{Tr}_R\,(\widehat{V}\,\rho_R^{(\mathrm{eq})}) , \qquad \textbf{(2.1.19)}$$

and the operator $\widehat{v}$ as a perturbation.

In the energy representation, the operator $\widehat{H}$ is diagonal:

$$H_{nN,\,n'N'} = (\varepsilon_n + E_N)\,\delta_{nn'}\,\delta_{NN'} , \qquad \textbf{(2.1.20)}$$

where ε_n and E_N are the eigenvalues of the Hamiltonian $\widehat{h}$ and $\widehat{H}_R$, n and N are quantum numbers characterizing the states of the system S and the reservoir R, respectively. In this representation, the equilibrium statistical operator (6) is also diagonal:

$$(\rho_R^{(\mathrm{eq})})_{NN'} = W_N\,\delta_{NN'} , \qquad \textbf{(2.1.21)}$$

where

$$W_N = \nu \exp(-\beta\,E_N) \qquad \textbf{(2.1.22)}$$

is Gibbs distribution.

We write the operator equality (2.1.3) in matrix form:

$$\varrho_{nn'}(t) = \sum_N \rho_{nN,\,n'N'}(t) = w_n(t)\,\delta_{nn'} , \qquad \textbf{(2.1.23)}$$

where w_n is the probability to find a system S in the state n with energy ε_n at time t. We will interpret the Greek indices ν, μ, ... in the formula (2.1.12) as quantum numbers of the composite system $S + R$:

$$\nu = \{n, N\}, \qquad \mu = \{m, M\}\ , \ldots$$

Substituting the expression (2.1.12) in the right part of equality (2.1.23), we will have

$$\begin{gathered}
\varrho_{nn'}(t) = \mathrm{e}^{-\mathrm{i}\,\omega_{nn'}\,t}\ \{\ \varrho_{nn'}(0) - \\
-\mathrm{i}/\hbar \sum_{mMN}\ [\ v_{nN,mM}\ \rho_{mM,n'N}(0)\, f(t,\ \omega_{nm} + \Omega_{NM}) - \\
\rho_{nN,mM}(0)\ v_{mM,n'N}\, f(t,\ \omega_{mn'} + \Omega_{MN})\ ] - \\
-1/\hbar^2 \sum_{mm'MM'N'} \int_0^t\ [\ v_{nN,mM}\ v_{mM,m'M'}\ \rho_{m'M',n'N}(0) \cdot \\
\cdot F(t',\ \omega_{nm} + \Omega_{NM},\ \omega_{mm'} + \Omega_{MM'}) - \\
-v_{nN,mM}\ \rho_{mM,m'M'}(0)\ v_{m'M',n'N} \cdot \\
\cdot\ (\ \mathrm{F}(t',\ \omega_{nm} + \Omega_{NM},\ \omega_{m'n'} + \Omega_{M'N}) + F(t,\ \omega_{m'n'} + \Omega_{M'N},\ \omega_{nm} + \Omega_{NM})\ ) + \\
+\ \rho_{nN,mM}(0)\ v_{mM,m'M'}\ v_{m'M',n'N} \cdot \\
\cdot F(t',\ \omega_{mm'} + \Omega_{MM'},\ \omega_{m'n'} + \Omega_{M'N})\ )\ ] + \ldots\ \}\ ,
\end{gathered} \tag{2.1.24}$$

where

$$\omega_{nm} = \mathrm{i}/\hbar\ (\ \varepsilon_n - \varepsilon_m)\ , \qquad \Omega_{NM} = \mathrm{i}/\hbar\ (\ E_N - E_M)\ .$$

Using matrix notation, we write the initial condition (2.1.5) and condition (2.1.16) as

$$\rho_{nN,n'N'}(0) = \varrho_{nn'}(0)\ W_N\ \delta_{NN'} \tag{2.1.25}$$

$$\sum_N v_{nN,n'N}\ W_N \equiv 0\ . \tag{2.1.26}$$

Functions (2.1.13) and (14) have the following properties

$$\begin{gathered}
f(t,\ 0) = t\ , \\
\lim_{t\to\infty} [\ f(t,\omega)/t] = 0
\end{gathered}$$

at $\omega \neq 0$,

$$\begin{gathered}
F(t,\ 0,\ 0) = t^2/2\ , \\
\lim_{t\to\infty} [\ F(t,\omega_1,\omega_2)/t\ ] = 0
\end{gathered}$$

at $\omega_1 \neq 0$ or $\omega_1 \neq 0$,

$$\lim_{\omega_\alpha \to \pm\infty} F(t, \omega_1, \omega_2) = 0 ,$$

where α =1, 2;

$$\int_{-\infty}^{\infty} F(t, \omega_1, \omega_2)\, d\omega_1 = 0 ,$$
$$\int_{-\infty}^{\infty} F(t, \omega_1, \omega_2)\, d\omega_2 = \pi f(t, \omega_1) .$$

Given these properties, we can approximate the function (2.1.14) for large values of $\omega_1 t$ and $\omega_2 t$ by the following expression:

$$F(t, \omega_1, \omega_2) \cong \pi\, \Delta(\omega_1)\, \delta_{\omega_2}\, t , \qquad \textbf{(2.1.27)}$$

where

$$\Delta(\omega) = \begin{cases} 1 & \text{at} \quad \omega = 0, \\ 0 & \text{at} \quad \omega \neq 0; \end{cases}$$

δ_ω is the Dirac delta-function.

Functions (2.1.14), included in the right part of equality (2.1.24), as arguments contain frequencies Ω_{NM}, which are proportional to the difference $E_N - E_M$ of energies of different States of the reservoir. As the size of the tank increases, its energy increases, and the energy spectrum becomes quasi-continuous. On this basis, it can be argued that the approximate equation (2.1.27) becomes accurate in the thermodynamic limit.

Using formulas (2.1.25) – (2.1.27), after differentiating both parts of equality (2.1.24) by t, we come to the equation

$$\dot{\varrho}_{nn'} = -\mathrm{i}\, \omega_{nn'}\, \varrho_{nn'} + \sum_{mm'} \gamma_{nm,m'n'}\, \varrho_{mm'} - 1/2 \sum_{m} (\gamma_{nm}\, \varrho_{mn'} + \varrho_{nm}\, \gamma_{mn'}), \qquad \textbf{(2.1.28)}$$

where

$$\gamma_{nm,m'n'} = 2\,\pi/\hbar \sum_{NM} v_{nN,mM}\, v^*_{n'N,m'M}\, W_M \,\cdot$$
$$\cdot\, \delta(\varepsilon_n - \varepsilon_m + E_N - E_M)\, \Delta(\varepsilon_n - \varepsilon_m + \varepsilon_{m'} - \varepsilon_{n'}) . \qquad \textbf{(2.1.29)}$$

Due to the fact that the energy spectrum of the reservoir is quasi-continuous, the summation of the quantum numbers N and M in formula (2.1.29) should, strictly speaking, be replaced by integration. Equation (2.1.28) is approximate due to the fact that the third and higher-order terms of perturbation v are discarded in its right part. In the second-order summands, the initial density matrix is replaced by an approximate expression

$$\varrho_{nn'}(0) \cong \mathrm{e}^{\,\mathrm{i}\,\omega_{nn'}\,t}\,\varrho_{nn'}(t)\,.$$

It is easy to see that equation (2.1.28) is equation (7) written in the energy representation.

The equilibrium density matrix

$$\varrho_{nn'} = w_n(t)\,\delta_{nn'}\,, \tag{2.1.30}$$

where

$$w_n(t) = \mathrm{e}^{-\beta\,\varepsilon_n(t)}, \tag{2.1.31}$$

is the solution of equation (2.1.28). This can be easily seen by directly substituting (2.1.30) into (2.1.28), which leads to equality

$$\dot{w}_n = \Sigma_m\,(p_{nm}\,w_m - p_{mn}\,w_n)\,, \tag{2.1.32}$$

where

$$p_{nm} = \gamma_{nm,mn} = 2\,\pi/\hbar\;\Sigma_{NM}\;|\;v_{nN,mM}\;|^2\;W_M\;\delta(\,\varepsilon_n - \varepsilon_m + E_N - E_M) \tag{2.1.33}$$

there is a probability of transition of the system per unit time from the state m to the state n. Formula (2.1.33) is a direct consequence of the Golden rule of Fermi. It follows from the definition (33) that the transition probabilities are subject to the detailed equilibrium principle

$$p_{nm}\,\mathrm{e}^{-\beta\,\varepsilon_m} = p_{mn}\,\mathrm{e}^{-\beta\,\varepsilon_n}. \tag{2.1.34}$$

Under conditions (2.1.31) and (2.1.34) equality (2.1.32) turns into identity.

2.1.4. Unitary Transformation

We proved that the equation

$$\dot{\varrho}_{nn'} = -\mathrm{i}\,\omega_{nn'}\,\varrho_{nn'} + \sum_{mm'} \gamma_{nm,m'n'}\,\varrho_{mm'} - 1/2 \sum_m (\gamma_{nm}\,\varrho_{mn'} + \varrho_{nm}\,\gamma_{mn'}) \qquad \textbf{(2.1.35)}$$

goes into the kinetic equation when the density matrix is diagonal, *i.e.* has the form

$$\varrho_{nn'} = w_n(t)\,\delta_{nn'}\,. \qquad \textbf{(2.1.36)}$$

We write the equation (2.1.7), replacing the state index from n to α. Get this equation in α –representation

$$\mathrm{i}\,\hbar\,\dot{\varrho}_{\alpha\alpha'} = \sum_{\alpha_1} (\,h_{\alpha\alpha_1}\,\varrho_{\alpha_1\alpha'} - \varrho_{\alpha\alpha_1}\,h_{\alpha_1\alpha'}\,) + \mathrm{i}\,\hbar\,[\,\sum_{\alpha_1\alpha'_1} \gamma_{\alpha\alpha_1,\alpha'_1\alpha'}\,\varrho_{\alpha_1\alpha'_1} - 1/2 \sum_{\alpha_1} (\,\gamma_{\alpha\alpha_1}\,\varrho_{\alpha_1\alpha'} + \varrho_{\alpha\alpha_1}\,\gamma_{\alpha_1\alpha'}\,)\,]\,. \qquad \textbf{(2.1.37)}$$

But the fact is that at the beginning of the problem we know the matrix elements of the Hamiltonian $h_{\alpha\alpha'}$ in the α-representation and do not know the eigenvalues of the energy ε_n, which set matrix

$$\omega_{nm} = \mathrm{i}/\hbar\,(\,\varepsilon_n - \varepsilon_m)$$

in the *n*- representation. Since we need to find the kinetic equation and coefficients transitions p_{nm}, it is necessary to establish the relationship between ideas. This relationship is established by a unitary matrix $U_{\alpha n}$.

2.1.5. Conclusion

In this paper, we propose a method for obtaining a quantum kinetic equation for the evolution of the system *S*, from the Liouville – von Neumann equation, which describes the evolution of the composite system $S + R$. The interaction of the system *S* and the reservoir *R* is divided into two terms, the first of which is the averaged value of this Hamiltonian, and the second – the deviation from the mean – is considered as a small perturbation $\hat{v}$. The obtained quantum Markov kinetic equation is valid only in the second approximation of the perturbation theory and does not take into account the influence of the system on the state of the reservoir. The calculation formulas for the coefficients of this equation are established, the construction of which is such that over time any solution of the kinetic equation relaxes to the equilibrium density matrix.

REFERENCES

[1] B.V. Bondarev, "Derivation of the quantum kinetic equation from the equation Liouville – von Neumann," *Theor. Mat. Phys.*, vol. 100, no. 1, pp. 33-43, 1994.
[2] G. Lindblad, "On the generators of quantum dynamical semigroups," *Commun. Math. Phys.*, vol. 48, pp. 119, 1976.

CHAPTER 3

New Theory of Superconductivity

3.1. HISTORY OF SUPERCONDUCTIVITY

3.1.1. Discovery of Superconductivity

Kamerlingh Onnes discovered the phenomenon of superconductivity at the Leiden Laboratory, Holland, in 1991 [1]. While studying the dependence of Hg resistance on temperature, he found out that when the material is cooled down to the temperature of about 4K the resistance drops abruptly to zero. This phenomenon was called **superconductivity**. Soon after that, other elements with similar properties were discovered. Fig. (**3.1.1**) demonstrates the scheme of measurement of superconductor resistance.

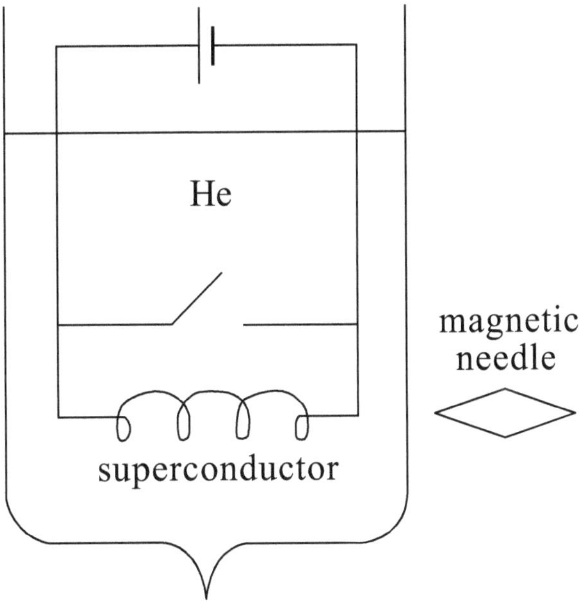

Fig. (3.1.1). The magnetic needle detects a supercurrent-induced magnetic field.

A superconductor is immersed in liquid helium. Initially, a weak current is supplied from a battery, then temperature is reduced. When the temperature falls below a certain value, the superconductor circuit is shortened. The current in the superconductor circuit can be sustained for a long time. A magnetic needle provided as a detector indicates the magnetic field produced by the current in the solenoid.

Boris V. Bondarev

Fig. (**3.1.2**) shows the dependence of resistivity ρ on temperature T in a superconductor. Temperature T_c is called **critical** temperature. This means that we cannot measure the resistance of the superconductor at $T < T_c$. At the same time, we cannot say that the resistivity ρ is equal to zero. The superconductor has a property that makes it impossible to measure the resistivity.

3.1.2. Meissner – Ochsenfeld Effect Silsbee Effect

It was discovered that superconductivity disappears when a test piece is placed in a relatively weak magnetic field. This phenomenon was discovered by Meissner and Ochsenfeld [2]. Value H_m of the magnetic field strength in which superconductivity is disrupted is called a **critical** field. The temperature dependence of the critical field is described by the following empirical formula:

$$H_m(T) = H_m(0)\,[\,1 - (T/T_c)^2]\ , \tag{3.1.1}$$

where $H_m(0)$ is a critical field produced at absolute zero of temperature energy gap $T = 0$. Dependence (1.1) is shown in Fig. (**3.1.3**). Plane (H, T) represents a phase diagram of the superconductive state. The substance in the superconductive state S is shown below the curve (3.1.1) and this substance in the normal state N is above the curve. The superconductor that demonstrates such states is called the type-I superconductor. Superconductivity is disrupted when the current in the substance exceeds a certain critical value (The Silsbee effect).

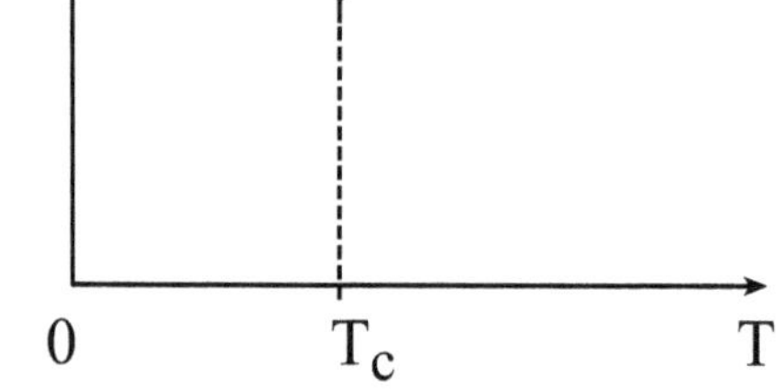

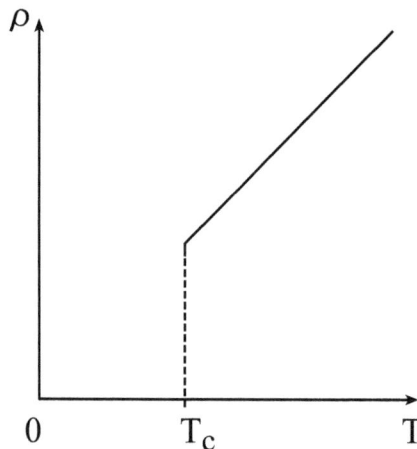

Fig. (3.1.2). The dependence of the resistivity on temperature.

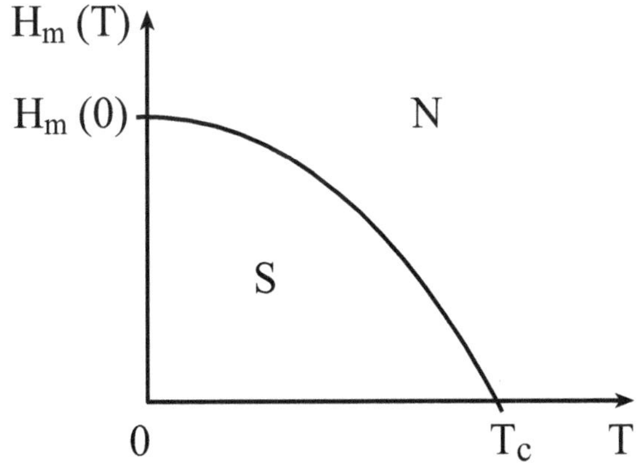

Fig. (3.1.3). Phase diagram of the type-I superconductive state at coordinates (*H*, *T*).

There are two magnetic fields in the superconductor. One magnetic field is created by the supercurrent and another external field is induced from other sources. The compass needle shown in Fig. (**3.1.3**) responds to the supercurrent-induced field. Let us denote such strength of field by parameter $\boldsymbol{H}^{(\mathrm{super})}$ and name this field the superconductor self-generated magnetic field. We shall denote the strength of other magnetic fields by parameter $\boldsymbol{H}^{(\mathrm{exter})}$. This is an external magnetic field. Let the strength of the external magnetic field on the surface of the superconductor be equal to $H^{(\mathrm{exter})} = H_0$. The Meissner – Ochsenfeld effect can be expressed by the following inequality. Superconductivity is generated in metal when its temperature T drops down below the critical temperature T_c:

$$T < T_c , \tag{3.1.2}$$

where the strength of the external magnetic field at the surface of the superconductor is less than that of the critical field:

$$H_0 < H_m(T) . \tag{3.1.3}$$

In other cases, the superconductor will show ordinary metal properties.

There are superconductors of type II, for which the phase diagram has the form shown in Fig. (**3.1.4**). The state of the superconductor, which lies between the normal state N and superconducting state S is called **mixed**.

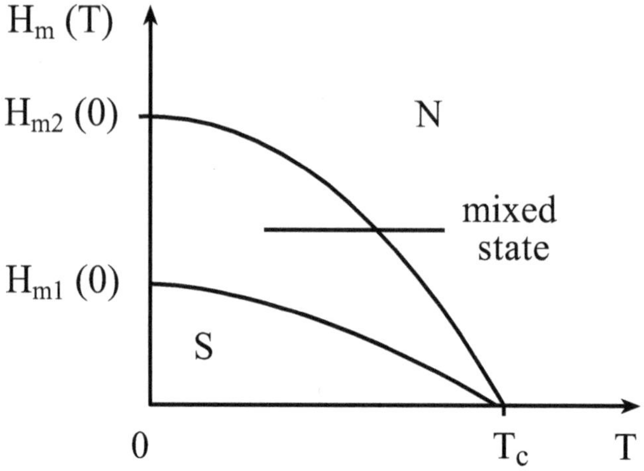

Fig. (3.1.4). Phase diagram of the superconducting state of type II.

3.1.3. Energy Gap

The next important step was finding out the existence of an energy gap. Experimental dependence of the gap width $\Delta = \Delta\ (T)$ on temperature is shown in Fig. (**3.1.5**). The dependence of the heat capacity on the superconducting and normal states, which corresponds to the width of the energy gap $\Delta = \Delta\ (T)$, is shown in Fig. (**3.1.6**).

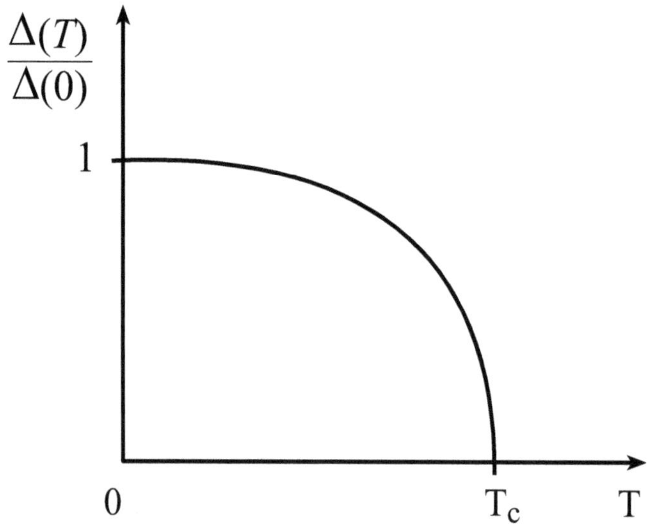

Fig. (3.1.5). The width of the energy gap.

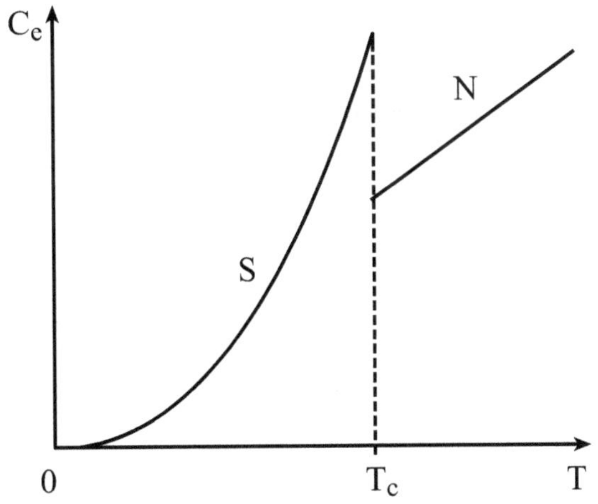

Fig. (3.1.6). The heat capacity of the superconducting and normal states.

In 1950, superconductivity has received its theoretical explanation on the phenomenological level in the Ginzburg – Landau theory [3] and 1957 on the microscopic level in the Bardeen – Cooper – Schrieffer theory [4]. The latter theory was based on a preposition as described in electronical second quantization theory.

3.2. DENSITY MATRIX METHOD VARIATIONAL PRINCIPLE FOR EQUILIBRIUM FERMIONS SYSTEM

3.2.1. Lagrange Method

The thermodynamic potential Ω of a system of particles is related to the average internal energy *E*, entropy *S*, the absolute temperature *T*, chemical potential μ and the average number of particles *N* by formula

$$\Omega = E - S\,T - \mu\,N. \tag{3.2.1}$$

The functions $y = y(x)$, describing the equilibrium state of the system can be found by the Lagrange method. To do this, the thermodynamic potential should be expressed in terms of the function $\Omega = \Omega(y)$. Then, the thermodynamic potential should be differentiated with respect to those functions and the derivatives should be equated to zero:

$$d\Omega/dy = 0. \tag{3.2.2}$$

As a result, we obtain the equations for the equilibrium functions.

3.2.2. The Hierarchy of Density Matrices

Consider a system consisting of N identical particles. Let α be a set of quantum numbers, which reciprocally uniquely determines the state of one particle. The microstate of the entire system will be fully defined if A is given, where A is a set of α_i (i = 1, 2, ..., N), each of which determines the state of one particle:

$$A \equiv \{ \alpha_1, \alpha_2, \dots , \alpha_N \} ,$$

i.e. when it is known that one of the particles is in the state α_1, another is in the state α_2, *etc*. Thus, the distribution function W_A, describing the microstates of the system is a function of multi-dimensional variable A:

$$W_A = W(\alpha_1, \ \alpha_2, \dots , \alpha_N) \equiv \ W_{12\dots N} \,. \tag{3.2.3}$$

This function is the probability that one of the particles of the system is in the state α_1, another is in state α_2, and so on. The probability satisfies the normalization condition

$$\sum_A W_A = N! \ \text{ or } \ \sum_{\alpha_1} \sum_{\alpha_2} \cdots \sum_{\alpha_N} W_{12\dots N} = N! \tag{3.2.4}$$

General statistical description of the evolution of a many-particle system in quantum mechanics is carried out by means of the density matrix [5]

$$\varrho_{AA'} = \varrho(\alpha_1, \alpha_2, \dots , \alpha_N; \ \alpha_1', \ \alpha_2', \dots , \alpha_N') \equiv$$

$$\equiv \varrho_{\alpha_1,\alpha_2,\dots,\alpha_N; \ \alpha_1',\alpha_2',\dots,\alpha_N'} \equiv \varrho_{12\dots N,1'\,2'\dots N'} \,. \tag{3.2.5}$$

A diagonal element ϱ_{AA} of this matrix is equal to the probability W_A of the many-particle system being in state A:

$$\varrho_{AA} = W_A \quad \text{or} \quad \varrho_{12\ldots N,12\ldots N} = W_{12\ldots N}\,.$$

(3.2.6)

The density matrix must be antisymmetric, *i.e.* its sign must change after the permutation of any two elements either among the first N arguments $\alpha_1,\ \alpha_2,\ \ldots,\ \alpha_N$, or among the following arguments $\alpha'_1,\ \alpha'_2,\ \ldots,\ \alpha'_N$:

$$\varrho_{\ldots i\ldots j\ldots;\ldots k'\ldots l'\ldots} = -\varrho_{\ldots j\ldots i\ldots;\ldots k'\ldots l'\ldots} = -\varrho_{\ldots i\ldots j\ldots;\ldots l'\ldots k'\ldots} =$$

$$= \varrho_{\ldots j\ldots i\ldots;\ldots l'\ldots k'\ldots}\,.$$

(3.2.7)

Particles, the set of which is described by the antisymmetric density matrix are called fermions.

Let us write the condition (2.4) for the density matrix

$$\textstyle\sum_A \varrho_{AA} = N! \quad \text{or} \quad \sum_{\alpha_1}\sum_{\alpha_2}\cdots\sum_{\alpha_N} \varrho_{12\ldots N,\ 12\ldots N} = N!$$

(3.2.8)

We define the one-particle density matrix:

$$\varrho_{\alpha_1\alpha'_1} \equiv \varrho_{11'} = \textstyle\sum_{\alpha_2}\cdots\sum_{\alpha_N} \varrho_{12\ldots N,1'2\ldots N}\,.$$

(3.2.9)

The diagonal element $\varrho_{\alpha\alpha}$ of this matrix satisfies the condition

$$\textstyle\sum_\alpha \varrho_{\alpha\alpha} = N.$$

(3.2.10)

Matrix elements of the two-particle density matrix

$$\varrho_{\alpha_1\alpha_2,\alpha_1'\alpha_2'} \equiv \varrho_{12,1'2'} = \sum_{\alpha_3} \cdots \sum_{\alpha_N} \varrho_{123\ldots N,1'2'3\ldots N}\,.$$

(3.2.11)

satisfy the normalization condition

$$\sum_{\alpha_1}\sum_{\alpha_2} \varrho_{12,12} = N(N-1)\,.$$

(3.2.12)

The diagonal element $\varrho_{12,1'2'}$ of the two-particle density matrix is antisymmetric:

$$\varrho_{12,1'2'} = -\,\varrho_{21,1'2'} = -\,\varrho_{12,2'1'} = \varrho_{21,2'1'}\,.$$

(3.2.13)

Densities (3.2.9) and (3.2.11) are related by

$$\sum_{\alpha_2} \varrho_{12,1'2} = (N-1)\,\varrho_{11'}\,.$$

(3.2.14)

3.2.3. Introduction of the Occupation Numbers

Let us associate a number j to each single-particle state. The value of α corresponding to the state with number j we denote as $\alpha^{(j)}$. Let n_1 be the number of particles in the state $\alpha^{(1)}$, n_1-number of particles in the state of $\alpha^{(2)}$, ···, n_j-the number of particles in the state $\alpha^{(j)}$, *etc*. Numbers n_1, n_2, ···, n_j, ··· are called the numbers of occupancy of states by particles. For electrons due to the Pauli principle, each of these numbers can take only two values of 0 and 1.

Let us introduce a multi-dimensional value

$$v \equiv \{\,\{n_1,\alpha^{(1)}\},\ \{n_2,\alpha^{(2)}\},\ \ldots,\ \{n_j,\alpha^{(j)}\},\ \ldots\,\}\,.$$

If the numbering of single-particle states remains unchanged, the values of $\alpha^{(1)}$, $\alpha^{(2)}$,... , $\alpha^{(j)}$, ... are also permanent. The change of the microstate of the system will only change the numbers of occupancy of the states by particles. And one and only one value of v will correspond to each microstate of the system. Therefore, it is convenient to consider the distribution function of a system of fermions on microstates as a function of the argument v. In other words, instead of the function $W_{12...N} \equiv W_A$ one should use the function

$$W_v = W_{\alpha^{(1)}\,\alpha^{(2)}...\,\alpha^{(j)}\,...}^{(n_1\,n_2...\,n_j...)} \equiv W_{1\,2\,...}^{(n_1\,n_2...\,)}.$$

(3.2.15)

This function is such that

$$W_v \equiv W_A.$$

(3.2.16)

The distribution function W_v enables the statistical description of the system of identical particles in the representation of the numbers of the occupancy by the particles of their states. If the number of single-particle states is infinitely large, the number of arguments in this function will also be infinitely large W_v is the probability that in the state $\alpha^{(1)}$ there are n_1 particles, in the state $\alpha^{(2)}$ there are n_2 particles, *etc.*

According to the definition of v all values $\alpha^{(1)}$, $\alpha^{(2)}$, ... are different. The permutation of any pair $\{n_j,\ \alpha^{(j)}\}$ does not change the meaning of the value, because such permutation is equivalent to merely changing the order of numbering one-particle states.

Regardless of the arrangement of the pairs $\{n_j,\ \alpha^{(j)}\}$ with fixed values of $\alpha^{(1)}$, $\alpha^{(2)}$, ... the value v reciprocally uniquely determines the state of the particle system. However, number 1 can be given to any state of the particle. Therefore, $\alpha^{(1)}$ can be considered as a variable that can take values from the set α . The same can be said about the value $\alpha^{(2)}$ with the only difference that $\alpha^{(2)} \neq \alpha^{(1)}$. Generally speaking all $\alpha^{(1)}$, $\alpha^{(2)}$, ... in some cases can be considered as independent variables, provided only that they all take different values.

Since v reciprocally uniquely determines the microstate of a system of particles, the normalization condition can be written as

$$\sum_v W_v = 1 \text{or} \ \sum_{n_1}\sum_{n_2} \cdots \sum_{n_s} \cdots W_{\alpha^{(1)}\,\alpha^{(2)}...\,\alpha^{(s)}\,...}^{(n_1\,n_2...\,n_s...)} = 1,$$

(3.2.17)

where the summation over n_s with fixed values $n_1, \ldots, n_{s-1}, n_{s+1}, \ldots$ is made in the range from 0 to 1.

Let us define one-particle and two-particle probability as

$$W_1^{(n_1)} = \sum_{n_2=0}^{1} \sum_{n_3=0}^{1} \cdots W_{1\,2\,3\ldots}^{(n_1\,n_2\,n_3\,\ldots)},$$

(3.2.18)

$$W_{1\,2}^{(n_1 n_2)} = \sum_{n_3=0}^{1} \cdots W_{1\,2\,3\ldots}^{(n_1\,n_2\,n_3\,\ldots)}.$$

(3.2.19)

From the normalization condition (3.2.17) we have

$$\sum_{n_1=0}^{1} W_1^{(n_1)} = 1,$$

(3.2.20)

$$\sum_{n_2=0}^{1} W_{1\,2}^{(n_1 n_2)} = W_1^{(n_1)},$$

(3.2.21)

$$\sum_{n_1=0}^{1} \sum_{n_2=0}^{1} W_{1\,2}^{(n_1 n_2)} = 1.$$

(3.2.22)

In greater detail, these equations can be written as:

$$W_1^{(0)} + W_1^{(1)} = 1,$$

(3.2.23)

$$W_{12}^{(0\,0)} + W_{12}^{(0\,1)} = W_1^{(0)},$$

(3.2.24)

$$W_{12}^{(10)} + W_{12}^{(11)} = W_{1}^{(1)},$$

(3.2.25)

$$W_{12}^{(00)} + W_{12}^{(01)} + W_{12}^{(10)} + W_{12}^{(11)} = 1.$$

It is easy to show that

$$\sum_{\alpha} W_{\alpha}^{(1)} = N,$$

(3.2.26)

$$\sum_{\alpha_1} \sum_{\alpha_2 \neq \alpha_1} W_{\alpha_1 \alpha_2}^{(11)} = \overline{N(N-1)}.$$

(3.2.27)

Let us define the correlation function

$$\xi_{12} \equiv \xi(\alpha_1, \alpha_2)$$

(3.2.28)

by the formula

$$W_{12}^{(11)} = W_{1}^{(1)} W_{2}^{(1)} u_{12} + \xi_{12},$$

(3.2.29)

where

$$u_{12} \equiv u(\alpha_1, \alpha_2) = 1 - \delta_{\alpha_1 \alpha_2},$$

$\delta_{\alpha_1 \alpha_2}$ is Kronecker symbol. The correlation function is symmetrical, *i.e.*

$$\xi_{12} = \xi_{21},$$

(3.2.30)

and has the property

$$\xi_{12} = 0.$$

(3.2.31)

Let us write the Equation (3.2.23) as follows:

$$W_1^{(0)} = 1 - W_1^{(1)}.$$

(3.2.32)

We substitute expressions (3.2.29) and (3.2.32) into (3.2.24) and (2.25) and resolve the resulting equations with respect to therobability $W_{12}^{(1\,0)}$ and $W_{12}^{(0\,0)}$. We then derive the following formulas

$$W_{12}^{(1\,0)} = W_1^{(1)} \left(1 - W_2^{(1)}\right) u_{12} - \xi_{12},$$

(3.2.33)

$$W_{12}^{(0\,0)} = \left(1 - W_1^{(1)}\right)\left(1 - W_2^{(1)}\right) u_{12} + \xi_{12}.$$

(3.2.34)

A sit can be seen from (3.2.29) (3.2.32)-(3.2.34), all the functions $W_1^{(n_1)}$ and $W_{12}^{(n_1 n_2)}$ can be expressed in terms of the probability $W_1^{(1)}$ and the correlation function ξ_{12}.

3.2.4. The Unitary Transformation

The connection between the single-particle density matrices $\varrho_{\alpha\alpha'}$ and $\varrho_{\kappa\kappa'}$ in various representations is performed by means of a unitary transformation:

$$\varrho_{\alpha\alpha'} = \sum_{\kappa} \sum_{\kappa'} U_{\alpha\kappa}\, \varrho_{\kappa\kappa'}\, U_{\alpha'\kappa'}^{*},$$

(3.2.35)

where $U_{\alpha\kappa}$ is a unitary matrix,

$$\sum_{\alpha} U_{\alpha\kappa}\, U^{*}_{\alpha\kappa'} = \delta_{\kappa\kappa'}\,, \qquad \sum_{\kappa} U^{*}_{\alpha\kappa}\, U_{\alpha'\kappa} = \delta_{\alpha\alpha'}.$$

(3.2.36)

Two-particle density matrix during the transition from one representation to another is transformed according to the law

$$\varrho_{\alpha_1\alpha_2,\alpha_1'\alpha_2'} = \sum_{\kappa_1} \sum_{\kappa_2} \sum_{\kappa_1'} \sum_{\kappa_2'} U_{\alpha_1\alpha_2\,\kappa_1\kappa_2}\, \varrho_{\kappa_1\kappa_2,\kappa_1'\kappa_2'}\, U^{*}_{\alpha_1'\alpha_2'\,\kappa_1'\kappa_2'},$$

(3.2.37)

where $U_{\alpha_1\alpha_2\,\kappa_1\kappa_2}$ is a unitary matrix satisfying the normalization conditions

$$\sum_{\kappa_1} \sum_{\kappa_2} U_{\alpha_1\alpha_2\,\kappa_1\kappa_2}\, U^{*}_{\alpha_1'\alpha_2'\kappa_1\kappa_2} = (\,\delta_{\alpha_1\alpha_1'}\,\delta_{\alpha_2\alpha_2'} - \delta_{\alpha_1\alpha_2'}\,\delta_{\alpha_2\alpha_1'}\,)/2.$$

(3.2.38)

Unitary transformations (3.2.35) and (3.2.37) do not change all the properties of the density matrix provided that the unitary matrix $U_{\alpha_1\alpha_2\,\kappa_1\kappa_2}$ is antisymmetric

$$U_{\alpha_1\alpha_2\,\kappa_1\kappa_2} = -\,U_{\alpha_2\alpha_1\,\kappa_1\kappa_2} = -\,U_{\alpha_1\alpha_2\,\kappa_2\kappa_1} = U_{\alpha_2\alpha_1\,\kappa_2\kappa_1}$$

and satisfies

$$\sum_{\alpha_2} U_{\alpha_1\alpha_2\,\kappa_1\kappa_2}\, U^{*}_{\alpha_1'\alpha_2'\kappa_1'\kappa_2'} = (\,U_{\alpha_1\kappa_1}\, U^{*}_{\alpha_1'\kappa_1'}\,\delta_{\kappa_2\kappa_2'} - U_{\alpha_1\kappa_2}\, U^{*}_{\alpha_1'\kappa_1'}\,\delta_{\kappa_1\kappa_2'} -$$

$$-\,U_{\alpha_2\kappa_1}\, U^{*}_{\alpha_1'\kappa_2'}\,\delta_{\kappa_2\kappa_1'} + U_{\alpha_1\kappa_2}\, U^{*}_{\alpha_1'\kappa_2'}\,\delta_{\kappa_1\kappa_1'}\,)/4\,.$$

In the particular case we can put

$$U_{\alpha_1\alpha_2\,\kappa_1\kappa_2} = (\,U_{\alpha_1\kappa_1} U_{\alpha_2\,\kappa_2} - U_{\alpha_1\kappa_2} U_{\alpha_2\,\kappa_1}\,)/2\,.$$

(3.2.39)

The Equation (3.2.37) becomes

$$\varrho_{\alpha_1\alpha_2,\alpha_1'\alpha_2'} = U_{\alpha_1\kappa_1} U_{\alpha_2\kappa_2}\, \varrho_{\kappa_1\kappa_2,\kappa_1'\kappa_2'}\, U^*_{\alpha_1'\kappa_1'} U^*_{\alpha_2'\kappa_2'}.$$

(3.2.40)

3.2.5. Internal Energy of Fermions System

Let us write the well-known expression for the internal energy of a system of fermions

$$E = \sum_{\alpha}\sum_{\alpha'} H_{\alpha\alpha'}\, \varrho_{\alpha'\alpha} + 1/2 \sum_{\{\alpha\}} H_{\alpha_1\alpha_2,\alpha_1'\alpha_2'}\, \varrho_{\alpha_1'\alpha_2',\alpha_1\alpha_2},$$

(3.2.41)

where $\{\alpha\} = \alpha_1, \alpha_2, \alpha_1', \alpha_2'$. The matrix elements of the Hamiltonian of the interaction of fermions included in this expression $H_{\alpha_1\alpha_2,\alpha_1'\alpha_2'} \equiv H_{12,1'2'}$, must be antisymmetric:

$$H_{12,1'2'} = -H_{21,1'2'} = -H_{12,2'1'} = H_{21,2'1'}.$$

(3.2.42)

Suppose that in a κ-representation the density matrix are diagonal, *i.e.* they have the form

$$\varrho_{\kappa\kappa'} = W^{(1)}_{\kappa}\, \delta_{\kappa\kappa'},$$

(3.2.43)

$$\varrho_{\kappa_1\kappa_2,\kappa_1'\kappa_2'} = W^{(1\,1)}_{\kappa_1\kappa_2}\,(\delta_{\kappa_1\kappa_1'}\,\delta_{\kappa_2\kappa_2'} - \delta_{\kappa_1\kappa_2'}\,\delta_{\kappa_2\kappa_1'}).$$

(3.2.44)

The probabilities of $W^{(1)}_{\kappa}$ and $W^{(1\,1)}_{\kappa_1\kappa_2}$ satisfy the conditions

$$\sum_{\kappa} W^{(1)}_{\kappa} = N, \quad \sum_{\kappa_1}\sum_{\kappa_2 \neq \kappa_1} W^{(1\,1)}_{\kappa_1\kappa_2} = \overline{N(N-1)}.$$

(3.2.45)

Substituting expressions (3.2.43) and (3.2.44) into (3.2.35) and (3.2.40), we transform the expression (3.2.41) to the form

$$E = \sum_{\kappa} \varepsilon_{\kappa} W_{\kappa}^{(1)} + 1/2 \sum_{\kappa_1} \sum_{\kappa_2 \neq \kappa_1} \varepsilon_{\kappa_1 \kappa_2} W_{\kappa_1 \kappa_2}^{(1\,1)},$$

(3.2.46)

where

$$\varepsilon_{\kappa} = \sum_{\alpha} \sum_{\alpha'} U_{\alpha\kappa}^{*} H_{\alpha\alpha'} U_{\alpha'\kappa'}$$

(3.2.47)

is the energy of a particle in the state κ,

$$\varepsilon_{\kappa_1\kappa_2} = 2 \sum_{\{\alpha\}} U_{\alpha_1\kappa_1}^{*} U_{\alpha_2\kappa_2}^{*} H_{\alpha_1\alpha_2,\alpha_1'\alpha_2'} U_{\alpha_1'\kappa_1} U_{\alpha_2'\kappa_2}$$

(3.2.48)

is the energy of interaction between two particles, one of which is in the state κ_1, and the other is in the state κ_2;

$$\varepsilon_{\kappa_1\kappa_2} = \varepsilon_{\kappa_2\kappa_1}, \varepsilon_{\kappa\kappa} = 0 .$$

(3.2.49)

3.2.6. Entropy

The general definition of entropy in quantum mechanics is given by the formula

$$S = -\, k_{\mathrm{B}} \,\mathrm{Tr}\, (\hat{\varrho} \ln \hat{\varrho}\,) ,$$

(3.2.50)

which, however, is of little use for specific calculations. It is natural to assume that this formula takes the simplest form in the representation of the numbers of occupancy, in which the density matrix is diagonal, *i.e.* it looks like

$$\varrho_{\nu\nu'} = W_{\nu}\, \delta_{\nu\nu'}.$$

(3.2.51)

This expression is

$$S = -\, k_{\mathrm{B}} \sum_v W_v \ln W_v \,.$$

(3.2.52)

We write this expression in more detail, using the definition (3.2.15):

$$S = -\, k_{\mathrm{B}} \sum_{n_1} \sum_{n_2} \cdots \sum_{n_s} \cdots W_{1\;2\,\ldots}^{(n_1 n_2 \ldots)} \ln W_{1\;2\,\ldots}^{(n_1 n_2 \ldots)} .$$

(3.2.53)

This is a complex expression. Only in very rare cases it is possible to obtain an exact solution. One should mostly rely on approximate methods.

3.2.7. Fermi – Dirac Function

The simplest statistical description can be done for such a system of identical particles-fermions, where the particles do not interact with one another. Consider the system of non-interacting particles and assume that their number can change randomly with time. Due to the absence of the interaction between the particles their distributions on the states are statistically independent, *i.e.* the number n_j of particles in the state κ_j can take the values 0 and 1, regardless of how many particles there are in other states:

$$W_v = W_{\kappa_1 \kappa_2 \ldots}^{(n_1 n_2 \ldots)} = \prod_j W_j^{(n_j)} ,$$

(3.2.54)

where $W_j^{(n_j)}$ is the probability that n_j particles are in the state κ_j. Recall that the average number n_j of particles in the state κ_j is equal to the probability $W_j^{(1)}$ of occupancy of this state by a particle:

$$W_j^{(1)} = N_j \,.$$

(3.2.55)

The number of particles in the system and their energy are given by the expressions

$$N = \sum_j N_j \, ,$$

(3.2.56)

$$E = \sum_j \varepsilon_j N_j \, ,$$

(3.2.57)

where ε_j is the energy of a fermion in the state κ_j.

Let us represent the entropy as a functional of the probability $W_j^{(n_j)}$. To do this, we put the product (3.2.54) into the logarithm in (3.2.52). We get:

$$S = -k_B \sum_v W_v \ln W_v = -k_B \sum_v W_v \sum_j ln\, W_j^{(n_j)} =$$

$$= -k_B \sum_j \sum_{n=0}^{1} W_j^{(n)} \ln W_j^{(n)} = -k_B \sum_j \left(W_j^{(0)} \ln W_j^{(0)} + W_j^{(1)} \ln W_j^{(1)} \right).$$

Taking into account (3.2.23) and (3.2.55), this formula can be written as

$$S = -k_B \sum_j \left[(1 - N_j) \ln (1 - N_j) + N_j \ln N_j \right].$$

(3.2.58)

Let us now write the expression for the thermodynamic potential:

$$\Omega = \sum_j \{ (\varepsilon_j - \mu) - N_j + \vartheta \left[(1 - N_j) \ln (1 - N_j) + N_j \ln N_j \right] \},$$

(3.2.59)

where $\vartheta = k_B T$. It is evident that the potential depends only on N_j. We calculate the derivative:

$$\partial \Omega / \partial N_j = 0.$$

This formula leads to the following dependence of the average number of fermions in one state on the energy of this state:

$$N_j = 1/[\, e^{\beta(\varepsilon_j - \mu)} + 1\,] ,$$

(3.2.60)

where $\beta = 1/(\,k_B T)$ is inverse temperature. This function is called the Fermi- Dirac distribution. The graph of this function is shown in Fig. (**3.2.1**).

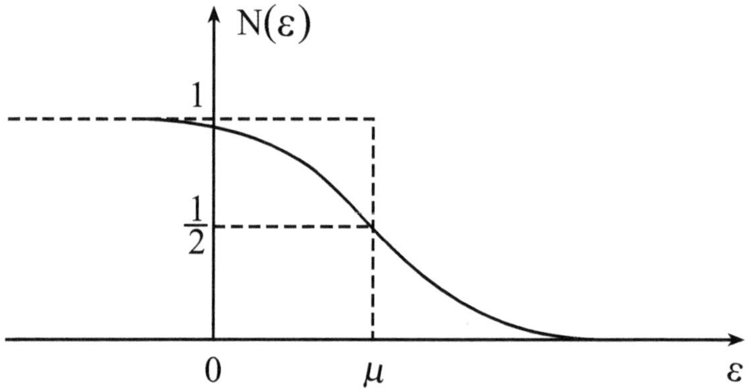

Fig. (3.2.1). The function of Fermi – Dirac.

3.2.8. Mean-field Approximation for Fermions System

Suppose that (3.2.54) is approximately true for a system of interacting particles. In this case, we have approximate equality

$$W^{(1\,1)}_{\kappa_1 \kappa_2} = W_{\kappa_1} W_{\kappa_2} ,$$

(3.2.61)

where

$$W^{(1)}_{\kappa} \equiv W_{\kappa} .$$

This equation allows us to represent the internal energy (3.2.46) as a functional of probability W_κ of occupancy of the state κ by a particle:

$$E = \sum_{\kappa} \varepsilon_{\kappa}\, W_{\kappa} + 1/2 \sum_{\kappa_1} \sum_{\kappa_2} \varepsilon_{\kappa_1\kappa_2}\, W_{\kappa_1} W_{\kappa_2}.$$

(3.2.62)

The transition to representation, in which the density matrix is diagonal, is necessary because only in this representation, a simple approximate expression for the entropy of a system of fermions can be obtained. Distribution (3.2.54) leads to the formula

$$S = -\, k_{\mathrm{B}} \sum_{j} \, [\, (\, 1 - W_{\kappa}) \ln (\, 1 - W_{\kappa}) + W_{\kappa} \ln W_{\kappa}] \, .$$

(3.2.63)

Since the smallest value of the thermodynamic potential corresponds to the equilibrium state of the system with fixed temperature and volume, the functions W_{κ} can be found by minimizing Ω. Thus, we approach the problem of conditional extremum. To solve it by the Lagrange method, we construct a functional Ω. The necessary condition for the extremum of the functional Ω

$$\partial \Omega / \partial\, W_{\kappa} = 0$$

leads to the equation for the distribution function W_{κ}:

$$\ln\,[(1 - \ W_{\kappa})/\ W_{\kappa}] = \beta\ (\ \bar{\varepsilon}_{\kappa} - \mu\ )\, ,$$

(3.2.64)

where $\bar{\varepsilon}_{\kappa}$ is the average energy of a particle in the state κ, with the account of its interaction with other particles. According to the formula

$$\bar{\varepsilon}_{\kappa} = \partial E / \partial\, W_{\kappa}$$

this energy is equal to

$$\bar{\varepsilon}_{\kappa} = \varepsilon_{\kappa} + \sum_{\kappa} \varepsilon_{\kappa\kappa'}\, W_{\kappa'}\, .$$

(3.2.65)

According to this formula, function $\bar{\varepsilon}_{\kappa}$ determines the dependence on the probability W_{κ} of occupancy of the state κ. Thus, the movement of one particle is defined by the distribution of other particles on the states.

3.2.9. Multiplicative Approach of Second Order

Grand canonical Gibbs distribution, which describes the distribution on microstates of an equilibrium system with variable number of particles, taking into account the formula for the system energy and the formula for the number of particles in it can be written as follows:

$$W_\nu = 1/Z \prod_\kappa e^{-\beta(\varepsilon_\kappa - \mu) n_\kappa} \prod_{\kappa_1} \prod_{\kappa_2 \neq \kappa_1} e^{-\frac{1}{2}\beta\, \varepsilon_{\kappa_1\kappa_2} n_{\kappa_1} n_{\kappa_2}},$$

(3.2.66)

where n_κ is the number of particles in the state κ, Z is a statistical sum. Difficulties associated with this expression occur during the calculation of statistical sum. It is even more difficult to find the probability $W_\kappa^{(n)}$ in general case.

Let us assume for the probability W_ν the following approximate formula, which is similar by its form to the expression (3.2.66)

$$W_\nu = \prod_\kappa W_\kappa^{(n_\kappa)} \prod_{\kappa_1} \prod_{\kappa_2 \neq \kappa_1} \left(\mu_{\kappa_1\kappa_2}^{(n_{\kappa_1} n_{\kappa_2})}\right)^{1/2},$$

(3.2.67)

where correlation factors $\mu_{\kappa_1\kappa_2}^{(n_1 n_2)}$ are defined by the relation

$$\mu_{\kappa_1\kappa_2}^{(n_1 n_2)} = W_{\kappa_1\kappa_2}^{(n_1 n_2)} / [\, W_{\kappa_1}^{(n_1)} W_{\kappa_2}^{(n_2)} \,].$$

(3.2.68)

If the two-particle probability $W_{\kappa_1\kappa_2}^{(n_1 n_2)}$ is equal to the product of single-particle probabilities (2.61), then $\mu_{\kappa_1\kappa_2}^{(n_1 n_2)}$ is equal to one and the formula (3.2.67) coincides with probability (2.54) of non-interacting particles.

We obtain an approximate expression for the entropy of a system of fermions. To do this, we substitute the expression (2.67) under the logarithm sign in (3.2.52):

$$S = -\, k_B \sum_\nu W_\nu \left[\sum_\kappa \ln W_\kappa^{(n_\kappa)} + 1/2 \prod_{\kappa_1} \prod_{\kappa_2 \neq \kappa_1} \ln \mu_{\kappa_1\kappa_2}^{(n_{\kappa_1} n_{\kappa_2})} \right].$$

After simple transformations we get the formula

$$S = -\, k_{\text{B}} \,[\, \sum_{\kappa} \sum_{n=0}^{1} W_{\kappa}^{(n)} \ln W_{\kappa}^{(n)} +$$

$$+ 1/2 \, \sum_{\kappa_1} \sum_{\kappa_2 \neq \kappa_1} \sum_{n_1=0}^{1} \sum_{n_2=0}^{1} W_{\kappa_1\kappa_2}^{(n_1 n_2)} \ln \mu_{\kappa_1\kappa_2}^{(n_1 n_2)} \,] \,.$$

(3.2.69)

The probability $W_{\kappa}^{(n)}$ that there are n particles in the state κ satisfies the normalization condition

$$\sum_{n=0}^{1} W_{\kappa}^{(n)} = 1$$

(3.2.70)

and it is related to the probability $W_{\kappa_1\kappa_2}^{(n_1 n_2)}$ by formula

$$\sum_{n_2=0}^{1} W_{\kappa_1\kappa_2}^{(n_1 n_2)} = W_{\kappa_1}^{(n_1)} .$$

(3.2.71)

Due to these equations, not all probabilities $W_{\kappa}^{(n)}$ and $W_{\kappa_1\kappa_2}^{(n_1 n_2)}$ are independent of each other. It is convenient to assume that the probability

$$W_{\kappa} \equiv W_{\kappa}^{(1)}$$

and the correlation function $\xi_{\kappa_1\kappa_2}$ are independent. From the normalization condition (3.2.70) it follows that

$$W_{\kappa}^{(0)} = 1 - W_{\kappa} \,.$$

(3.2.72)

We define the correlation function $\xi_{\kappa_1\kappa_2}$ by the formula

$$W_{\kappa_1\kappa_2}^{(1\,1)} = W_{\kappa_1}^{(1)} \, W_{\kappa_2}^{(1)} \, u_{\kappa_1\kappa_2} + \xi_{\kappa_1\kappa_2} \,,$$

(3.2.73)

where

$$u_{\kappa_1\kappa_2} = 1 - \delta_{\kappa_1\kappa_2} .$$

This definition implies the following properties of the correlation function:

$$\xi_{\kappa_1\kappa_2} = \xi_{\kappa_2\kappa_1}, \qquad \xi_{\kappa_1\kappa_1} = 0.$$

(3.2.74)

From (3.2.71) taking into account (3.2.73) we find that

$$W^{(1\,0)}_{\kappa_1\kappa_2} = W^{(1)}_{\kappa_1} W^{(0)}_{\kappa_2} u_{\kappa_1\kappa_2} - \xi_{\kappa_1\kappa_2},$$

(3.2.75)

$$W^{(0\,0)}_{\kappa_1\kappa_2} = W^{(0)}_{\kappa_1} W^{(0)}_{\kappa_2} u_{\kappa_1\kappa_2} + \xi_{\kappa_1\kappa_2}.$$

(3.2.76)

Thus, the thermodynamic potential is a functional depending on the probability W_κ of a particle filling the state κ and the correlation function $\xi_{\kappa_1\kappa_2}$.

With the help of formulas (3.2.46) and (3.2.70) we can write the following expression for the thermodynamic potential:

$$\Omega = \sum_\kappa (\varepsilon_\kappa - \mu) W^{(1)}_\kappa + 1/2 \sum_{\kappa_1} \sum_{\kappa_2 \neq \kappa_1} \varepsilon_{\kappa_1\kappa_2} W_{\kappa_1\kappa_2} +$$

$$+ \vartheta \sum_\kappa (W^{(0)}_\kappa \ln W^{(0)}_\kappa + W^{(1)}_\kappa \ln W^{(1)}_\kappa) +$$

$$+ 1/2\, \vartheta \sum_{\kappa_1} \sum_{\kappa_2 \neq \kappa_1} [W^{(0\,0)}_{\kappa_1\kappa_2} \ln (W^{(0\,0)}_{\kappa_1\kappa_2} / W^{(0)}_{\kappa_1} W^{(0)}_{\kappa_2}) +$$

$$+ 2\, W^{(1\,0)}_{\kappa_1\kappa_2} \ln (W^{(1\,0)}_{\kappa_1\kappa_2} / W^{(1)}_{\kappa_1} W^{(0)}_{\kappa_2}) +$$

$$+ W^{(1\,1)}_{\kappa_1\kappa_2} \ln (W^{(1\,1)}_{\kappa_1\kappa_2} / W^{(1)}_{\kappa_1} W^{(1)}_{\kappa_2})] .$$

(3.2.77)

Necessary conditions for the extremum of functional

$$\partial\Omega/\partial W_\kappa = 0 \text{ and } \partial\Omega/\partial\xi_{\kappa\kappa'} = 0$$

leads to the following equations:

$$\varepsilon_\kappa + \sum_{\kappa'} \varepsilon_{\kappa\kappa'} W_{\kappa'}^{(1)} + \vartheta \ln (W_\kappa^{(1)} / W_\kappa^{(0)}) +$$
$$+ \vartheta \sum_{\kappa'} u_{\kappa\kappa'} [W_{\kappa\kappa'}^{(1\,1)} \ln (W_{\kappa\kappa'}^{(1\,1)} W_{\kappa\kappa'}^{(0\,0)} / W_{\kappa\kappa'}^{(1\,0)} W_{\kappa\kappa'}^{(0\,1)}) +$$
$$+ \ln (W_{\kappa\kappa'}^{(1\,0)} W_\kappa^{(0)} / W_\kappa^{(1)} W_{\kappa\kappa'}^{(0\,0)})] - \mu = 0 , \tag{3.2.78}$$

$$\varepsilon_{\kappa\kappa'} + \vartheta \ln (W_{\kappa\kappa'}^{(1\,1)} W_{\kappa\kappa'}^{(0\,0)} / W_{\kappa\kappa'}^{(1\,0)} W_{\kappa\kappa'}^{(0\,1)}) = 0 . \tag{3.2.79}$$

If in the Equation (3.2.78) we put $\xi_{\kappa\kappa'} \equiv 0$, then we get the Equation (3.2.64). As it can be seen from the Equation (3.2.79), the correlation function $\xi_{\kappa\kappa'}$ is equal to zero only for those values of κ and κ', for which the interaction energy $\varepsilon_{\kappa\kappa'}$ of the particles in these states is zero.

Using (3.2.77), the Equation (3.2.78) can be transformed to the form

$$\beta (\varepsilon_\kappa - \mu) + \ln (W_\kappa^{(1)} / W_\kappa^{(0)}) + \sum_{\kappa'} u_{\kappa\kappa'} \ln (W_{\kappa\kappa'}^{(1\,0)} W_\kappa^{(0)} / W_\kappa^{(1)} W_{\kappa\kappa'}^{(0\,0)}) . \tag{3.2.80}$$

The Equation (3.2.79) can be written as:

$$\ln (W_{\kappa\kappa'}^{(1\,1)} W_{\kappa\kappa'}^{(0\,0)} / W_{\kappa\kappa'}^{(1\,0)} W_{\kappa\kappa'}^{(0\,1)}) = - \beta\, \varepsilon_{\kappa\kappa'} . \tag{3.2.81}$$

The proposed variational methods can be used to study a variety of many-particle systems in which there are cooperative phenomena.

Now let us reconsider our actions, which will give us the necessary result. The first thing we should do is to find a unitary transformation (2.35), which will lead us to the representation, where the energy of the system (2.46) and its entropy (2.52) have the simplest view. Here, another problem arises. It is necessary to find the expressions for the energies ε_κ and $\varepsilon_{\kappa\kappa'}$ simple enough to enable solving the equation to find the probability W_κ in the mean-field approximation or in the second-order approximation of the probability W_κ and correlation function $\xi_{\kappa\kappa'}$. The next chapter is dedicated to this problem.

3.3. ENERGY OF ELECTRONS IN CRYSTAL LATTICE

3.3.1. Unitary Transformation

The arrangement of atoms in a crystal can be described by specifying its underlying Brava is lattice and specifying the separate position atoms in the unit cell. We will determine the position of one of the atoms in the unit cell at assistance to the vector $\boldsymbol{R}$, and the location of all other atoms in the cell relative to the first – with the help of $\boldsymbol{a}$. Let the set of quantum numbers characterize the wave function of one of the states of the electron, localized in the vicinity of the atom, whose position is determined by the vector $\boldsymbol{R} + \boldsymbol{a}$. using input notation, we write orthonormal wave functions describing localized states electron, in the form of

$$\varphi_\alpha(q) \equiv \varphi_s(\boldsymbol{r} - \boldsymbol{R} - \boldsymbol{a}\,)\,\chi_\sigma(\xi)\,, \tag{3.3.1}$$

where $q = \{\,\boldsymbol{r}, \xi\}$, $\varphi_s(\boldsymbol{r} - \boldsymbol{R} - \boldsymbol{a})$ is the Wannier function at numbers, $\chi_\sigma(\xi)$ - the spin function,

$\alpha = \{\,\boldsymbol{R} + \boldsymbol{a}\,, s, \sigma\}$ – set of quantum numbers that determine the state of an electron in a crystal lattice. Function (3.3.1) satisfies the normalization conditions

$$\int \varphi_\alpha^*(q)\,\varphi_{\alpha'}(q)\,\mathrm{d}q = \delta_{\alpha\alpha'}\,.$$

Assume for simplicity that the crystal lattice is simple, which means that there is a unit cell only one atom: $\boldsymbol{a} = 0$. The function (3.3.1) has the form

$$\varphi_\alpha(q) \equiv \varphi_s(\boldsymbol{r} - \boldsymbol{R})\,\chi_\sigma(\xi)\,, \tag{3.3.2}$$

where $\alpha = \{\,\boldsymbol{R}, s, \sigma\}$. The normalization condition can be written as:

$$\int \varphi_s^*(\boldsymbol{r} - \boldsymbol{R})\,\varphi_{s'}(\boldsymbol{r} - \boldsymbol{R}')\,\mathrm{d}\boldsymbol{r} = \delta_{ss'}\,\delta_{\boldsymbol{R}\boldsymbol{R}'}, \sum_\xi \chi_\sigma^*(\xi)\,\chi_{\sigma'}(\xi) = \delta_{\sigma\sigma'}.$$

When the system of electrons in the metal is in a state of thermodynamic equilibrium, any of the conditions $\varphi_s(\boldsymbol{r} - \boldsymbol{R})$ in the node $\boldsymbol{R}$ can be occupied by electrons with equal probability, and the electrons are distributed over the nodes lattice uniformly.

In this case the density matrix has the form

$$\varrho_{\alpha\alpha'} = \varrho_{\boldsymbol{R} - \boldsymbol{R}'}\, \delta_{ss'}\, \delta_{\sigma\sigma'} .$$

(3.3.3)

Probability W_α filled electronic state $\varphi_\alpha(q)$ is associated with diagonal elements of the density matrix ratio

$$W_\alpha = \varrho_{\alpha\alpha} .$$

In this case

$$W_\alpha \equiv W_{\boldsymbol{R}\, s\, \sigma} = W_{\boldsymbol{R}} = \text{const} ,$$

(3.3.4)

where $W_{\boldsymbol{R}}$ – filling one state probability in node ***R***.

Suppose that a number of electron are localized in the ***R*** site, which describes the functions of the Vanier $\varphi_s(\boldsymbol{r} - \boldsymbol{R})$, of course: $s = 1, 2, \ldots, n_s$. In this case, since $\alpha = \{\boldsymbol{R}, s, \sigma\}$, normalization condition

$$\sum_\alpha W_\alpha = N ,$$

it leads to the formula

$$2\, n_s\, N_L\, W_{\boldsymbol{R}} = N, \qquad W_{\boldsymbol{R}} = \nu,$$

(3.3.5)

where N_L – number of nodes in the lattice,

$$\nu = N/(G\, N_L)$$

(3.3.6)

– the degree of filling of the states,

$$G = 2\, n_s \tag{3.3.7}$$

– number of states in a single node.

Imagine density matrix $\varrho_{\boldsymbol{R}-\boldsymbol{R}'}$ in the form of a complex Fourier series:

$$\varrho_{\boldsymbol{R}-\boldsymbol{R}'} = 1/\,N_L \sum_{\boldsymbol{k}} w_{\boldsymbol{k}}\, e^{\,\mathrm{i}\,\boldsymbol{k}\,(\boldsymbol{R}-\boldsymbol{R}')}, \tag{3.3.8}$$

where the summation is over the wave the vectors $\boldsymbol{k}$, belonging to the first Brillouin zone; $w_{\boldsymbol{k}}$ – function the electron distribution over the wave vectors satisfying the normalization condition

$$G \sum_{\boldsymbol{k}} w_{\boldsymbol{k}} = N, \qquad \text{or } 1/N_L \sum_{\boldsymbol{k}} w_{\boldsymbol{k}} = \nu\,. \tag{3.3.9}$$

From (3.3) it follows that the transformation of the density matrix $\varrho_{\alpha\alpha'}$ is carried out in a diagonal for musing unitary matrix

$$U_{\alpha\kappa'} \equiv U_{\boldsymbol{R}\,s\,\sigma,\ \boldsymbol{k}\,s'\sigma'} = 1/\sqrt{N_L}\, e^{\,\mathrm{i}\,\boldsymbol{k}\boldsymbol{R}}\, \delta_{ss'}\, \delta_{\sigma\sigma'}. \tag{3.3.10}$$

Thus, the density matrix $\varrho_{\alpha\alpha'}$ goes into the matrix

$$\varrho_{\kappa\kappa'} = w_{\boldsymbol{k}}\, \delta_{\boldsymbol{k}\boldsymbol{k}'}\, \delta_{ss'}\, \delta_{\sigma\sigma'}, \tag{3.3.11}$$

where $\kappa = \{\boldsymbol{k}, s, \sigma\}$. The probability $W_\kappa = \varrho_{\kappa\kappa}$ then the state of κ, *i.e.* state described by the wave function $\psi_{\boldsymbol{k}\,s}(\boldsymbol{r})\,\chi_\sigma(\xi)$, will be

$$W_\kappa = w_{\boldsymbol{k}}\,. \tag{3.3.12}$$

Two-electron density matrix can also be diagonal, *i.e.* represented as

$$\varrho_{\kappa_1\kappa_2,\kappa_1'\kappa_2'} = W^{(1\,1)}_{\kappa_1\kappa_2}\,(\,\delta_{\kappa_1\kappa_1'}\,\delta_{\kappa_2\kappa_2'} - \delta_{\kappa_1\kappa_2'}\,\delta_{\kappa_2\kappa_1'}\,),$$

where in

$$W^{(1\,1)}_{\kappa_1\kappa_2} = w^{(1\,1)}_{\boldsymbol{k}_1\boldsymbol{k}_2}\,(1 - \delta_{\boldsymbol{k}_1\boldsymbol{k}_2}\,\delta_{s_1\,s_2}\,\delta_{\sigma_1\sigma_2}\,).$$

(3.3.13)

Binary probability $W^{(1\,1)}_{\kappa_1\kappa_2}$ satisfies the normalization condition

$$\sum_{\kappa_1}\sum_{\kappa_2} W^{(1\,1)}_{\kappa_1\kappa_2} = \overline{N\,(N-1)}\,.$$

(3.3.14)

We found a unitary transformation. This is perhaps the only case in which a unitary transformation can be set immediately.

3.3.2. Hamiltonian of Fermions System

The coordinate representation to which the Hamiltonians are commonly assigned passes to the certain α-representation through the use of the system of the orthonormal wave functions $\varphi_\alpha(q)$, where $q = \{\boldsymbol{r},\ \xi\}$, ξ is a particle spin. If we know these functions, we can calculate matrix elements for Hamiltonians by using the known formulas:

$$H_{\alpha\alpha'} = \int \varphi^*_\alpha\,\widehat{H}^{(1)}\varphi_{\alpha'}\,dq\,,$$

(3.3.15)

$$H_{12,1'2'} = \int \Phi^*_{12}\,\widehat{H}^{(2)}\Phi_{1'2'}\,dq_1dq_2\,,$$

(3.3.16)

where the above sign of integration is a symbol of both the coordinate integration and spin variable summation. $\widehat{H}^{(1)}$ – single particle Hamiltonian,

$$\hat{H}^{(2)} = U(q_1, q_2),$$

$U(q_1, q_2)$ – Coulomb interaction energy of two fermions. Elements $H_{12,1'2'}$, as well as elements $\varrho_{12,1'2'}$, must be anti-symmetric:

$$H_{12,1'2'} = -H_{21,1'2'} = -H_{12,2'1'} = H_{21,2'1'}.$$

Therefore, we will use Slater two-particle wave function Φ_{12}:

$$\Phi_{12} = 1/2\ \{\varphi_1(q_1)\,\varphi_2(q_2) - \varphi_1(q_2)\,\varphi_2(q_1)\}.$$

(3.3.17)

On substituting such function in the formula (3.3.16), we can obtain an antisymmetric matrix as follows:

$$H_{12,1'2'} = 1/4\left(V_{12,1'2'} - V_{21,1'2'} - V_{12,2'1'} + V_{21,2'1'}\right),$$

(3.3.18)

where

$$V_{12,1'2'} = \int \varphi_1^*(q_1)\,\varphi_2^*(q_2)\,U(q_1, q_2)\,\varphi_{1'}(q_1)\,\varphi_{2'}(q_2)\,dq_1 dq_2.$$

(3.3.19)

3.3.3. The Energy of the Electrons in the Crystal Lattice

Let each of the states at the node $\boldsymbol{R}$ correspond to the same value of electron energy. In this case, the single-particle matrix elements of the Hamiltonian can be written as:

$$H_{\alpha\alpha'} = \varepsilon_{\boldsymbol{R}-\boldsymbol{R}'}\,\delta_{ss'},$$

(3.3.20)

Here $\varepsilon_{\boldsymbol{R}-\boldsymbol{R}'}$ – the energy of the particles in the transition from node $\boldsymbol{R}$ to node $\boldsymbol{R}'$

We adopt the following expression for the

$$V_{12,1'2'} = V_{\boldsymbol{R}_1\boldsymbol{R}_2,\boldsymbol{R}'_1\boldsymbol{R}'_2}\, \delta_{s_1 s'_1}\, \delta_{s_2 s'_2},$$

(3.3.21)

where

$$V_{\boldsymbol{R}_1\boldsymbol{R}_2,\boldsymbol{R}'_1\boldsymbol{R}'_2} = \int \varphi(\boldsymbol{r}_1)\, \varphi(\boldsymbol{r}_1 + \boldsymbol{R}_1 - \boldsymbol{R}'_1)\, U(\boldsymbol{r}_1 - \boldsymbol{r}_2 + \boldsymbol{R}_1 - \boldsymbol{R}_2)\cdot$$

$$\cdot\, \varphi(\boldsymbol{r}_2)\, \varphi(\boldsymbol{r}_2 + \boldsymbol{R}_2 - \boldsymbol{R}'_2)\, \mathrm{d}\boldsymbol{r}_1\, \mathrm{d}\boldsymbol{r}_2\,.$$

(3.3.22)

$\varphi(\boldsymbol{r} - \boldsymbol{R})$ is averaged wave function describing an electron localized in the vicinity of the site $\boldsymbol{R}$; $U(r_1 - r_2)$ is the potential energy of Coulomb repulsion between electrons.

Using the approximate formula

$$\varrho_{\alpha_1\alpha_2,\alpha'_1\alpha'_2} \cong \varrho_{\alpha_1\alpha'_1}\, \varrho_{\alpha_2\alpha'_2} - \varrho_{\alpha_1\alpha'_2}\, \varrho_{\alpha_2\alpha'_1}\,.$$

(3.3.23)

and formula (3.3.20) – (3.3.22), after simple transformations of formula (3.2.41) we arrive at the following expression for the energy of the electrons in the mean-field approximation

$$E = G\,\{\, \textstyle\sum_{\boldsymbol{R}\boldsymbol{R}'}\, \varepsilon_{\,\boldsymbol{R}-\boldsymbol{R}'}\, \varrho_{\boldsymbol{R}'\boldsymbol{R}} + \sum_{\{\boldsymbol{R}\}} H_{\boldsymbol{R}_1\boldsymbol{R}_2,\boldsymbol{R}'_1\boldsymbol{R}'_2}\, \varrho_{\boldsymbol{R}'_1\boldsymbol{R}_1} \varrho_{\boldsymbol{R}'_2\boldsymbol{R}_2} \}\,,$$

(3.3.24)

where $\{\boldsymbol{R}\} = \boldsymbol{R}_1,\ \boldsymbol{R}_2,\ \boldsymbol{R}'_1,\ \boldsymbol{R}'_2$;

$$H_{\boldsymbol{R}_1\boldsymbol{R}_2\,\boldsymbol{R}'_1\boldsymbol{R}'_2} = 1/4\,\{\, G\,(V_{\boldsymbol{R}_1\boldsymbol{R}_2,\boldsymbol{R}'_1\boldsymbol{R}'_2} + V_{\boldsymbol{R}_2\boldsymbol{R}_1,\boldsymbol{R}'_2\boldsymbol{R}'_1}) - V_{\boldsymbol{R}_2\boldsymbol{R}_1,\boldsymbol{R}'_1\boldsymbol{R}'_2} - V_{\boldsymbol{R}_1\boldsymbol{R}_2,\boldsymbol{R}'_2\boldsymbol{R}'_1}\,\}.$$

(3.3.25)

Substitution of expression (3.3.8) in the formula (3.3.23) gives

$$E = G \{ \sum_k \varepsilon_k w_k + 1/2 \sum_{kk'} \varepsilon_{kk'} w_k w_{k'} \} ,$$

(3.3.26)

where ε_k is the kinetic energy of an electron:

$$\varepsilon_k = \sum_R \varepsilon_R e^{-ikR};$$

(3.3.27)

$\varepsilon_{kk'}$ – energy of the interacting of two electrons with wave vectors $\boldsymbol{k}$ and $\boldsymbol{k}'$;

$$\varepsilon_{kk'} = 2/N_L^2 \sum_{\{R\}} H_{R_1R_2,R_1'R_2'} e^{\{ik(R_1'-R_1)+ik'(R_2'-R_2)\}}.$$

(3.3.28)

To determine the structure of the kernel $\varepsilon_{kk'}$ in functional (3.3.26), refer to the formula (3.3.22). Taking into account that among matrix elements (3.3.22) the greatest are diagonal elements that correspond $\boldsymbol{R}_1' - \boldsymbol{R}_1$ and $\boldsymbol{R}_2' - \boldsymbol{R}_2$, we write the approximate formula

$$V_{R_1R_2,R_1'R_2'} = U_{R_1-R_2}\delta_{R_1R_1'}\delta_{R_2R_2'} + U^{(o)}_{R_1-R_2}\delta(\boldsymbol{R}_1 - \boldsymbol{R}_1' - \boldsymbol{R}_2 + \boldsymbol{R}_2'),$$

(3.3.29)

where $U_{R_1-R_2}$ is the average energy of Coulomb interaction of two electrons localized at the sites $\boldsymbol{R}_1 - \boldsymbol{R}_2$ and the second term approximates the off-diagonal elements.

Strictly speaking, the function $U^{(o)}_{R_1-R_1'}$ in the formula (3.3.29) should depend not only on $\boldsymbol{R}_1 - \boldsymbol{R}_2$, but also on $\boldsymbol{R}_1 - \boldsymbol{R}_1'$. We obtain the following expression for the interaction energy of the electrons

$$E_{\text{int}} = G/2 \sum_{kk'} \varepsilon_{kk'} w_k w_{k'} .$$

Using formula (3.3.25), (3.3.26) and (3.3.28), (3.29), we obtain the following approximate expression for the interaction energy of electrons:

$$E_{\text{int}} = G/2\,(\,v\,U_o\,N - \textstyle\sum_{kk'} J_{k-k'}\, w_k\, w_{k'} \;+\; \sum_k I_k\, w_k\, w_{-k}\,)\,,$$

(3.3.30)

where

$$J_{k-k'} = 1/\,N_L\,\textstyle\sum_R U_R\, e^{\,\mathrm{i}\,(k-k')\,R}\,;$$

(3.3.31)

$$I_k = \textstyle\sum_R U_R^{(o)}\,(\,G - e^{-2\,\mathrm{i}\,k\,R}\,)\,.$$

(3.3.32)

In the formula (3.3.30), the first term is the energy of the direct Coulomb interaction of electrons, which does not depend on the distribution function w_k. The following component of the sum is the exchange energy of electrons. The kernel $J_{k-k'}$ in this sum is a positive function that takes the largest value at $\boldsymbol{k}' = \boldsymbol{k}$ and rapidly decreases as $|\,\boldsymbol{k} - \boldsymbol{k}'|$ increases because the Coulomb interaction is long- range. Because the exchange energy is negative, the behavior of the function $J_{k-k'}$ determines the effective attraction between electrons with close values of the wave vectors. While the positive terms in the formula (3..3.12) containing values I_k determine effective repulsion between electrons with wave vectors $\boldsymbol{k}$ and $\boldsymbol{k}' = -\,\boldsymbol{k}$.

Mean energy of an electron, corresponding to (3.3.30), excluding the constant $v\,U_o\,N$, will have the form

$$\bar{\varepsilon}_k = \varepsilon_k - \textstyle\sum_{k'} J_{k-k'}\, w_{k'} \;+\; I_k\, w_{-k}\,.$$

(3.3.33)

In the model of free electrons moving in a field of positive charge ions evenly distributed in space, the energy of an electron is described by the formula

$$\bar{\varepsilon}_k = (\hbar\, k)^2/(2\,m) \;-\; e^2/(2\pi^2)\int w_{k'}\,/|\,\boldsymbol{k}' - \boldsymbol{k}\,|^2\, \mathrm{d}\boldsymbol{k}',$$

(3.3.34)

where m and e are the mass and charge of an electron. Both in the formula (3.3.33) and in the formula (3.3.34) kernel $\bar{\varepsilon}_k$ has one common feature. It takes the smallest value $\boldsymbol{k}' = \boldsymbol{k}$, and the largest, with $|\, \boldsymbol{k}' = -\, \boldsymbol{k}\, | \to \infty$. .

Unfortunately, using the formula (3.3.33), it is not only impossible to analytically solve Equation (3.3.3), but even in any depth to research it. Therefore, we approximate the function (3.3.31) by expression

$$J_{k-k'} = J\,\delta_{kk'},$$

(3.3.35)

where J is a positive constant, $\delta_{kk'}$ is Kronecker symbol; and the value of (3.3.32) we assume not depending on the wave vector

$$I_k = I > 0\,.$$

(3.3.36)

In this case the formula (3.33) has the form

$$\bar{\varepsilon}_k = \varepsilon_k + I\,w_{-k} - J w_k\,,$$

(3.3.37)

and any interaction energy would be

$$\varepsilon_{kk'} = I\,\delta_{k+k'} - J\,\delta_{k-k'}\,.$$

(3.3.38)

Since in such approximation, the property $\bar{\varepsilon}_k$ described above is preserved, we can assume that associated with this property, features of the distribution function w_k will not change significantly.

Function (3.3.37) is the model Hamiltonian of the interaction of electrons. Putting in Equation (4.19), the wave vector $\boldsymbol{k}'$ equal in magnitude and opposite in direction to the vector $\boldsymbol{k}$, we see that the electron energy $I\,w_{-k}$ will decrease, when w_{-k} decreases. This means that the electron in the state $\boldsymbol{k}$ displaces an electron from the state $\boldsymbol{k}' = -\,\boldsymbol{k}$.

With the help of formula (3.3.38) Expression (3.3.26) for the energy electrons in the mean-field approximation can be written as:

$$E = G/2 \ \sum_k (2\, \varepsilon_k \, w_k + I \, w_{k} w_{-k} - J w_k^2) .$$

(3.3.39)

Using the model Hamiltonian (3.3.38), we can write an exact expression for the average energy of the electrons:

$$E = G/2 \sum_k (2\, \varepsilon_k \, w_k + I \, w_{k,-k}^{(11)} - J w_{k\,k}^{(11)}) .$$

(3.3.40)

3.3.4. Fermi – Dirac Function and Distribution of Electrons Over the Wave Vectors

Equality (3.3.12) essentially is a unitary transformation, diagonalizing the density matrix. The equation (2.2.63) takes the form

$$S = - G\, k_B \sum_k [\, w_k \ln w_k + (1 - w_k) \ln (1 - w_k)] .$$

(3.3.41)

Minimizing the thermodynamic potential Ω, taking into account the energy (3.3.39), the entropy (3.3.41) and the normalization condition (3.9) we come to the integral equation for finding the distribution function w_k of conduction electrons over the wave vectors

$$\ln [(1 - w_k) / w_k] = \beta (\varepsilon_k + I \, w_{-k} - J \, w_k - \mu) ,$$

(3.3.42)

The solution of this book will be dedicating this equation.

3.3.5. We Also Need to Find the Thermodynamic Functions

Knowing the electron distribution function w_k, we find its appropriate thermodynamic functions: the energy E, the entropy S, the chemical potential μ, thermodynamic potential Ω and heat capacity C_V. Firstly, we introduce the approximate formula

$$\varepsilon_k = (\hbar\, k)^2/(2\, m)\ .$$

Then, we find the chemical potential μ from the normalization condition (3.3.27): $G\ \sum_k w_k\ = N$. We receive an average energy of the electrons by the formula (3.3.39):

$$E = G/2 \sum_k (\, 2\ \varepsilon_k\ w_k\ + I\, w_{k\_} w_{-\,k}\ - J\, w_k^2\ )\ .$$

Entropy by the formula (3.3.41):

$$S = -\ G\, k_{\rm B} \sum_k\ [\ w_k \ln w_k\ + (\ 1 - w_k\ ) \ln\ (\ 1\ -\ w_k\ )\ ]\ .$$

The thermodynamic potential Ω by the formula (3.2.1):

$$\Omega = E - S\, T - \mu\, N.$$

Finally, the heat capacity C_V by the formula

$$C_V = (\partial E/\partial T)_V\ .$$

3.4. ANISOTROPY AND SUPERCONDUCTIVITY

3.4.1. Equation for Probability and Model Hamiltonian

As discussed before, we start from the equation for the Fermi – Dirac function in the mean-field approximation as follows:

$$\ln\ [\ (\ 1 -\ w_k\ )/w_k\ ] =\ \beta\ (\ \bar{\varepsilon}_k - \mu\ )\ .$$

(3.4.1)

where w_k is the probability that a state described by a wave function $\psi_{ks}(\boldsymbol{r})\,\chi_\sigma(\xi)$, occupied one of the electron systems in equilibrium with the wave vector $\boldsymbol{k}$,

$$\bar{\varepsilon}_k\ =\ \varepsilon_k + \sum_k\ \varepsilon_{kk'}\ \ w_{k'}.$$

(3.4.2)

is the average energy of an electron in one of the states with Bloch wave vector $\boldsymbol{k}$, ε_k is electron energy without taking into account its interaction with other electrons, and $\varepsilon_{kk'}$ is electronical energy interaction with wave vectors $\boldsymbol{k}$ and $\boldsymbol{k}'$.

Now we will look at the model Hamiltonian of the form

$$\varepsilon_{kk'} = I\,\delta_{k+k'} - J\,\delta_{k-k'}, \tag{3.4.3}$$

which consists of two components. The first term says that two electrons repel each other if their wave vectors are equal in magnitude and opposite in direction. The second term says that two electrons are attracted if the wave vectors are equal in magnitude and have the same direction.

Substituting in the formula (3.4.2) model Hamiltonian (3.4.3) leads to the dependence of the average energy of an electron species

$$\bar{\varepsilon}_k = \varepsilon_k + I\,w_{-k} - J\,w_k\,. \tag{3.4.4}$$

This formula shows that the average energy of an electron is less than the lower value of w_{-k}, and greater than the value of w_k.

3.4.2. Anisotropy

Let us now consider the anisotropic distribution. We have prescribed function $f = f(\boldsymbol{a})$, *i.e.* value f depends on vector $\boldsymbol{a}$. If value f depends on modulus a of this vector only, the distribution concerned is called isotropic, *i.e.* it may be formulated as $f = f(\boldsymbol{a})$. This kind of isotropy may be represented graphically (Fig. **3.4.1**). We will plot a sphere of radius $\boldsymbol{a}$ centered in the origin of coordinates. Therefore, the value f will remain equal at any point of this sphere, provided that $f = f(a)$ is the isotropic function. Any other $f = f(\boldsymbol{a})$ function will be referred to as the anisotropy function [6].

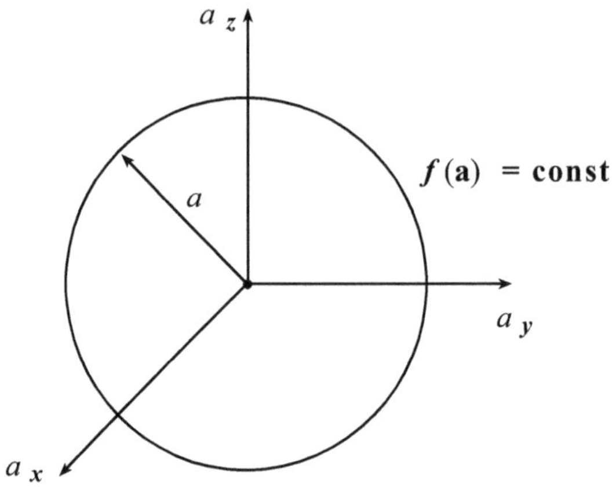

Fig. (3.4.1). Isotropic function.

Now, we will consider the example of the anisotropic function. We will plot two vectors. One of them will be an arbitrary Vector ***a*** and the other one will be rated as equal, but opposite in its direction − ***a***. Two such vectors are shown in Fig. (**3.4.2**). It appears that function values fail to match in the points concerned, *i.e.* $f(\boldsymbol{a}) \neq f(-\boldsymbol{a})$, this function will be called the anisotropic function. Some exhaustive examples of the anisotropic function may be additionally described, but the information provided is sufficient for understanding.

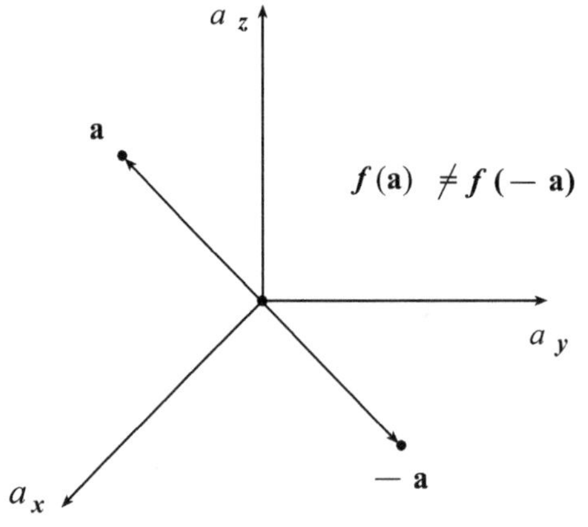

Fig. (3.4.2). Example of anisotropic function.

3.4.3. Mean-field Approximation

The formula (4.4) can be used for transforming Equation (3.3.4.1) as follows:

$$\ln\left[(1-w_k)/w_k\right]=\beta\left(\varepsilon_k+I\,w_{-k}-J\,w_k-\mu\right).$$

(3.4.5)

In this equation, we substitute ***k*** for – ***k***. Since

$$\varepsilon_{-k}=\varepsilon_k\,,$$

we can obtain the following expression:

$$\ln\left[(1-w_{-k})/w_{-k}\right]=\beta\left(\varepsilon_k+I\,w_k-J\,w_{-k}-\mu\right).$$

(3.4.6)

Equations (3.4.5) and (3.4.6) lead to the following system of equations for the two values w_k and w_{-k} of the electron distribution function:

$$\begin{cases}\ln\left[(1-w_k)/w_k\right]=\beta\left(\varepsilon_k+I\,w_{-k}-J\,w_k-\mu\right),\\ \ln\left[(1-w_{-k})/w_{-k}\right]=\beta\left(\varepsilon_k+I\,w_k-J\,w_{-k}-\mu\right).\end{cases}\qquad\textbf{(3.4.7)}$$

We can demonstrate that this system admits two types of solutions. One of them describes isotropic wave vectors distribution of electrons and other anisotropic wave vectors distribution of electrons. If

$$w_{-k}=w_k\,,$$

(3.4.8)

then each of the Equations (3.4.7) transformed can be formulated as follows:

$$\ln\left[(1-w_k)/w_k\right]=\beta\left[\varepsilon_k+(I-J)\,w_k-\mu\right].$$

(3.4.9)

If $I = J$, this equation is solvable as the Fermi – Dirac function.

The unknown functions w_k and w_{-k}, expressed *via* Equations (3.4.7), admit as combined functions where the kinetic electron energy ε_k acts as an intervening variable:

$$w_{-k}=w_1(\varepsilon_k)\text{and}w_k=w_2(\varepsilon_k)\,.$$

The functions

$$w_1 = w_1(\varepsilon) \text{and} \quad w_2 = w_2(\varepsilon)$$

are solvable as follows:

$$\begin{cases} \ln[(1 - w_1)/w_1] = \\ = 2[2\epsilon + (1 - f)w_2 - (1 + f)w_1]/\tau, \\ \ln[(1 - w_2)/w_2] = \\ = 2[2\epsilon + (1 - f)w_1 - (1 + f)w_2]/\tau, \end{cases}$$

(3.4.10)

where

$$\epsilon = (\varepsilon - \mu)/(J + I), \qquad \tau = 4\vartheta/(J + I)$$

(3.4.11)

energy ratio I and J defined by the parameter

$$f = (J - I)/(J + I).$$

(3.4.12)

In this chapter, we will study the case when the parameter $J = 3\,I$; in this case $f = 1/2$.

3.4.4. Isotropic Distribution of Electrons

We express the combined Equations (3.4.10) regarding the case when no anisotropic condition is available, *i.e.*

$$w_1 = w_2 = w_0.$$

Now we can obtain the following equation:

$$\ln\left[(1 - w_o)/w_o\right] = 4(\epsilon - f\,w_o)/\tau\,. \tag{3.4.13}$$

This function is graphically represented in Fig. (**3.4.3**).

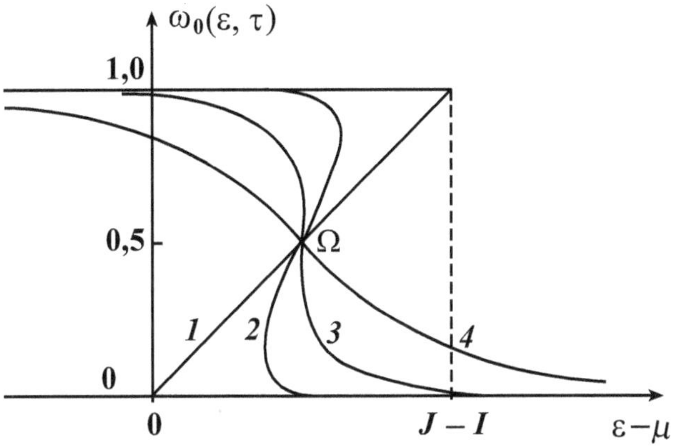

Fig. (3.4.3). Isotropic energy distribution of electrons regarding the case when $J = 3I$ and at various temperature values τ: $1 - \tau = 0$; $2 - \tau = 0.25$; $3 - \tau = 0.5$; $4 - \tau = 1$.

3.4.5. Anisotropic Distribution of Electrons

With the electrons being distributed over wave vectors in an anisotropic manner, we will introduce new variables d and s using the following relations:

$$w_2 - w_1 = d\,,\;\; w_1 + w_2 = 1 + s\,. \tag{3.4.14}$$

Without loss of generality, the difference d of two values w_1 and w_2 of the distribution function can be considered as a nonnegative value: $d \geq 0$,

$$w_1(\epsilon) \leq w_2(\epsilon)\,. \tag{3.4.15}$$

In this case, the largest value d equals to one: $d \in [0,1]$. Value s can take on those to be ranged from -1 to 1: $s \in [-1,1]$. Now, we solve Equations (6.14) relative to the probabilities w_1 and w_2:

$$w_1 = (1 + s - d)/2, \qquad w_2 = (1 + s + d)/2.$$

(3.4.16)

Using the formulae (3.4.16), we transform the combined Equations (3.4.10). For this purpose, we at first subtract one equation from the other and then sum them up. As a result, we can obtain the following combination:

$$[(1+d)^2 - s^2]/[(1-d)^2 - s^2] = \mathrm{e}^{4d/\tau},$$

$$\epsilon = (\tau/8) \ln \{[(1-s)^2 - d^2]/[(1+s)^2 - d^2]\} + (1+s) f/2.$$

(3.4.17)

The first equation of the above combination is easy to solve in relation to s:

$$s(d) = \pm\sqrt{[(1-d)^2 \mathrm{e}^{4d/\tau} - (1+d)^2]/(\mathrm{e}^{4d/\tau} - 1)}.$$

(3.4.18)

As provided by the relations (3.4.12) hereinabove, the probabilities w_1 and w_2 can be considered as the functions of parameter d: $w_1 = w_1(d)$, $w_2 = w_2(d)$. The second combined Equation (3.4.10) makes it possible to express the electron energy ϵ by using parameter d. As based on the obtained dependencies, it is easy enough to plot the function graphs

$$w_1 = w_1(\epsilon) \text{ and } w_2 = w_2(\epsilon)$$

for various temperature values. Such anisotropic curve graphs are demonstrated in Fig. (**3.4.4**). Many-valuedness of the function $w = w(\epsilon)$ proves that various equilibrium states of conduction electrons in metals are possible at the same temperature. These macro-states are different from those represented by the Bloch electron distribution function. As a matter of the fact, it is the electron minimum energy macro-state that can be implemented provided that this kind of state is rather stable and it is kept out of any disruption under external effects.

The plots demonstrated in Figs. (**3.4.3** and **3.4.4**) provide particular insight into electron state distribution pattern shaped up under various metal temperatures. When temperature $\tau \geq 1$, the isotropic electron wave vector distribution only is possible that is described by the function $w_k = w_0(\varepsilon_k)$. The function graph curve $w_0 = w_0(\epsilon)$ stated against all temperature values passes point Ω at the coordinates of $\varepsilon = \mu + (J - I)/2$ and $w = 0.5$. When the temperature falls down $(\tau < 1)$, the curve slope at this point is increased. When the temperature drops down to rather low values, the dependency curve $w_0 = w_0(\epsilon)$ is bent so that it looks like letter *Z*.

A closed anisotropic curve originates at point Ω of the curve $w_0 = w_0(\epsilon)$ at $\tau = 1$ and its dimensions increase against the falling temperature. The shape of the curve also changes. The following critical temperature corresponds to value $\tau = 1$:

$$T_c = (I + J)/(4\, k_B)\,. \tag{3.4.19}$$

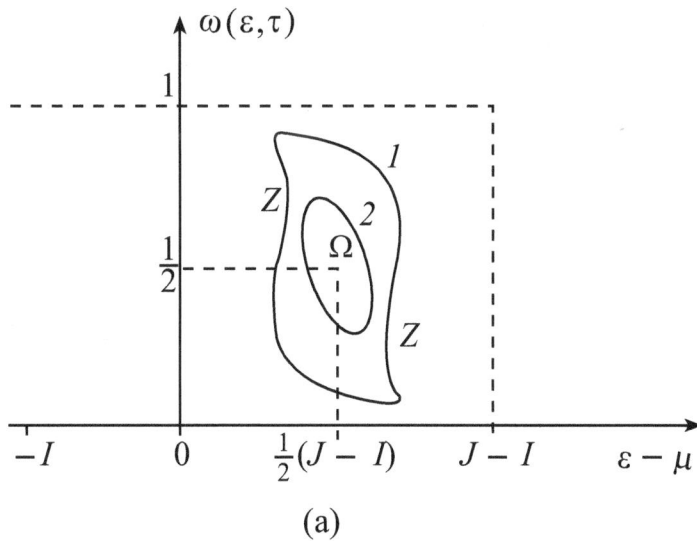

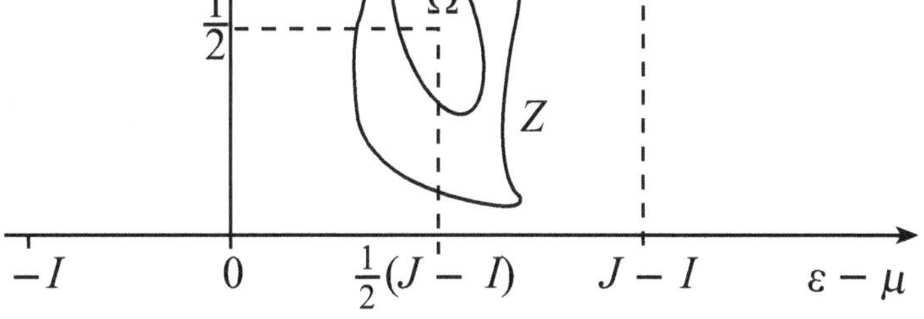

(a)

(Fig. 3.4.4) contd.....

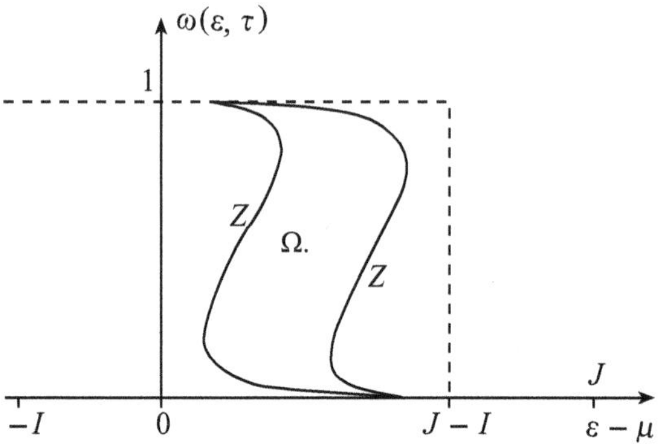

(b)

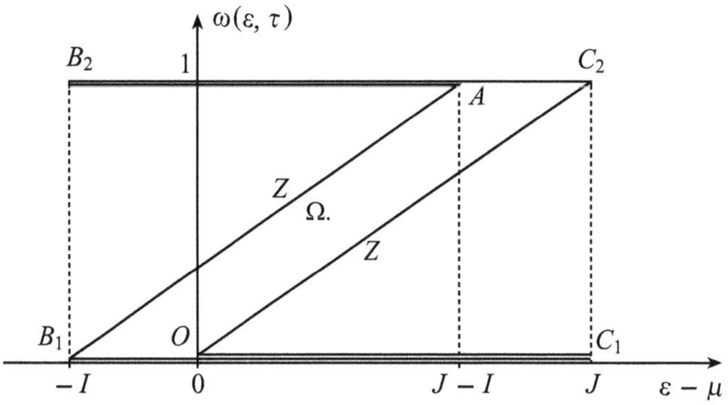

(c)

Fig. (3.4.4). Anisotropic energy distribution of conduction electrons regarding the cases. when: $J = 3I$, at various temperatures τ: (a) $1 - \tau = 0.75$, $2 - \tau = 0.95$; (b) $\tau = 0.5$; (c) $\tau = 0$.

If values $\tau \in (\tau', 1)$ are used, where τ' is a certain critical value, the vertical straight line meets an anisotropic curve maximum at two points (curve 2 in Fig. **3.4.4a**). If $\tau < \tau'$, the anisotropic curve Z bends so that the vertical straight line cuts it at four points (curve 1 in Figs. **3.4.4a** and **3.4.4b**). The anisotropic curve Z transforms into a polygon at $\tau \to 0$. As a result, the $AB_1C_1OC_2B_2A$ polygon line resembles the letter Z. This polygonal line is shown in Fig. (**4.4c**).

3.4.6. Electron Distribution at $T = 0$

Now, we consider the electron distribution function at $T = 0$ in detail. Within the range of $\tau \to 0$, the isotropic distribution solvable by the Equation (3.4.1) is expressed as follows:

$$w_0(\varepsilon) = \begin{cases} 1 & \text{at} \quad \varepsilon \le \mu + J - I, \\ (\varepsilon - \mu)/(J - I) & \text{at } \mu \le \varepsilon \le \mu + J - I, \\ 0 & \text{at} \quad \varepsilon \ge \mu. \end{cases}$$

w **(3.4.20)**

This dependence is graphically represented in Figs. (**3.4.3** and **3.4.4c**) by the dotted broken line B_2AOC_1.

Within the range of $\tau \to 0$, as stated by Equation (3.4.10), the occupation probability dependence w of the kinetic el electron energy ε that describes anisotropic electron wave vector distribution can be formulated as:

$$w(\varepsilon) = \begin{cases} 1 & \text{at} \quad \varepsilon \le \mu - I, \\ w_i(\varepsilon) & \text{at } \mu - I \le \varepsilon \le \mu + J, \\ 0 & \text{at} \quad \varepsilon \ge \mu + J, \end{cases}$$

(3.4.21)

where $i = 1$ or 2. So, the values of functions $w_1 = w_1(\varepsilon)$ and $w_2 = w_2(\varepsilon)$ produce the following pairs:

$$w_1(\varepsilon) = 0 \quad \text{and} \quad w_2(\varepsilon) = 1$$

(3.4.22)

at $\mu - I \le \varepsilon \le \mu + J,$

or $w_1(\varepsilon) = (\varepsilon - \mu + I)/J$ and $w_2(\varepsilon) = 1$

(3.4.23)

at $\mu - I \le \varepsilon \le \mu + J - I,$

or $w_1(\varepsilon) = 0$ and $w_2(\varepsilon) = (\varepsilon - \mu)/J$

(3.4.24)

$$\text{at } 0 \le \varepsilon \le \mu + J.$$

As demonstrated in Fig. (**3.4.4c**), the AB_1C_1O polygonal line meets the relationship $w_1 = w_1(\varepsilon)$ and the AB_2C_2O polygonal line the relationship $w_2 = w_2(\varepsilon)$.

3.4.7. Superconductivity

In the scope of normalization conditions, the average itinerant electron velocity may be defined by the formula as follows:

$$v = G\,\hbar/(m\,N)\,\Sigma_k k\; w_k\,, \tag{3.4.25}$$

where $G = 2\,n_s$ is a number of states in one node and N-mean number of conductivity electrons in a crystal. If the distribution function is isotropic, the mean electron velocity v gets equal to zero. Formula (3.4.22) may assign specific nonzero electron ordered motion velocity values to some anisotropic distribution functions, *i.e.* these distribution functions are applicable for defining electric current. If there are steady-state currents to exist with no external fields available then such itinerant electron system states shall be considered as the superconductive ones.

We will assume that the state of the electron gas is described by the anisotropic distribution function (3.4.21). In this case, mean electron ordered motion velocity modulus v may assign any value rated from zero to certain $v_{\max}$. The mean velocity will be equal to zero, providing that free pairs and those occupied by wave vectors $\boldsymbol{k}$ and $-\boldsymbol{k}$ are chaotically distributed within layer S.

In other words, if $T < T_c$ in the conductor, there will be no superconducting current. For example, even at $T = 0$ the probability of $w_{k_1} = 1$. Then, if the vector k_1 is in

$$S \in \mu - I < \varepsilon_k < \mu\,,$$

then the probability of $w_{-k_1} = 0$. The same can be said about the probabilities $w_{k_2} = 1$ and $w_{-k_2} = 0$. When the vectors $\boldsymbol{k}_1$ and $\boldsymbol{k}_2$ are distributed in S randomly, the average velocity of the electrons is zero: $v_{\max} = 0$. Fig. (**3.4.5a**) shows the wave vectors $\boldsymbol{k}_i$ and $-\boldsymbol{k}_i$, distributed in S chaotically, for which the probability of $w_{k_i} = 1$ and $w_{-k_i} = 0$.

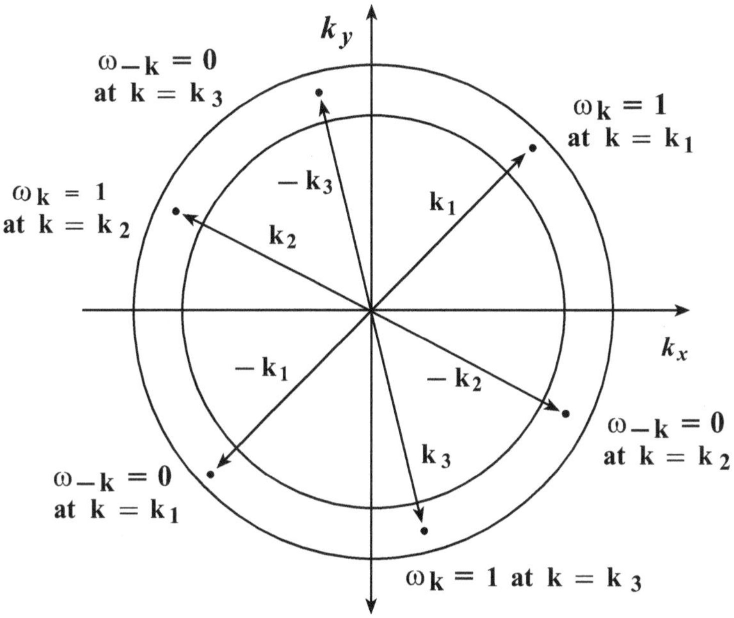

Fig. (3.4.5a). The average velocity of the electrons is zero when the wave vectors k_i and $-k_i$ are distributed randomly in the area of $\mu - I < \varepsilon_k < \mu$.

If all the states concerned are occupied in one half of the layer (this is to say at $k_x > 0$) and the other half of the layer is free (at $k_x < 0$), the electrons will gain their maximum ordered motion velocity. The value assigned by the mean electron velocity is defined by the nature of the initial electron gas state. If the pattern of anisotropic wave vector electron distribution is rather steady with respect to small environment variations, the electron velocity value will survive for ages. This means that the concerned metal will be able to gain its specific superconductive characteristics. Anisotropic distribution pattern is represented in Fig. **(3.4.5b)**.

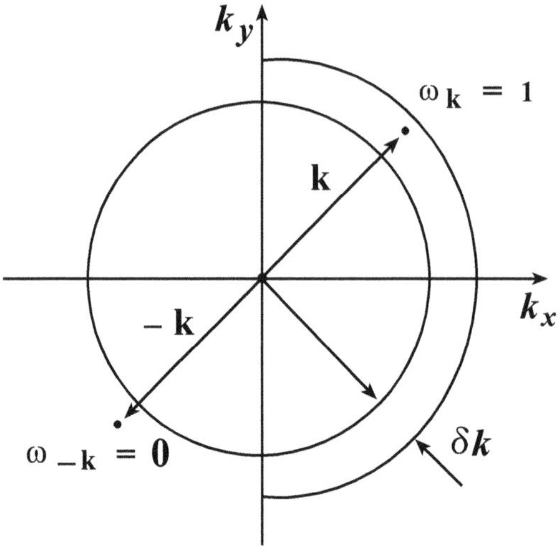

Fig. (3.4.5b). Anisotropic distribution function at $T = 0$. The average electron velocity is maximum. Metal has become a superconductor.

The interaction of electrons with wave vectors $\boldsymbol{k}$ and $\boldsymbol{k'}$ is carried out by the formula for energy (3.3.39)

$$E = G/2 \ \sum_k (2 \ \varepsilon_k \ w_k + I \ w_{k\_} w_{-k} \ - J \ w_k^2) \ .$$

As we can see, only those electrons can interact, which can either fly towards each other or fly together in the same direction. Figs. (**3.4.6a** and **3.4.6b**) show such electron-waves.

Fig. (**3.4.6a**) shows two waves flying towards each other with wave vectors $\boldsymbol{k}$ and – $\boldsymbol{k}$. These electrons are characterized by wave functions $\psi_{\boldsymbol{k}\, s}(\boldsymbol{r}) \ \chi_{\sigma}(\xi_1)$ and $\psi_{-\boldsymbol{k}\, s}(\boldsymbol{r}) \ \chi_{\sigma}(\xi_2)$. Since the wave vectors of these functions are different, the spins of ξ_1 and ξ_2 of these electrons can be different. The energy of these electrons will be proportional

$$I \ w_k \ w_{-k} \ .$$

It can ve observed that for this energy to be minimal, one of the probabilities w_k or w_{-k} must be zero when the temperature $T = 0$. The other probability is not zero. **This indicates the presence of superconductivity**.

Fig. (**3.4.6b**) shows two waves that fly in the same direction with equal wave vectors. These electrons are characterized by wave functions $\psi_{\boldsymbol{k}\, s}(\boldsymbol{r}) \ \chi_{\sigma}(\xi_1)$ and $\psi_{\boldsymbol{k}\, s}(\boldsymbol{r}) \ \chi_{\sigma}(\xi_2)$. Since the wave vectors of these functions are equal to each other, the spins of ξ_1 and ξ_2 of these electrons can be different. The energy of these electrons will be proportional to

$$- J\, w_k^2 \,.$$

The greater the probability of w_k the minimal will be this energy. **This indicates the presence of paired electrons.**

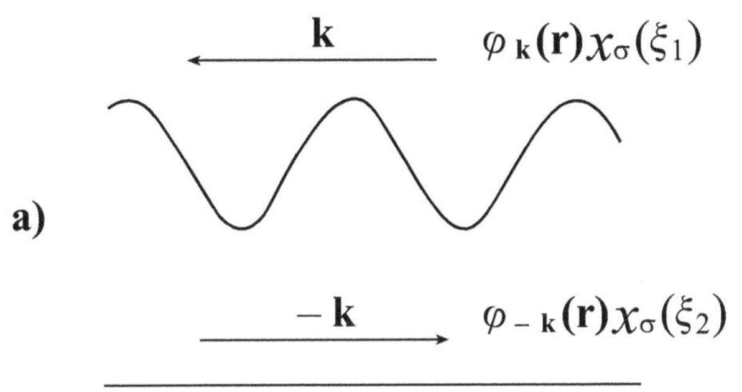

Fig. (3.4.6a). The energy $I\, w_k\, w_{-k}$ of these electrons must be the smallest. This is possible when the probabilities w_k or w_{-k} superconductivity are zero. **This Indicates the Presence of Superconductivity.**

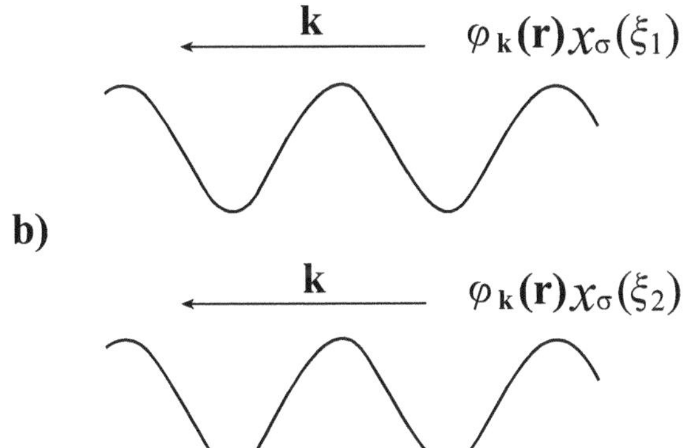

Fig. (3.4.6b). The energy $- J\, w_k^2$ of these electrons will be the smallest, when the probability of w_k is not zero. **This indicates the presence of paired electrons.**

3.4.7. Normalization Condition and Electron Energy

Now, we will calculate the energy that the isotropic and anisotropic distribution electrons exhibit. We will apply the normalization condition formulated as follows:

$$G \sum_k w_k = N.$$

(3.4.26)

Mean-field approximation electron energy takes on the following form:

$$E = G/2 \sum_k (2\,\varepsilon_k w_k + I\, w_k w_{-k} - J\, w_k^2).$$

(3.4.27)

We will approximate dependence of electron kinetic energy ε_k from wave vector ***k*** by applying the formula as follows:

$$\varepsilon_k = \hbar^2 k^2/(2\,m),$$

(3.4.28)

where m is the effective itinerant electron mass. As provided by this formula, any electron kinetic energy shall be counted from the band bottom to be also called "bottom of conduction band" *i.e.* $\varepsilon_{k=0} = 0$.

To simplify calculations, instead of ***k*** summing we will produce integration by ε electron kinetic energy. By applying the dependence (3.4.28)

$$d\varepsilon = \hbar^2 k\, dk/m$$

we will obtain the following symbolic equation:

$$G \sum_k \ldots = A\, N \int_0^\infty \ldots \sqrt{\varepsilon}\, d\varepsilon,$$

(3.4.29)

where

$$A = G\, m \sqrt{2\,m}\, V/(2\,\pi^2 \hbar^3 N).$$

The upper integration limit may be equal to ∞ since the occupational probability of states in which energy ε is specified at the ceiling of the conduction band is actually equal to zero. Now, we will formulate the normalization condition (3.4.26) by the method as follows:

$$A \int_0^{\infty} w(\varepsilon)\ \sqrt{\varepsilon}\ d\varepsilon = 1.$$

(3.4.30)

Knowing the distribution function w = $w()$, energy of electrons (3.4.27) can be obtained from the formula:

$$E = 1/2\ A\,N \int_0^{\infty} [2\,\varepsilon + I\,w_1(\varepsilon) - J\ w_2(\varepsilon)]\,w_2(\varepsilon)\ \sqrt{\varepsilon}\ d\varepsilon\,.$$

(3.4.31)

3.4.9. Electron Energy Calculation at $T = 0$

Consider some of the distribution functions $w_k = w(\varepsilon_k)$, describing the different equilibrium macrostate of the system of conduction electrons at temperature $T =$ 0. The simplest function describing the isotropic distribution of electron wave vector has the form

$$w_k = \begin{cases} 1 \text{ at} & \varepsilon_k < \mu\,, \\ 0 \text{ at} & \varepsilon_k > \mu\,. \end{cases}$$

(3.4.32)

This distribution will be called normal. The normalization condition (3.4.30) and formula (3.4.20) for the electron energy can now be written as:

$$A \int_0^{\mu} \sqrt{\varepsilon}\ d\varepsilon = 1\,, \quad E_0^{(n)} = A\,N \int_0^{\mu} [\ \varepsilon - (J - I\,)/2\ ]\,\sqrt{\varepsilon}\ d\varepsilon\,.$$

These equations lead to the following expressions:

$$\mu = \varepsilon_F\,, \quad E_0^{(n)} = N[\ 3\ \varepsilon_F/5 - (\,J - I\,)/2\ ].$$

(3.4.33)

where the energy $\bar{\varepsilon}_k$ of one electron because of its interaction with other electrons will be

$$\bar{\varepsilon}_k = \begin{cases} \varepsilon_k - J + I & \text{at } \varepsilon_k \leq \mu\,, \\ \varepsilon_k & \text{at } \varepsilon_k > \mu\,. \end{cases}$$

As can be seen from this formula, the energy spectrum of the conduction electrons has a gap width of $J - I$, which is located on the surface of $\varepsilon_k = \mu$.

Let the electrons distributed over the states in accordance with the formula (3.4.34) are isotropic so that the probability filling of the states w_k depends on the kinetic energy of the electrons as follows:

$$w_k = \begin{cases} 1 & \text{at } \varepsilon_k \leq \mu\,, \\ (\varepsilon_k - \mu)/(J - I) & \text{at } \mu < \varepsilon_k < \mu + J - I\,, \\ 0 & \text{at } \varepsilon_k \geq \mu + J - I\,. \end{cases}$$

(3.4.34)

In this case, the normalization condition and the formula for the energy of the electrons take the form

$$A \int_0^{\mu} \sqrt{\varepsilon}\, d\varepsilon + A \int_{\mu}^{\mu + J - I} (\varepsilon - \mu)/(J - I)\, \sqrt{\varepsilon}\, d\varepsilon = 1\,,$$

$$E_o^{(i)} = A\,N \int_0^{\mu} [\,\varepsilon - (J - I)/2\,]\sqrt{\varepsilon}\, d\varepsilon + A\,N \int_{\mu}^{\mu + J - I} \{(\varepsilon^2 - \mu^2)/[2\,(J - I)]\}\sqrt{\varepsilon}\, d\varepsilon\,.$$

Due to the small ratio I/ε_F and J/ε_F :

$$\mu = \varepsilon_F\,[\,1 - (J - I)/(2\,\varepsilon_F) - 5\,(J - I)^2/(48\,\varepsilon_F^2) + \ldots\,]\,,$$

$$E_o^{(i)} = N\,\varepsilon_F\,[\,3/5 - (J - I)/(2\,\varepsilon_F) - 7\,(J - I)^2/(16\,\varepsilon_F^2) + \ldots\,]\,.$$

(3.4.35)

The energy $E_0^{(i)}$ of electrons in the state (3.4.34) is greater than the energy $E_0^{(n)}$ of electrons in the normal state by the amount

$$E_0^{(i)} - E_0^{(n)} = 7\,N\,(J - I)^2\,/(\,16\,\varepsilon_F)\,.$$

(3.4.36)

Suppose that the conduction electrons distributed in Bloch states in accordance with the formula (3.4.32) are anisotropic, the probability of filling is given as:

$$w_k = \begin{cases} 1 \text{ at } \varepsilon_k \le \mu - I\,, \\ 1 \text{ at } k \in S\,,\ k \in S^*, \\ 0 \text{ at } k \in S\,,\ k \ni S^*, \\ 0 \text{ at } \varepsilon_k \ge \mu\,, \end{cases}$$

(3.4.37)

where S is a set of wave vectors for which the inequality,

$$\mu - I < \varepsilon_k < \mu\,,$$

(3.4.38)

S^* is the arbitrary subset of the wave vectors $\boldsymbol{k} \in S$, which contains half the vectors than the set of S. Among the vectors $\boldsymbol{k} \in S^*$ no two vectors have the sum equal tozero. The graph of $w = w(\varepsilon)$ corresponding to such distribution is shown in Fig. (**3.4.7**).

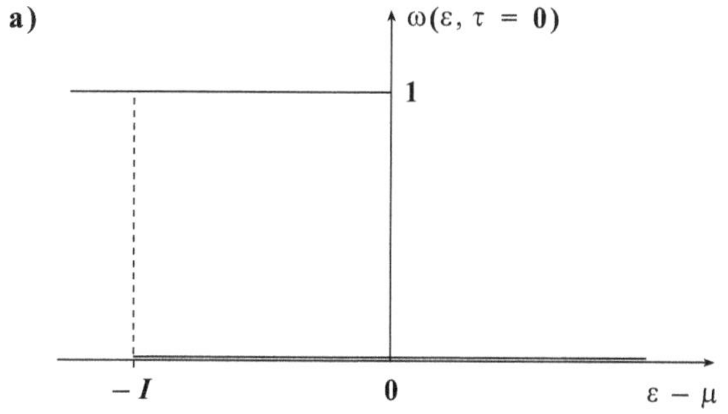

Fig. (3.4.7). One of the functions that describes the equilibrium of the conduction electron energy distribution at $\tau = 0$.

As can be seen from the formulas (3.4.37), the Bloch condition wave vectors $\boldsymbol{k} \in S$ form pairs. Each pair consists of a state filled with an electron with wave vector $\boldsymbol{k} \in S^*$, and the free state with wave vector $-\boldsymbol{k}$.

The substitution of function (3.4.37) to the normalization condition (3.4.30) and formula (4.25) leads to the following equations:

$$A \int_0^{\mu - I} \sqrt{\varepsilon}\, d\varepsilon + 1/2\, A \int_{\mu - I}^{\mu} \sqrt{\varepsilon}\, d\varepsilon = 1 ,$$

$$E_0^{(s)} = A\, N \int_0^{\mu - I} [\, \varepsilon - (J - I)/2\,]\, \sqrt{\varepsilon}\, d\varepsilon + 1/2\, A\, N \int_{\mu - I}^{\mu} (\, \varepsilon - J/2\,)\, \sqrt{\varepsilon}\, d\varepsilon .$$

From these equations, we find the chemical potential energy of the electrons in an anisotropic condition (3.4.37)

$$\mu = \varepsilon_F\, [\, 1 + I/(\, 2\, \varepsilon_F\,) \; - \; I^2/(\, 16\, \varepsilon_F^2) + \ldots\,] ,$$

$$E_0^{(s)} = N\, \varepsilon_F\, [\, 3/5 - (J - I\,)/(2\, \varepsilon_F\,) \; - \; 3\, I^2/(16\, \varepsilon_F^2) + \ldots\,] .$$

(3.4.39)

This energy value is less than (3.4.33) in the normal state of electrons by the amount

$$E_0^{(n)} - E_0^{(s)} = 3\, N\, I^2/(16\, \varepsilon_F) .$$

(3.4.40)

Thus, it is proved that the superconducting state is thermodynamically stable. A subset S^* of wave vectors $k \in S$, which contains half the vectors $\boldsymbol{k}$ than the set S. Moreover, for any two vectors $\boldsymbol{k}_1$ and $\boldsymbol{k}_2$ of S^* condition $\boldsymbol{k}_1 + \boldsymbol{k}_2 \neq 0$. According to (3.4.37) in the k-space, a layer S, defined by the inequalities (3.4.38), in which electrons are distributed over the wave vectors so that two states with wave vectors $\boldsymbol{k}$ and $-\boldsymbol{k}$ are present in this layer, one is free electron and the other one is occupied electron. Such a distribution of the electrons causes the superconducting state of the electron conductivity system in the metal.

3.4.10. Real Distribution Function

Suppose that at any temperature $T < T_c$, electrons are distributed in Bloch states so that the edges of the "pit" in the graph of $\bar{\varepsilon} = \bar{\varepsilon}(\varepsilon)$ are on the same level. Following this principle, we can establish that equilibrium distribution function describing the macro-state conduction electrons of the lowest energy of the system will have the form:

$$w(\varepsilon_k) = \begin{cases} w_0(\varepsilon_k) \text{ at } \varepsilon_k \le \varepsilon_1, \ \varepsilon_k \ge \varepsilon_2\,; \\ w_1(\varepsilon_k) \text{ at } \ k \in S, \ k \in S^*; \\ w_2(\varepsilon_k) \text{ at } \ k \in S, \ k \ni S^*, \end{cases}$$

(3.4.41)

where S is set of wave vectors for which the inequality

$$\varepsilon_1 < \varepsilon_k < \varepsilon_2\,.$$

(3.4.42)

The value of ε_1 is the smallest of the electron kinetic energies ε , with functions $w_1(\varepsilon)$ and $w_2(\varepsilon)$. The value of ε_2 satisfies

$$\bar{\varepsilon}(\varepsilon_1) = \bar{\varepsilon}(\varepsilon_2)\,,$$

(3.4.43)

where by the edges of the "pit" in the graph of $\bar{\varepsilon} = \bar{\varepsilon}(\varepsilon)$ are on the same level. Schedule distribution function $w = w(\varepsilon)$, depending on the kinetic energy of the electron ε , corresponding to the formula (3.4.43), for temperatures $\tau = 0.75$ is shown in Fig. (**3.4.8**).

You must now determine the distribution function. The isotropic distribution function should be unambiguous. This means that one value of energy corresponds to one value of the distribution function. The anisotropic distribution function is two-valued. It defines just two values of the wave vector k and $-k$, which correspond to two values of the distribution function $w_{-k} = w_1(\varepsilon_k)$ and $w_k = w_2(\varepsilon_k)$. But there is is a possibility that these vectors correspond to the values $w_{-k} = w_2(\varepsilon_k)$ and $w_k = w_1(\varepsilon_k)$. Here lies the phenomenon of superconductivity.

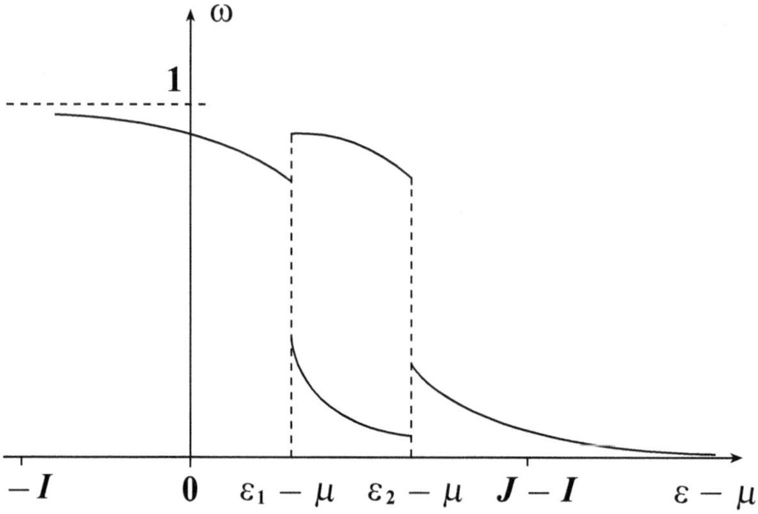

Fig. (3.4.8). Real-valued conduction electron equilibrium distribution function at $\tau = 0.75$. Electron interaction energy values I and J are coupled by the $J = 3I$ relation.

The real equilibrium distribution function is equal to the lowest energy of the electrons. Energy calculation of electrons for different distributions leads to a real equilibrium distribution function. Graphic arts of this function is shown in Figure 4.8, which corresponds to the temperature $\tau = 0.75$.

3.4.11. Electron Mean Energy

Dependence $\bar{\varepsilon} = \bar{\varepsilon}(\varepsilon)$ of the electron mean energy $\bar{\varepsilon}$ against kinetic energy ε can be found by the formula (3.4.4):

$$\bar{\varepsilon}(\varepsilon) = \begin{cases} \varepsilon - (J - I)\, w_0(\varepsilon) & \text{at } \varepsilon \leq \varepsilon_1,\ \varepsilon \geq \varepsilon_2\,; \\ \varepsilon + I\, w_1(\varepsilon) - J\, w_2(\varepsilon) & \text{at } \quad \varepsilon_1 < \varepsilon < \varepsilon_2\,. \end{cases}$$

(3.4.44)

This dependence, as rated at various temperatures τ, is graphically represented in Fig. (**3.4.9**).

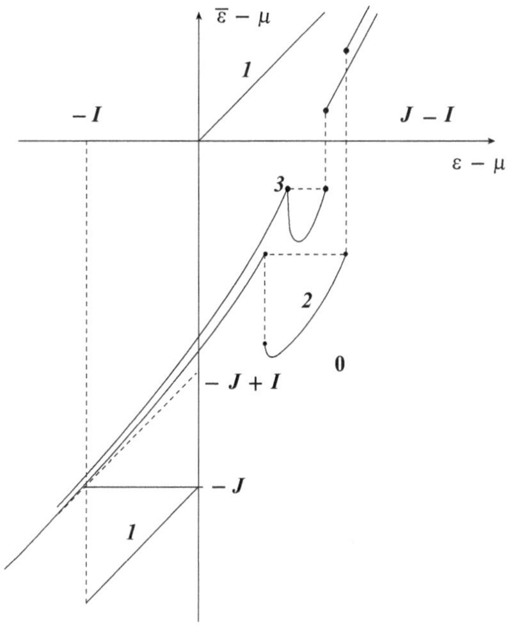

Fig. (3.4.9). Dependence of the mean electron energy $\bar{\varepsilon}$ of the kinetic energy ε at various temperature values τ: 1 – τ = 0; 2 – τ = 0,75; 3 – τ = 0,95.

The graphs have the following characteristics. Each of the curves $\bar{\varepsilon}=\bar{\varepsilon}(\varepsilon)$ at $T<T_c$ has a "pit", which corresponds to the values of the kinetic energy of the ε_1, satisfies (3.4.44). At the right edge of the hole feature discontinuity. This means that the spectrum of electron energies $\bar{\varepsilon}(\varepsilon)$ has a "slot". The slit width Δ is increased from zero to a value of *J*, and the temperature decreases the value of T_c to zero. The width of the pit

$$\delta\varepsilon=\varepsilon_2 -- \varepsilon_1$$

(3.4.45)

is lowered and temperature also increases from zero to a value of *I* at $T = 0$.

3.4.12. Type-I and Type-II Superconductors

For characterizing the type of a superconductor, the following parameter value is involved:

$$\zeta = \sqrt{I/(I+J)}\,.$$

(3.4.46)

We express parameter ζ in terms of parameter f. Now, we can obtain the following formula:

$$\zeta = \sqrt{(1-f)/2}\,.$$

(3.4.47)

This function is graphically represented in Fig. (**3.4.10**). Using known inequalities, we can write the superconductor type condition. The condition

□<1/2

indicates a type-I superconductor. The condition

$$\zeta > 1/\sqrt{2}$$

indicates a type-II superconductor. These conditions have been originally gained by A. A. Abrikosov.

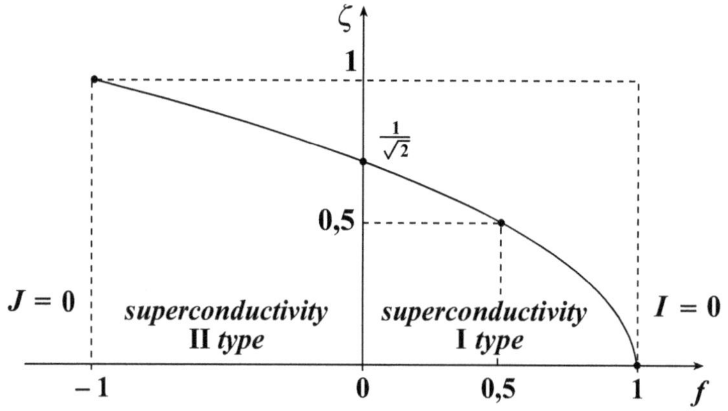

Fig. (3.4.10). Graph of function $\zeta = \zeta(f)$.

Parameter $f = -1$ shows that the value of $J = 0$, that characterizes the gap width Δ, does not produce any pair. Consequently, the coherent length ξ (*i.e.* electron pair interaction length) is equal to zero. As applied in this case, the condition ξ

$< \lambda$ is satisfied where λ is the superconductor magnetic field penetration depth. This condition shows that type-II superconductor is in the range of the f parameter defined values.

3.4.13. Density Matrix

Now, when we find the probability w_k, it is possible in principle to find the density matrix. Using a unitary transformation, we have

$$\rho_{RR'} = (1/N_L) \sum_k w_k \, e^{\,i k (R - R')} \, . \tag{3.4.48}$$

Unfortunately, the calculations carried out at arbitrary temperatures are very complex. The density matrix can be calculated only for the probability at the temperature $T = 0$:

$$\varrho_{RR'} = 1/N_L \, \{ \sum_{k,\ 0 < \varepsilon_k < \mu - I} e^{\,i k (R - R')} + \\ + \sum_{k,\ k_x > 0,\ \mu - I < \varepsilon_k < \mu} e^{\,i k (R - R')} \} \, . \tag{3.4.49}$$

3.4.14. Silsby Effect

We apply an external voltage V to the superconductor. If the conductor was superconducting current j, with increasing voltage, this current will decrease and then returns to zero. This can be explained by the formula given as:

$$j(V) = j(0) - \psi(V) \, ,$$

where $j(0)$ is the initial density of the superconducting current when the voltage was zero, $j(V)$ – the current density under the voltage, $\psi(V)$ is the reduction of the current density under the applied voltage. When the voltage reaches some critical value V_c, the superconducting current vanishes

(Silsby effect):

$$\psi(V_c) = j(0) \, .$$

The dependence of the superconducting current density on the applied voltage is shown in Fig. (**3.4.11**).

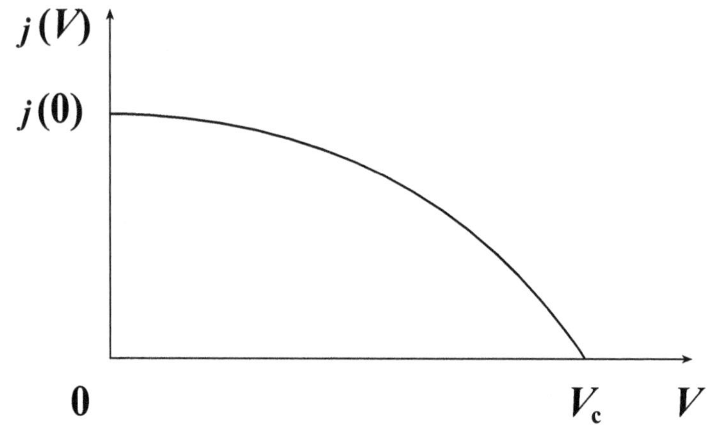

Fig. (3.4.11). The dependence of the superconducting current density on the applied voltage.

We do not know the function $\psi(V)$. Therefore, we consider an approximate solution to this problem. In the theory under consideration, superconductivity is a consequence of the interaction of electrons, whose wave vectors $\boldsymbol{k}$ and $-\boldsymbol{k}$ belong to the spherical layer S. The radius of this layer is equal to the Fermi energy k_{F}. In this layer, superconductivity sometimes occurs.

Fig. (**3.4.12**) shows the area A, which belongs to the conduction band. This zone is obtained by applying voltage. The distance OO' is proportional to this voltage V. When the voltage is zero, the region A contains a layer S. Once the voltage has been applied, a superconducting current is generated in the S layer. Vector $\boldsymbol{j}$, the density of this current, is directed in the direction opposite to the applied voltage V. As you increase the voltage V, layer S is shifted tothe border region S. This reduces the superconducting current. Inotropnoy, the distribution of electrons over the wave vectors, and superconductivity do not increase when the offset $OO'$of area A is so large that S which is anisotropy is outside the area A.

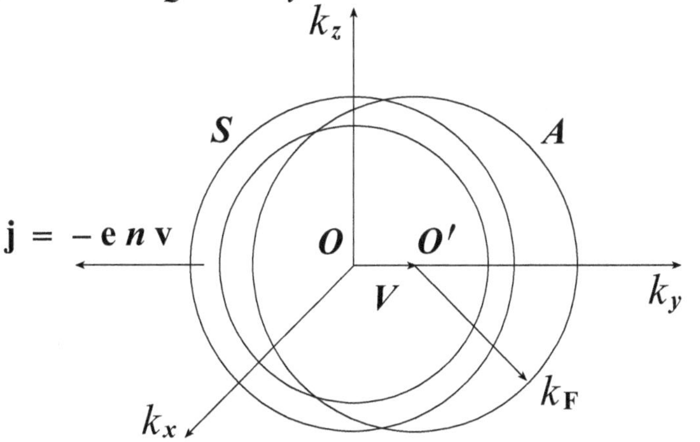

Fig. (3.4.12). Displacement of the Fermi sphere under the action of an electric field.

3.4.15. Conclusion

The model for electrons in a metal, described in this paper, can be assumed as a basis of an alternative theory of superconductivity. This model significantly differs from those which have been applied in the contemporary theory of superconductivity. Superconductivity, described herein, is caused by repulsion of wave vector k and $-k$ electrons, but electron pairs and energy gap in the spectrum – by attraction between the electrons of equal wave vectors. Electrons are statistically described in terms of the density matrix formalism, represented in a simple way by specific formulae and physical content. The problem examined demonstrates the advantages of the density matrix method.

3.5. SUPERCONDUCTIVITY DISAPPEARS

3.5.1. The Mean-field Approximation for $I = 0$

As before, we start from equation (3.3.42) in the mean-field approximation

$$\ln\left[(1 - w_k)/w_k\right] = \beta\left(\varepsilon_k + I\,w_{-k} - J\,w_k - \mu\right), \quad \textbf{(3.5.1)}$$

Only now put

$$I = 0 . \quad \textbf{(3.5.2)}$$

In this case, the interaction energy of two electrons with the wave vectors k ⟦and k⟧ ^ 'will be equal to

$$\varepsilon_{kk'} = - J\,\delta_{k-k'} \,.$$

(3.5.3)

This formula gives the average energy of one electron in the form

$$\bar{\varepsilon}_k = \varepsilon_k - J\,w_k ,$$

(3.5.4)

Substituting equation (3.5.2) into equation (3.5.1) leads to

$$\ln\,[\,(1 - w_k)/w_k] = \beta\,(\,\varepsilon_k - J\,w_k - \mu\,) ,$$

(3.5.5)

The solution of equation (5.4) can be written as:

$$w_k = w(x_k \,, \tau) ,$$

(3.5.6)

where

$$x_k = (\,\varepsilon_k - \mu\,)/J \,, \qquad \tau = 4\,\vartheta/J.$$

(3.5.7)

The dependence of

$$w = w(x,\, \tau)$$

(3.5.8)

is given by:

$$x = w + (\tau/4) \ln [\,(1 - w)/w\,].$$

(3.5.9)

Plots of function (3.5.8) for different values of the temperature τ are shown in Fig. **(3.5.1)**.

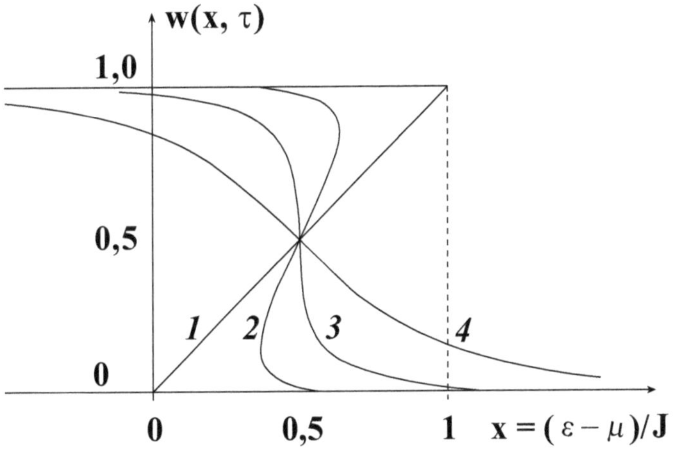

Fig. (3.5.1). Electron energy distribution function at various temperature values τ: $1 - \tau = 0$; $2 - \tau = 0.5$; $3 - \tau = 1$; $4 - \tau = 2$.

At temperatures T above critical point:

$$T_c = J/(4\,k_B)$$

(3.5.10)

($\tau \geq 1$) distribution function $w = w(x, \tau)$ has a monotonic decreasing quantity. At a particular range of kinetic electron energies $(\varepsilon_1, \varepsilon_2)$ $T < T_c$ and the distribution function gets three-valued. Naturally, the probability $w(\varepsilon, \tau)$ of filling with electrons may gain one of the three possible values only. Actually, electrons are distributed by states provided that electron energy takes on the least value.

3.5.2. Real Distribution of the Electrons

Now, we consider different electron equilibrium states at $T = 0$. One of the distribution functions, which describes the state of the electron system at $T = 0$ takes the form as follows:

$$w(\varepsilon, 0) = \begin{cases} 1 \text{ at } \varepsilon < \mu + J/2 \,, \\ 0 \text{ at } \varepsilon > \mu + J/2 \,. \end{cases} \tag{3.5.11}$$

Other electron system distribution function gets the state described by the following equation:

$$w(\varepsilon, 0) = \begin{cases} 1 & \text{at} \quad \varepsilon < \mu \,, \\ (\varepsilon - \mu)/J & \text{at} \quad \mu < \varepsilon < \mu + J \,, \\ \mathbf{0} & \text{at} \quad \varepsilon > \mu + J \,. \end{cases} \tag{3.5.12}$$

Energy E_0 of electrons in the state (3.5.11) is less than energy E_1 of electrons in the state (3.5.12) by the following quantity:

$$E_1 - E_0 = 7\, N J^2/(16\, \varepsilon_F)\,, \tag{3.5.13}$$

where N is the number of electrons, and ε_F is Fermi energy. Thus, state (3.5.11) is a basic one of the electron systems, *i.e.* the least energy state.

At temperatures $T < T_c$, the state of the electron system which matches the least energy E is described by the distribution function shown in Fig. (**3.5.2**). This function gets nonremovable discontinuity at the kinetic energy of electrons as follows:

$$\varepsilon = \mu + J/2\,. \tag{3.5.14}$$

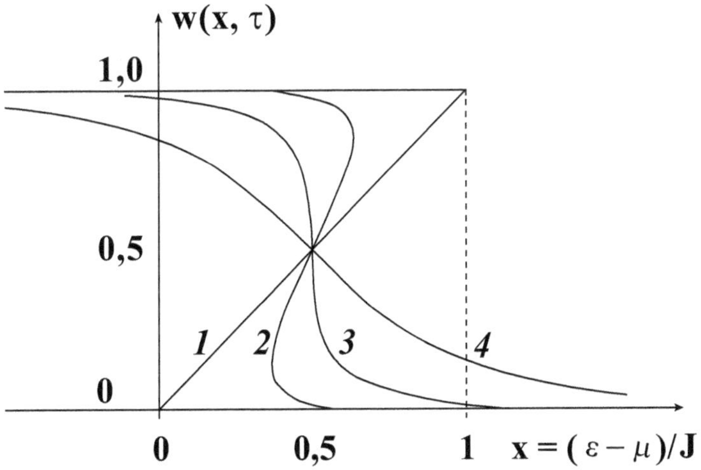

Fig. (3.5.2). Real electron energy distribution function at $\tau = 0.85$.

3.5.3. Energy Gap

Gap width Δw of distribution function can be determined by equation (3.5.9) using $x = 1/2$. On making simple calculations, we obtain the following equation:

$$\ln [(1 + \Delta w)/(1 - \Delta w)] = 2 \Delta w/\tau , \quad \textbf{(3.5.15)}$$

which describes the relation of gap width Δw to temperature. Fig. (**3.5.3**) shows the relation diagram.

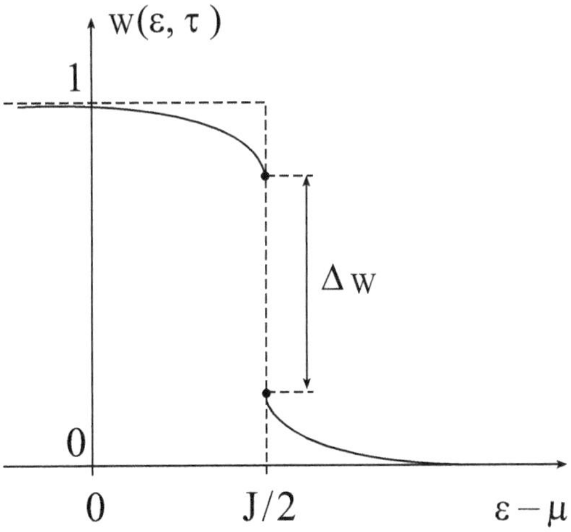

Fig. (3.5.3). Real distribution function gap width Δw to temperature τ.

3.5.4. Medium Electron Energy

The medium electron energy is determined by equation (3.5.4). As provided by this formula, electron medium energy $\bar{\varepsilon}_k$, like distribution function w_k, is a wave $\boldsymbol{k}$ vector composed function, where kinetic electron energy ε_k is applied as an intervening variable (3.5.5). Fig. (**3.5.4**) shows the relation diagram at $\tau = 0.85$.

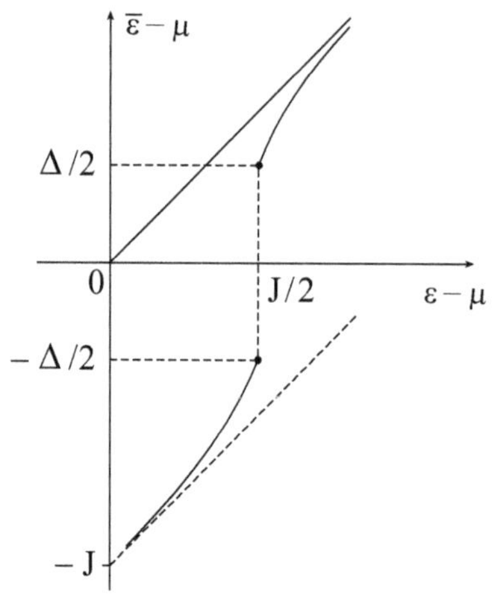

Fig. (3.5.4). Relation of medium electron energy $\bar{\varepsilon}$ to its kinetic energy ε at temperature $\tau = 0.85$.

As evident from Fig. (**3.5.4**), the electron energy spectrum has a "discontinuity" width $2\,\Delta$ which depends on gap width Δw of the distribution function and determined by the ratio as follows:

$$2\,\Delta = J\,\Delta w. \tag{3.5.16}$$

At temperature $T = 0$ the energy gap acquires the largest width. As the temperature goes up, gap width monotonically decreases. At $T = T_c$, the gap will disappear, since distribution function $w = w(\varepsilon, \tau)$ becomes continuous.

3.6. SUPERCONDUCTIVITY TYPE-II

3.6.1. Mean-field Approximation for $J = 0$

As discussed earlier, we start from equation (3.3.42) in the mean-field approximation

$$\ln\left[(1 - w_k)/w_k\right] = \beta(\varepsilon_k + I\, w_{-k} - J\, w_k - \mu)\,.$$

(3.6.1)

Only superconductivity will be presented here

$$J = 0\,.$$

(3.6.2)

In this case, the interaction energy of two electrons with wave vectors k and k ′ will be equal to

$$\varepsilon_{kk'} = I\,\delta_{k+k'}\,.$$

(3.6.3)

This formula gives the average energy of one electron in the form

$$\bar{\varepsilon}_k = \varepsilon_k + I\, w_{-k}\,,$$

(3.6.4)

Substituting in equation (3.6.1) условие $J = 0$. In this case, the equation is represented as

$$\ln\left[(1 - w_k)/w_k\right] = \beta(\varepsilon_k + I\, w_{-k} - \mu)\,,$$

(3.6.5)

where, w_k is the probability that the state described by a wave function $\psi_{ks}(\boldsymbol{r})$ $\chi_\sigma(\xi)$, occupied one of the electron systems in equilibrium with the wave vector $\boldsymbol{k}$.

Replace vector $\boldsymbol{k}$ with vector $-\boldsymbol{k}$. and consider that kinetic energy is the isotropic function, *i.e.*

$$-\varepsilon_k = \varepsilon_k\,,$$

we will formulate the following equation:

$$\ln [(1 - w_{-k})/w_{-k}] = \beta (\varepsilon_k + I w_k - \mu) .$$

(3.6.6)

Thus, the equation (6.5) and (6.6) containing two function values w_k and w_{-k} is produced. At the same time, it is clear enough that the probability w_k is the composite ***k*** vector function, where electron kinetic energy is applied as an intervening variable ε_k:

$$w_k = w(\varepsilon_k).$$

Equations (3.6.5) and (6.6) have two types of solutions. The first is the isotropic function distribution, that for any wave vector values satisfy

$$w_{-k} = w_k .$$

(3.6.7)

The isotropic distribution function is a solution of

$$\ln [(1 - w_k)/w_k] = \beta (\varepsilon_k + I w_k - \mu) .$$

(3.6.8)

Some anisotropic distribution functions fall out of formula (3.6.5) when specific wave vector values are applied:

$$w_{-k} \neq w_k .$$

(3.6.9)

Such kind of electron state distribution anisotropy may occur even when no external field is available. While applying forms:

$$w_{-k} = w_1(\varepsilon_k) , \qquad w_k = w_2(\varepsilon_k) .$$

(3.6.10)

We may formulate Equations (3.6.5) and (6.6) by the method as follows:

$$\begin{cases} \ln[(1 - w_1)/w_1] = (4/\tau)(\epsilon + I\, w_2), \\ \ln[(1 - w_2)/w_2] = (4/\tau)(\epsilon + I\, w_1). \end{cases}$$

(3.6.11)

where

$$\epsilon = (\varepsilon - \mu)/I, \qquad \tau = 4\, \epsilon/I.$$

(3.6.12)

The following functions remain unknown in the combined Equations (3.6.11):

$$w_1 = w_1(\varepsilon) \text{ and } w_2 = w_2(\varepsilon).$$

(3.6.13)

If electrons have isotropic wave vector distribution, it is necessary to insert

$$w_1 = w_2 = w_0$$

in the combined Equation (3.6.8).

$$w_0 = w_0(\varepsilon)$$

In this case, the equation gained may be formulated by the method as follows:

$$\epsilon = (\tau/4) \ln[(1 - w_0)/w_0] - w_0.$$

(3.6.14)

This equation states specific dependence of $w_0 = w_0(\varepsilon)$ with various temperature values graphically represented in Fig. (**3.6.1**) in the form of monotonically decreasing curves.

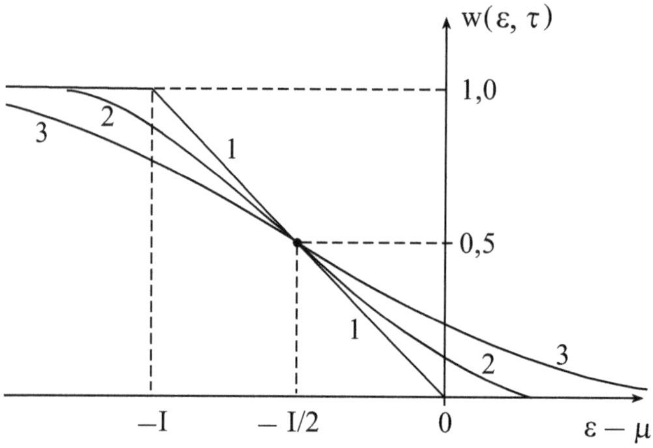

Fig. (3.6.1). Isotropic function of distribution of conductivity electron energy at various temperature values: $1-\tau=0$; $2-\tau=0.25$; $3-\tau=0.8$.

If electrons have anisotropic wave vector distribution, probabilities w_1 and w_2 in the combined Equations (6.11) shall be considered as various functions ϵ:

$$w_1 = w_1(\epsilon) \text{ and } w_2 = w_2(\epsilon)$$

subject to energy. To determine these dependencies, we will introduce new variables d and s applying the relations as follows:

$$w_2 - w_1 = d\,, \qquad w_1 + w_2 = 1 + s\,.$$

(3.6.15)

Without the loss of generality, we will assume that non-negative difference d of two distribution function values w_{1} and w_2 is $d > 0$. At the same time, d remains equal to $d \in [0, 1]$. Value s may possess the values within the range of -1 to 1: $s \in [-1, 1]$.

We will determine the equalities (3.6.15) regarding the probabilities w_1 and w_2:

$$w_1 = \frac{(1+s-d)}{2}, \qquad w_2 = (1+s+d)/2\,.$$

(3.6.16)

We will transform the combined Equations (3.6.11) by applying the formulas (3.6.16). We will firstly subtract the specific equation from one of the combined equations and then we will add the equations. As a result, we will obtain the following combined equations:

$$[\,(1+d)^2 - s^2\,]/[\,(1-d)^2 - s^2\,] = \mathrm{e}^{4d/\tau},$$

$$\epsilon = (\tau/8)\ \ln\{\,[\,(1-s)^2 - d^2\,]/[\,(1+s)^2 - d^2\,]\,\} - (1+s)/2\,.$$

(3.6.17)

The first equation of the combined ones may easily be resolved against *s*:

$$s(d) = \pm\sqrt{[\,(1-d)^2\mathrm{e}^{4d/\tau} - (1+d)^2]/[\,\mathrm{e}^{4d/\tau} - 1\,]}\,.$$

(3.6.18)

According to the relations (3.6.16), probabilities w_1 and w_2 may also be considered as *d* functions:

$$w_1 = w_1(d), \qquad w_2 = w_2(d)\,.$$

With the second equation of the combined ones (3.6.17) applied, we may express ϵ electron energy in terms of parameter *d*. Using the dependences produced specific graphs of functions

$$\mathrm{w}_1 = \mathrm{w}_1(\epsilon), \qquad \mathrm{w}_2 = \mathrm{w}_2(\epsilon)$$

may easily be plotted for various temperature values. For the plotted curve, see (Fig. **3.6.2**).

The pattern of distribution of electrons by their states depends on their relation between metal temperature *T* and critical temperature:

$$T_\mathrm{c} = I/(4\,k_\mathrm{B})\,.$$

(3.6.19)

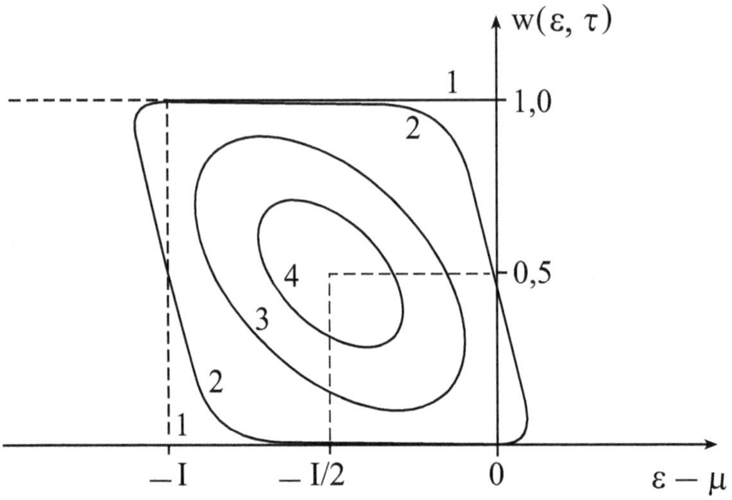

Fig. (3.6.2). Anisotropic function of the distribution of conductivity electron energy at various temperature values: 1 – τ = 0; 2 – τ = 0,25; 3 – τ = 0,8; 4 – τ = 0,95.

At the temperature of $T \geq T_c$ the distribution function $w = w(\epsilon)$ is single-valued and satisfies the condition (6.7), with respect to all the ϵ energy values. At the temperature of $T < T_c$ the energy is limited by (ϵ_1, ϵ_2) with function $w = w(\epsilon)$ possessing any of three values at every point of the limit, particularly $w_1 < w_o < w_2$. Being out of the aforesaid limit, the distribution function $w = w(\epsilon)$ possesses only a single value $w_o(\epsilon)$. Thus, Equation (6.14) is resolved by applying the function $w_k = w_o(\epsilon_k)$ to describe isotropic wave vector electron distribution.

At $T < T_c$ some kind of anisotropic wave vector electron distribution may occur in the narrow layer S under Fermi surface $\varepsilon_k = \mu$. This kind of distribution is formulated by

$$w_k = w_2(\epsilon_k), \qquad w_{-k} = w_1(\epsilon_k) \tag{3.6.20}$$

subject to $\epsilon \in (\epsilon_1, \epsilon_2)$.

3.6.2. Distribution of Electrons at $T = 0$

At $T = 0$ the isotropic distribution function is formulated as follows:

$$w_k = \begin{cases} 1 & \text{at} \quad \varepsilon_k \leq \mu - I, \\ (\varepsilon_k - \mu)/I & \text{at} \ \mu - I < \varepsilon_k < \mu, \\ 0 & \text{at} \quad \varepsilon_k \geq \mu. \end{cases}$$

(3.6.21)

As for the anisotropic distribution, it is formulated as follows:

$$w_k = 1 \text{ at } \varepsilon_k \leq \mu - I,$$

$$w_k = 1, \ w_{-k} = 0 \text{ or } w_k = 0, \ w_{-k} = 1$$

(3.6.22)

$$\text{at } \mu - I < \varepsilon_k < \mu,$$

$$w_k = 0 \text{ at } \varepsilon_k \geq \mu.$$

(3.6.23)

As provided by the formula (4.20), layer S may be determined under Fermi surface by the inequality (3.6.21), in which the electrons have anisotropic wave vector distribution, *i.e.* one of both $\boldsymbol{k}$ and $-\boldsymbol{k}$ wave vector states in this layer is free and another one is occupied. For the function curves, see Fig. (**3.6.3**).

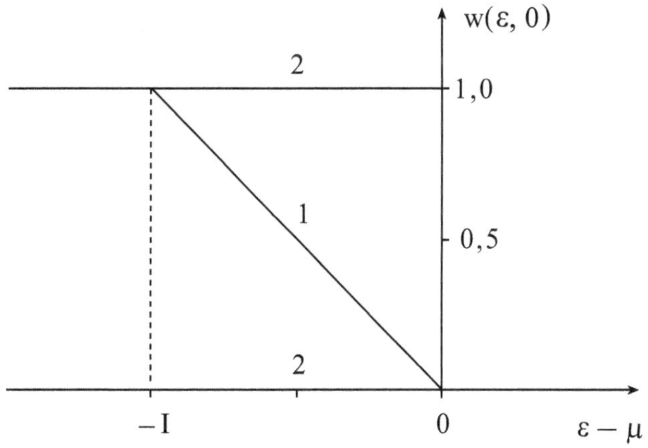

Fig. (3.6.3). Isotropic and anisotropic distribution of conductivity electrons depending on their kinetic energy at temperature τ = 0: 1 – isotropic distribution, 2 — anisotropic distribution.

The electron distribution function obtains its three *S* layer values. And what is the matter it stands for? The answer is in the value of energy, the isotropic or anisotropic distribution electrons exhibit. The electrons gain their steady-state when they have the lowest energy. It can be shown that the anisotropic distribution of the electrons has lower energy than the isotropic. So, in practice, isotropic distribution disappears, where it is replaced by anisotropic.

3.6.3. Superconductivity Energy of States

Now, we will calculate the energy, the isotropic and anisotropic distribution electrons exhibit. We will apply the normalization condition formulated as follows:

$$G \sum_k w_k = N. \tag{3.6.24}$$

Mean-field approximation electron energy takes on the following form:

$$E = G \sum_k \left(\varepsilon_k w_k + I\, w_k\, w_{-k}/2 \right). \tag{3.6.25}$$

Примем приближенные формулы

$$\varepsilon = \hbar^2 k^2/(2\,m), \qquad d\varepsilon = \hbar^2 k\, dk/m.$$

The upper integration limit may be equal to ∞, since the occupational probability of states in which energy ε is specified at the ceiling of the conduction band is equal to zero. Now, we will formulate the normalization condition (6.24) by the method as follows:

$$A \int_0^\infty w(\varepsilon)\, \sqrt{\varepsilon}\, d\varepsilon = 1. \tag{3.6.26}$$

As for the isotropic distribution electron energy, we will formulate the following formula:

$$E^{(i)} = A\,N \int_0^{\infty} (\varepsilon + I\; w(\varepsilon)/2)\; w(\varepsilon) \sqrt{\varepsilon}\; d\varepsilon\,.$$

(3.4.27)

If the isotropic electron distribution function applied at $T = 0$ is formulated according to (3.6.21), the Equations (3.6.26) and (3.6.27) take on the following form:

$$A \int_0^{\mu - I} \sqrt{\varepsilon}\; d\varepsilon + A \int_{\mu - I}^{\mu} (\mu - \varepsilon)/I \sqrt{\varepsilon}\; d\varepsilon = 1\,,$$

$$E_0^{(i)} = A\,N \int_0^{\mu - I} (\varepsilon + I/2) \sqrt{\varepsilon}\; d\varepsilon + A\,N/(2\,I) \int_{\mu - I}^{\mu} (\mu^2 - \varepsilon^2) \sqrt{\varepsilon}\; d\varepsilon\,.$$

Since we apply the small parameter

$$\lambda = I/\varepsilon_F\,,$$

where

$$\varepsilon_F = (3/2\,A)^{2/3}$$

refers to the Fermi energy, we will define that the chemical potential and isotropic wave vector distribution electron energy at $T = 0$ take on the forms as follows:

$$\mu_0 = \varepsilon_F\,(1 + \lambda/2 - \lambda^2/48 + \cdots)\,,$$

$$E_0^{(i)} = N\,\varepsilon_F\,(3/5 + \lambda/2 - \lambda^2/16 + \cdots)\,.$$

(3.4.28)

We will assume that the anisotropic wave vector electron distribution at $T = 0$ is defined by the function as follows:

$$w_{k} = \begin{cases} 1 \text{ at } \varepsilon_{k} \le \mu - I\,, \\ 1 \text{ at } \mu - I < \varepsilon_{k} < \mu\,, \quad k_x > 0\,, \\ 0 \text{ at } \mu - I < \varepsilon_{k} < \mu\,, \quad k_x < 0\,, \\ 0 \text{ at } \varepsilon_{k} \ge \mu\,, \end{cases}$$

(3.6.29)

As provided by the above formula, only one half of the $k_x > 0$ states may be referred to as the occupied ones to occur in layer S above the Fermi surface, in which thickness δk is proportional to the interaction parameter I. In this case, the average velocity of the electrons is maximal: $\bar{v} = v_{\max}$ (see Fig. **3.6.4.**).

Here, the normalization condition gives rise to the following equation:

$$A \int_0^{\mu - I} \sqrt{\varepsilon}\ \mathrm{d}\varepsilon + A/2 \int_{\mu - I}^{\mu} \sqrt{\varepsilon}\ \mathrm{d}\varepsilon = 1 \,,$$

As for electron energy, it may be calculated by the formula as follows:

$$E_0^{(s)} = A\,N \int_0^{\mu - I} (\varepsilon + I/2)\sqrt{\varepsilon}\ \mathrm{d}\varepsilon + A\,N/2 \int_{\mu - I}^{\mu} \varepsilon \sqrt{\varepsilon}\ \mathrm{d}\varepsilon \,.$$

As provided by the above calculation, the following formulation is obtained:

$$\mu_0 = \varepsilon_F \left(1 + \lambda/2 - \lambda^2/16 + \cdots \right),$$

$$E_0^{(s)} = N\,\varepsilon_F \left(3/5 + \lambda/2 - 3\,\lambda^2/16 + \cdots \right).$$

(3.6.30)

If occupied and free state pairs that match specific wave vectors $\boldsymbol{k}$ and $-\boldsymbol{k}$ will be distributed within layer S by any other way, the chemical potential and electron energy rating will remain the same. The difference in electron energy values (3.6.28) and (3.6.30) will be formulated by the equation as follows:

$$E_0^{(i)} - E_0^{(s)} \cong N\,I^2/(8\,\varepsilon_F) > 0 \,.$$

(3.6.31)

Thus, we get to the conclusion that the state of itinerant electrons described by the anisotropic distribution function is the primary one, *i.e.* the electron system specified in this condition is of the lowest energy. Considering the aforesaid about the anisotropic electron energy distribution, we have plotted the pattern of superconductive state, as shown in Figs. (**3.6.4** and **3.6.5**).

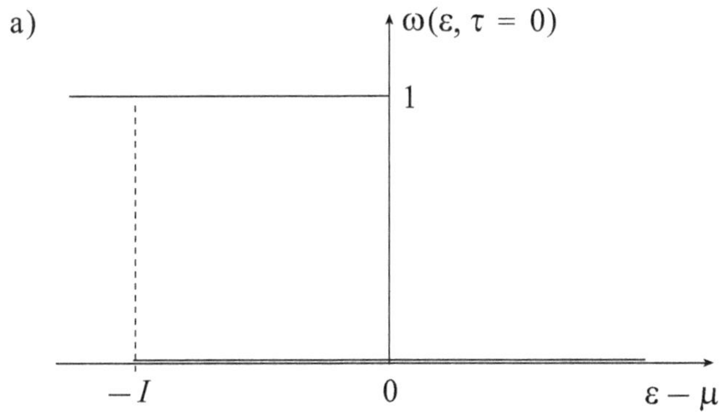

Fig. (3.6.4). Real anisotropic distribution of energy ε conductivity electrons subject to the lowest energy E at the temperature of $\tau = 0$.

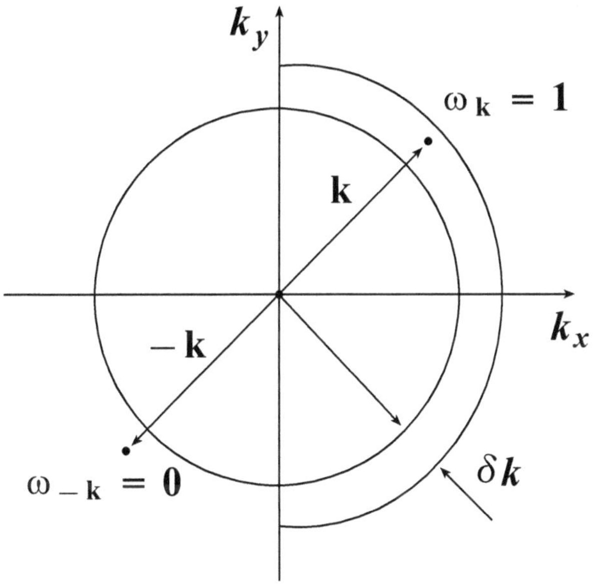

Fig. (3.6.5). Anisotropic distribution function at $T = 0$.

Considering the aforesaid about the anisotropic electron energy distribution, we will plot the pattern of superconductive state, as shown in Fig. (**3.6.6**).

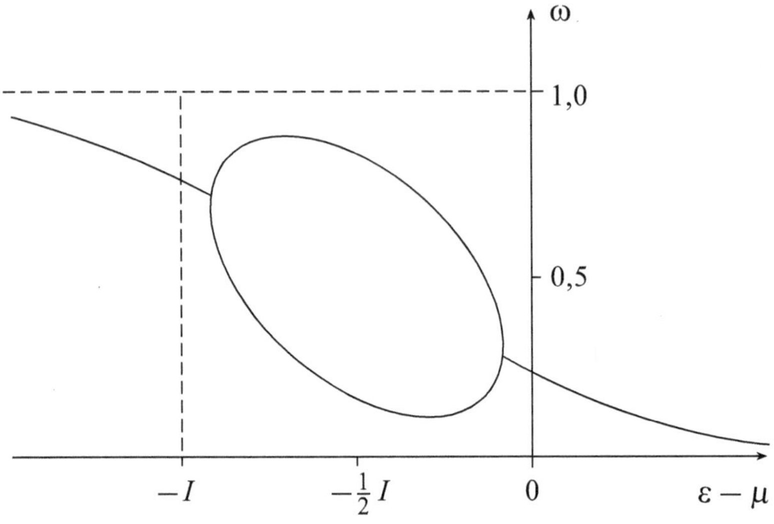

Fig. (3.6.6). Real anisotropic distribution of energy ε conductivity electrons subject to the lowest energy E at the temperature of $\tau = 0$.

3.6.4. Order Parameter

At $T < T_c$ in a narrow layer S under the Fermi surface $\varepsilon_k = \mu$ is the possible anisotropic electron distribution over the wave vectors. This distribution is described by the functions

$$w_k = w_2(\epsilon_k) \ \text{ and } \ w_{-k} = w_1(\epsilon_k)\,,$$

when $\epsilon \in (\epsilon_1, \epsilon_2)$. Difference $d = w_2 - w_1$ of two anisotropic distribution function has the largest value d_{max} at $\epsilon = 0{,}5$. This $w_0 = 0{,}5$ and $s = 0$. We will determine difference d_{max} from temperature τ by applying $s = 0$ in the first Equation (3.6.18):

$$2\, d_{max}/\tau = \ln\left[\,(1 + d_{max})/(1 - d_{max})\,\right]. \qquad \textbf{(3.6.32)}$$

For the dependence curve, see (Fig. **3.6.7**).

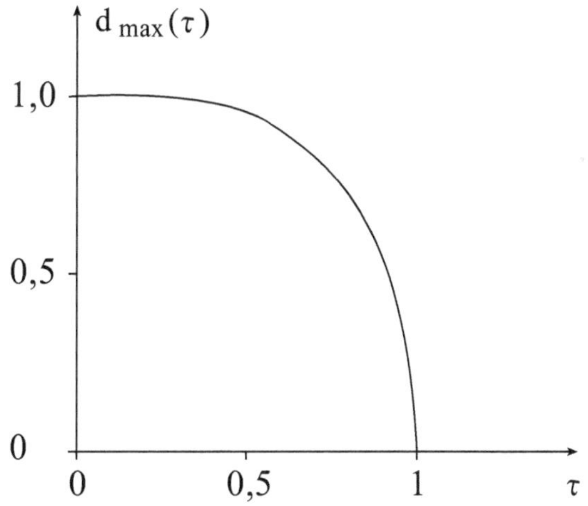

Fig. (3.6.7). Electron distribution anisotropy parameter $d_{\max}$, as τ temperature function.

3.6.5. Mean Energy Dependence of Single Electron

The dependence of the mean electron energy $\bar{\varepsilon}_{\boldsymbol{k}}$ of its kinetic energy $\varepsilon_{\boldsymbol{k}}$ is defined by the formula

$$\bar{\varepsilon}_{\boldsymbol{k}} = \varepsilon_{\boldsymbol{k}} + I\, w_{-\boldsymbol{k}}\,. \tag{3.6.33}$$

As may be inferred from the above formula, the electron energy with wave vector $\boldsymbol{k}$ depends on whether the $-\ \boldsymbol{k}$ wave vector state is free or occupied. Electron energy $\bar{\varepsilon}$ may be specified by the ε kinetic energy functional form as follows:

$$\bar{\varepsilon} = \varepsilon + I\, w_1(\varepsilon)\,. \tag{3.6.34}$$

For the pattern of this function at various temperatures, (see Fig. **3.6.8**). As can be seen from this graph, there is a potential well on the curve. The depth of the pit increases from zero at $T = T_c$ the value of I at $T = 0$.

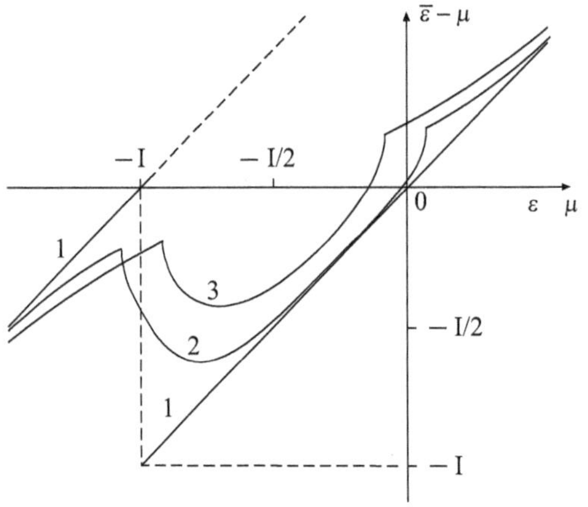

Fig. (3.6.8). Mean electron energy $\bar{\varepsilon}$ dependence of its kinetic energy ε at various temperature values: $1-\tau=0$, $2-\tau=0{,}5$, $3-\tau=0{,}8$.

The parameter $f=-1$ indicates that the superconductor belongs to type-II. We show how, judging by the graph of the dependence of the average electron energy $\bar{\varepsilon}$ on its kinetic energy ε, we can say that the critical magnetic field strength $H_{c1}(\tau)$ and $H_{c2}(\tau)$ are equal, which were discussed in the phase diagram in Fig. **(3.1.4)**. To do this, we construct a dependence $\bar{\varepsilon}=\bar{\varepsilon}(\varepsilon)$ for temperature $\tau=0{,}5$ (see Fig. **3.6.9**). On this curve, there is a potential well, where there's a steady anisotropic state. The right end of the pit corresponds to the kinetic energy ε_1. The critical field $H_{c1}(\tau)$ corresponds to this energy. Further, the pit is followed by not quite steady States, which are called mixed. These states end with the kinetic energy ε_2, which corresponds to the critical field $H_{c2}(\tau)$.

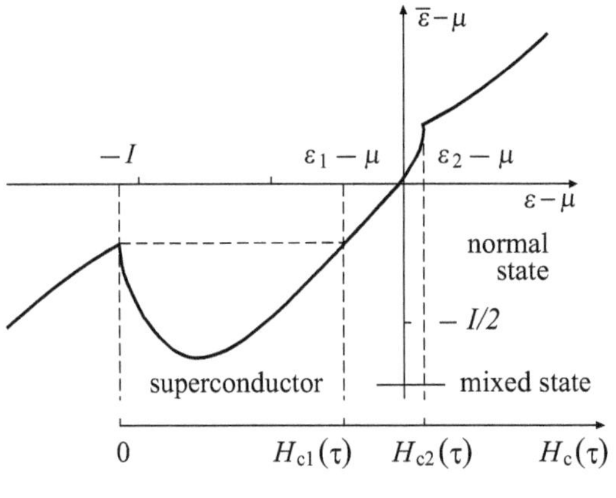

Fig. (3.6.9). The dependence of the average energy electron $\bar{\varepsilon}$ from its kinetic energy ε at temperature $\tau=0{,}5$.

3.6.6. Maximum Superconductivity Electron Velocity at $T = 0$

Now, we will find ordered electron motion velocity in the state described by the distribution function (3.6.29) at the temperature of $T = 0$. For this purpose, we will substitute the wave vector sum specified in the formula (3.6.22) for the following integral:

$$\mathbf{v} = G\,\hbar\,V/(8\,\pi^3\,m\,N)\ \int \mathbf{k}\ w_{\mathbf{k}}\ \mathrm{d}^3 k\,, \tag{3.6.35}$$

If $T = 0$, the inner and outer radii of layer S shall be respectively equal to as follows:

$$k_1 = 1/\hbar\,\sqrt{2\,m\,(\,\mu - I\,)}\,, \qquad k_2 = 1/\hbar\,\sqrt{2\,m\,\mu}\ .$$

When this thickness is

$$\delta k = k_2 - k_1 = I/\hbar\ \sqrt{m/2\,\varepsilon_{\mathrm{F}}}\,. \tag{3.6.36}$$

If the states are filled with electrons in one half of layer S, but another one remains free, the rate of ordered electron motion velocity exhibits its maximum value

$$v_{\max} = 3\,I/(\,4\sqrt{\,2\,m\,\varepsilon_{\mathrm{F}}}\,)\,. \tag{3.6.37}$$

3.7. MAGNETIC FIELD IN SUPERCONDUCTOR

3.7.1. Wave Function

Let us find the wave function $\varphi_\alpha(q)$ from the Schrodinger equation:

$$\widehat{H}^{(1)}\ \varphi_\alpha(q) = E^{(1)}\ \varphi_\alpha(q)\,, \tag{3.7.1}$$

where $\widehat{H}^{(1)}$ is a single-particle Hamiltonian that describes the motion of one electron less interaction with other electrons. As for us, we are interested in the motion of a valence electron in the metal.

Since an electron has a spin, the Hamiltonian $\widehat{H}^{(1)}$ in a magnetic field will be formulated as:

$$\widehat{H}^{(1)} = -\, \hbar^2/(2\, m)\, \nabla^2 + U(\boldsymbol{r}) - \mu\, H\,, \tag{3.7.2}$$

where $\mu = \mu_{\mathrm{B}}\, \xi$ is a spin magnetic moment of an electron; $\mu_{\mathrm{B}} = \mathrm{e}\, \hbar/(2\, m)$ is a Bohr magneton; $\xi = \pm 1/2$ is an electron spin; $U(\boldsymbol{r})$ is the potential electron energy; H is the magnetic-field strength.

Let the wave function is the product of

$$\varphi_\alpha(q) = \phi_s(\boldsymbol{r} - \boldsymbol{R})\, \chi_\sigma(\xi)\,, \tag{3.7.3}$$

where $q = \{\boldsymbol{r}, \xi\}$, $\phi_s(\boldsymbol{r} - \boldsymbol{R})$ is a coordinate function; $\boldsymbol{R}$ is a vector that defines an ion around which an electron moves; $\chi_\sigma(\xi)$ is a spin function, $\alpha = \{\boldsymbol{R}, s, \sigma\}$.

Let us write the Schrodinger equation for the wave function $\varphi_\alpha(q)$:

$$[\, -\hbar^2/(2\, m)\, \nabla^2 + U(\boldsymbol{r}) - \mu_B\, H\, \xi\,]\, \varphi_\alpha(q) = E^{(1)}\, \varphi_\alpha(q)\,, \tag{3.7.4}$$

and on substituting the product (3.7.3), we will obtain

$$\chi_\sigma(\xi)\, [\, -\hbar^2/(2\, m)\, \nabla^2 + U(\boldsymbol{r})\,]\, \phi_s(\boldsymbol{r} - \boldsymbol{R}) \; -$$

$$-\, \phi_s(\boldsymbol{r} - \boldsymbol{R})\, \mu_{\mathrm{B}}\, H\, \xi\, \chi_\sigma(\xi) = E^{(1)}\, \phi_s(\boldsymbol{r} - \boldsymbol{R})\, \chi_\sigma(\xi)\,.$$

If the coordinate function satisfies the equation

$$[-\hbar^2/(2m)\,\nabla^2 + U(\boldsymbol{r})\,]\,\phi_s(\boldsymbol{r}) = E_1^{(1)}\phi_s(\boldsymbol{r})\,,$$

(3.7.5)

then a spin function shall be used for solving the following equation:

$$-\,\mu_{\mathrm{B}}\,H\,\xi\,\chi_\sigma(\xi) = E_2^{(1)}\chi_\sigma(\xi)\,,$$

(3.7.6)

where

$$E^{(1)} = E_1^{(1)} + E_2^{(1)}.$$

(3.7.7)

As follows from the Equation (3.7.6), the proper value $E_2^{(1)}$ will take up:

$$E_2^{(1)} = -\,\mu_{\mathrm{B}}\,H.$$

(3.7.8)

As a result, we will obtain:

$$E^{(1)} = E_1^{(1)} - \mu_{\mathrm{B}}\,H.$$

(3.7.9)

Now, we will accept the functions $\phi_s(\boldsymbol{r})$ as equal to each other:

$$\phi_s(\boldsymbol{r}) = \phi(\boldsymbol{r})\,,$$

(3.7.10)

where $s = 1, 2, \ldots, \mathrm{n_S}$. The spin functions are different only. Therefore, we will have two functions:

$$\varphi_\alpha(q) = \phi(\boldsymbol{r} - \boldsymbol{R})\,\chi_\sigma(\xi)\,,$$

(3.7.11)

Let us consider the energies possessed by an electron when its spin is valued differently. As follows from the formula (3.7.9), when an electron spin is equal to − 1/2, its energy is taken as

$$E^{(1)}(-1/2) = E_1^{(1)} + \mu_B\, H/2\,,$$

and when an electron spin is equal to 1/2, its energy is taken as

$$E^{(1)}(1/2) = E_1^{(1)} - \mu_B\, H/2\,,$$

thus,

$$E^{(1)}(\,1/2\,) < E^{(1)}(-1/2\,)\,.$$

Therefore, the probability $|\,\chi_{1/2}(1/2)\,|^2$, when an electron has the spin taken as 1/2 exceeds the

probability $|\,\chi_{-1/2}(-1/2)\,|^2$, when an electron has the spin taken as − 1/2:

$$|\,\chi_{1/2}(1/2)\,|^2 > |\,\chi_{-1/2}(-1/2)\,|^2\,.$$

3.7.2. The Kinetic Energy of Electrons in the Crystal Lattice

Let us consider that two values of electron energy, as determined by the formula (3.7.9), corresponding to each of the states specified in site ***R***. So, matrix elements of the single-particle Hamiltonian may be expressed as:

$$H_{\alpha\alpha'} = \int \varphi_\alpha^*\, \hat{H}^{(1)}\, \varphi_{\alpha'}(q)\, \mathrm{d}q =$$

$$= \sum_\xi \int \phi_s(\boldsymbol{r} - \boldsymbol{R})\, \chi_\sigma(\xi)\, [\, -\hbar^2/(2\,m)\, \nabla^2 + U(\boldsymbol{r}) - \mu_B\, H\, \xi\,] \cdot$$

$$\cdot\; \phi_s(\boldsymbol{r} - \boldsymbol{R}')\, \chi_{\sigma'}(\xi)\, \mathrm{d}\boldsymbol{r}\, \delta_{ss'}\,.$$

(3.7.12)

As a result, we will obtain:

$$H_{\alpha\alpha'} = \sum_\xi \int \phi_s(\boldsymbol{r} - \boldsymbol{R})\, \chi_\sigma(\xi)\, [\, E_1^{(1)} - \mu_B\, H\, \xi\,]\, \phi_s(\boldsymbol{r} - \boldsymbol{R}')\, \chi_{\sigma'}(\xi)\, \mathrm{d}\boldsymbol{r}\, \delta_{ss'}\,.$$

Since

$$\int \phi_s(\boldsymbol{r} - \boldsymbol{R})\, E_1^{(1)} \phi_s(\boldsymbol{r} - \boldsymbol{R}')\, \mathrm{d}\boldsymbol{r} = \varepsilon_{\boldsymbol{R}-\boldsymbol{R}'} \,, \tag{3.7.13}$$

here $\varepsilon_{\boldsymbol{R}-\boldsymbol{R}'}$ is the particle energy produced when passing from site $\boldsymbol{R}$ to site $\boldsymbol{R}'$. As applies to spin functions, the normalizing condition will take up the following form:

$$\sum_\xi \chi_\sigma(\xi)\, \chi_{\sigma'}(\xi) = \delta_{\sigma\sigma'} \,. \tag{3.7.14}$$

In this case

$$\sum_\xi \int \phi_s(\boldsymbol{r} - \boldsymbol{R})\, \chi_\sigma(\xi)\, [-\mu_\mathrm{B} H \xi\,]\, \phi_s(\boldsymbol{r} - \boldsymbol{R}')\, \chi_{\sigma'}(\xi)\, \mathrm{d}\boldsymbol{r} =$$

$$= \int \phi_s(\boldsymbol{r} - \boldsymbol{R})\, \phi_s(\boldsymbol{r} - \boldsymbol{R}')\, \mathrm{d}\boldsymbol{r} \sum_\xi \int \chi_\sigma(\xi)\, [-\mu_B H \xi\,]\, \chi_{\sigma'}(\xi) =$$

$$= \delta_{\boldsymbol{R}\boldsymbol{R}'} \sum_\xi \chi_\sigma(\xi)\, [-\mu_\mathrm{B} H \xi\,]\, \chi_{\sigma'}(\xi) \,,$$

where

$$\sum_\xi \chi_\sigma(\xi)\, [-\mu_\mathrm{B} H \xi\,]\, \chi_{\sigma'}(\xi) =$$

$$= (1/2)\, \mu_\mathrm{B} H\, [\, \chi_\sigma(-1/2)\, \chi_{\sigma'}(-1/2) - \chi_\sigma(1/2)\, \chi_{\sigma'}(1/2)\,] \,.$$

Thus, the following formula is obtained:

$$H_{\alpha\alpha'} = \varepsilon_{\boldsymbol{R}-\boldsymbol{R}'}\, \delta_{\sigma\sigma'}\, \delta_{ss'} + (1/2)\, \mu_\mathrm{B} H\, \delta_{\boldsymbol{R}\boldsymbol{R}'}\, \delta_{ss'} \cdot$$

$$\cdot [\, \chi_\sigma(-1/2)\, \chi_{\sigma'}(-1/2) - \chi_\sigma(1/2)\, \chi_{\sigma'}(1/2)\,] \,. \tag{3.7.15}$$

3.7.3. The Magnetic Field-dependent Unitary Transformation

As applies to the coordinate function (3.7.11) the normalizing condition will take up the following form:

$$\sum_{ss'} \int \varphi^* (\boldsymbol{r} - \boldsymbol{R})\, \varphi(\boldsymbol{r} - \boldsymbol{R}')\, \mathrm{d}\boldsymbol{r} = \delta_{\boldsymbol{R}\boldsymbol{R}'}\, \delta_{ss'}.$$

(3.7.16)

When a system of electrons in a metal is at the state of thermodynamic equilibrium, any of the states $\varphi_s(\boldsymbol{r} - \boldsymbol{R})$ in site $\boldsymbol{R}$ may be occupied by an electron of equal probability and such electrons will nearly be uniformly distributed among lattice sites , which is not absolutely true. As follows from the formula (3.7.15), the probabilities, as referred to those intended to occupy any states with various spin orientation, are different from each other. But we will ignore such differences. As a matter of the fact, the density matrix will take the following form:

$$\varrho_{\alpha\alpha'} = \varrho_{\boldsymbol{R}-\boldsymbol{R}'}\, \delta_{ss'}\, \delta_{\sigma\sigma'}\,.$$

(3.7.17)

The probability W_α electron state $\varphi_\alpha(q)$ associated with diagonal elements of the density matrix ratio

$$W_\alpha = \varrho_{\alpha\alpha}\,.$$

In this case

$$W_\alpha \equiv W_{\boldsymbol{R}s\sigma} = W_{\boldsymbol{R}} = \text{constant}\,,$$

where $W_{\boldsymbol{R}}$ is the probability of occupation of one state in the site $\boldsymbol{R}$.

Suppose that the number of electron states localized at the site $\boldsymbol{R}$, which describes the functions of the Vanier $\varphi_s(\boldsymbol{r} - \boldsymbol{R})$, of course: $s = 1, 2, \ldots, n_s$. Thus, since $\alpha = \{\boldsymbol{R}, s, \sigma\}$, normalization condition

$$\sum_\alpha W_\alpha = N$$

it leads to the formula

$$2\, n_s\, N_L\, W_{\boldsymbol{R}} = N\,,$$

(3.7.18)

where N_L is the number of sites in the crystal lattice, $G = 2\,n_s$ is the number of valence states in a single node.

Imagine the density matrix $\varrho_{\boldsymbol{RR}'}$ in the form of a complex Fourier series:

$$\varrho_{\boldsymbol{RR}'} = (1/N_L) \sum_{\boldsymbol{k}} w_{\boldsymbol{k}}\, e^{\,\mathrm{i}\,\boldsymbol{k}\,(\boldsymbol{R}-\boldsymbol{R}')},$$

(3.7.19)

where the summation is over the wave vectors $\boldsymbol{k}$, belonging to the first Brillouin zone; $w_{\boldsymbol{k}}$ is electron distribution function over the wave vectors satisfying the normalization condition

$$G \sum_{\boldsymbol{k}} w_{\boldsymbol{k}} = N,$$

(3.7.20)

From equation (3.7.19) that the transformation of density matrices $\varrho_{\alpha\alpha'}$ in diagonal form is carried out by means of a unitary matrix

$$U_{\alpha\kappa} \equiv U_{\boldsymbol{R}\,s\,\sigma,\ \boldsymbol{k}\,s'\sigma'} = (\,1/\sqrt{N_L}\,)\, e^{\,\mathrm{i}\,\boldsymbol{k}\boldsymbol{R}}\, \delta_{ss'}\, \delta_{\sigma\sigma'}\,,$$

(3.7.21)

where $\kappa = \{\,k,\ s',\ \sigma'\,\}$. Thus, the density matrix $\varrho_{\alpha\alpha'}$ goes into the matrix

$$\varrho_{\kappa\kappa'} = w_{\boldsymbol{k}}\, \delta_{\boldsymbol{kk}'}\, \delta_{ss'}\, \delta_{\sigma\sigma'}\,,$$

(3.7.22)

where $\kappa = \{\,k,\ s,\ \sigma\,\}$. The probability

$$W_\kappa = \varrho_{\kappa\kappa}$$

that the state κ, *i.e.* a state described by a wave function $\psi_{\boldsymbol{k}s}(\boldsymbol{r})\,\chi_\sigma(\xi)$, busy, will

$$W_\kappa = w_{\boldsymbol{k}}\,.$$

(3.7.23)

3.7.4. Electrons Energy in Wave Vector Space

Let us substitute the formula (3.7.7) in the following expression for noninteracting electron energy:

$$E^{(1)} = \sum_{\alpha} \sum_{\alpha'} H_{\alpha\alpha'} \varrho_{\alpha'\alpha} =$$

$$= \sum_{\alpha} \sum_{\alpha'} \{ \varepsilon_{\boldsymbol{R}-\boldsymbol{R}'} \, \delta_{\sigma\sigma'} \, \delta_{ss'} + \mu_{\mathrm{B}}/2 \, H \, \delta_{\boldsymbol{R}\boldsymbol{R}'} \, [\, \chi_{\sigma}(-1/2) \, \chi_{\sigma'}(-1/2) - \chi_{\sigma}(1/2) \, \chi_{\sigma'}(1/2) \,] \cdot$$

$$\cdot \, \delta_{ss'} \, \} \, (1/N_L) \sum_{\boldsymbol{k}} w_{\boldsymbol{k}} \, e^{\, i \boldsymbol{k} (\boldsymbol{R}-\boldsymbol{R}')} \, \delta_{ss'} \, \delta_{\sigma\sigma'} \, .$$

Upon simple transformation, we can obtain the formula shown below:

$$E^{(1)} = G \sum_{\boldsymbol{k}} \varepsilon_{\boldsymbol{k}} \, w_{\boldsymbol{k}} - \; G/4 \, \mu_{\mathrm{B}} \, H \sum_{\boldsymbol{k}} w_{\boldsymbol{k}} \sum_{\sigma} [\, \chi_{\sigma}^{2}(1/2) - \chi_{\sigma}^{2}(-1/2)] \, . \tag{3.7.24}$$

Now, we write:

$$\Lambda = \mu_{\mathrm{B}}/4 \, H f(\tau) \, , \tag{3.7.25}$$

where

$$f(\tau) = \sum_{\sigma} [\, \chi_{\sigma}^{2}(1/2) - \chi_{\sigma}^{2}(-1/2)] \, . \tag{3.7.26}$$

Let us write the energy $E^{(1)}$ as follows:

$$E^{(1)} = G \sum_{\boldsymbol{k}} \;\; (\, \varepsilon_{\boldsymbol{k}} - \Lambda \,) \, w_{\boldsymbol{k}} \, . \tag{3.7.27}$$

The interacting electron energy will take up the below form (3.3.30):

$$E_{\mathrm{int}} = 1/2 \sum_{\boldsymbol{k}} \sum_{\boldsymbol{k}'} \varepsilon_{\boldsymbol{k}\boldsymbol{k}'} \, w_{\boldsymbol{k}} \, w_{\boldsymbol{k}'} \, , \tag{3.7.28}$$

where an appropriate electron interaction energy will be formulated as:

$$\varepsilon_{kk'} = I\,\delta_{k+k'} - J\,\delta_{k-k'}\,.$$

On summing up the energies (3.7.27) and (3.7.28), we will obtain the following electron energy expression:

$$E = G\sum_k [\,(\varepsilon_k - \Lambda)\,w_k + 1/2\,(I\,w_k\,w_{-k} - J\,w_k^2\,]\,.$$

(3.7.29)

3.7.5. Equation for Electron Wave Vector Distribution Function

On minimizing the thermodynamic potential Ω, as referred to the energy (7.29), we obtain the following nonlinear equation to be applied for finding the distribution function w_k of wave-vector conduction electrons:

$$\ln[\,(1 - w_k)/w_k\,] = \beta\,(\,\varepsilon_k - \Lambda + I\,w_{-k} - J\,w_k - \mu\,).$$

(3.7.30)

Let us consider

$$\varepsilon_k' = \varepsilon_k - \Lambda\,.$$

(3.7.31)

than the equation (3.7.29) will take up its previous solution:

$$\ln[\,(1 - w_k)/w_k\,] = \beta\,(\,\varepsilon_k' + I\,w_{-k} - J\,w_k - \mu\,)\,.$$

(3.7.32)

Let us consider, for example,

$$\varepsilon_k' = \mu.$$

As follows from (7.30), we will obtain the following expression

$$\varepsilon_k = \Lambda + \mu.$$

This means I and J that a curve of the electron wave-vector distribution function is displaced on its right by value Λ as compared with the graphical representation without a magnetic field. As follows from the equation (7.30), the critical temperature is equal to:

$$T_c = (I + J)/(4\,k_B)\ .$$

Calculations show that the equilibrium distribution function in a certain region of the wave vectors of ambiguity. To highlight the real distribution function, you need to find the minimum energy of the electrons.

The ratio (3.7.31) indicates that the solution of the equation (3.7.30) is shifted towards larger values of ε_k when the magnetic field is turned on. In this case, equation (3.7.32) says that the structure of this solution does not change.

Consider how the electron distribution function $w(\varepsilon, \tau)$ will behave at $T = 0$ when an external magnetic field acts on the metal. It reflects a stable anisotropic distribution of electrons in the metal, *i.e.* it indicates the existence of superconductivity. While the magnetic field is absent, the graph of the function has the form shown in Fig. **(3.4.5)** in section **3.4** or in Fig. **(3.4.7)** in section **3.6.**

When the magnetic field increases to the value

$$\Lambda = I\,,$$

then the superconductivity becomes zero (see Fig. **3.7.1**).

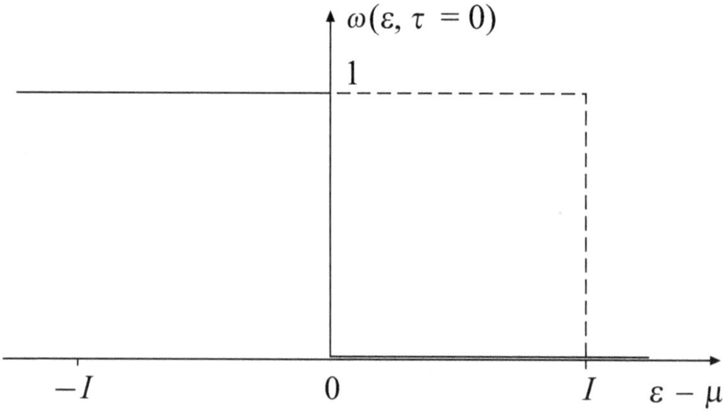

Fig. (3.7.1). Real distribution of conduction electrons by energy at temperature $\tau = 0$. External magnetic field enabled, when superconductivity disappears.

The reason why superconductivity disappears at $\Lambda = I$ (see Fig. **3.7.1**) is because the lower edge of the region where the electrons are distributed anisotropically corresponds to the Fermi energy at $\tau = 0$: $\varepsilon = \mu$. But there are no electrons in this region that can be distributed in any way.

The superconductivity disappears when the distribution function is taken as:

$$w_{\boldsymbol{k}} = \begin{cases} 1 \text{ at } \varepsilon_{\boldsymbol{k}} \leq \mu\,, \\ 0 \text{ at } \quad \varepsilon_{\boldsymbol{k}} > \mu\,. \end{cases}$$

(3.7.33)

As a result, we can obtain the following normalizing condition:

$$A \int_0^{\mu} \sqrt{\varepsilon}\, d\varepsilon = 1\,,$$

(3.7.34)

from which it follows that

$$\mu = \varepsilon_{\mathrm{F}}\,,$$

(3.7.35)

where ε_{F} is a Fermi energy.

Any valence electron energy specified at $J = 3I$ may be calculated as:

$$E^{(m)} = A\,N \int_0^{\mu} (\,\varepsilon - 2\,I\,) \sqrt{\varepsilon}\, d\varepsilon.$$

(3.7.36)

On calculating the above integral we will obtain:

$$E^{(m)} = A\,N\,[\,(2/5)\,\varepsilon^2\,\sqrt{\varepsilon} \;-\; (4/3)\,I\,\varepsilon\,\sqrt{\varepsilon}\,]\,\Big|_0^{\mu}.$$

As a result, the following formula is derived:

$$E^{(m)} = N\,\varepsilon_{\mathrm{F}}\,[\,(3/5) - (2\,I/\varepsilon_{\mathrm{F}})\,]\,. \tag{3.7.37}$$

But such superconducting state may in any way survive and the distribution function takes up the form as:

$$w_k = \begin{cases} 1 \text{ at } \varepsilon_k \le \mu\,, \\ 1 \text{ at } \mu < \varepsilon_k < \mu + I\,,\ k_x > 0\,, \\ 0 \text{ at } \mu < \varepsilon_k < \mu + I\,,\ k_x < 0\,, \\ 0 \text{ at } \varepsilon_k \ge \mu + I\,. \end{cases} \tag{3.7.38}$$

According to the normalizing condition, the equation takes up the form as:

$$A \int_0^{\mu} \sqrt{\varepsilon}\,\mathrm{d}\varepsilon + A/2 \int_{\mu}^{\mu+I} \sqrt{\varepsilon}\,\mathrm{d}\varepsilon = 1\,. \tag{3.7.39}$$

The equation gives the following chemical potential:

$$\mu = \varepsilon_{\mathrm{F}}\,[\,1 - (\,I/2\,\varepsilon_F\,) - I^2/(16\,\varepsilon_F^2) + \ldots\,]\,. \tag{3.7.40}$$

An electron energy may be calculated by the formula:

$$E^{(s)} = A\,N \int_0^{\mu} [\,\varepsilon - \Lambda - 1/2\,(\,J - I\,)\,]\,\sqrt{\varepsilon}\,\mathrm{d}\varepsilon +$$

$$+ 1/2\,A\,N \int_{\mu}^{\mu+I} (\,\varepsilon - \Lambda - J/2\,)\,\sqrt{\varepsilon}\,\mathrm{d}\varepsilon\,.$$

Let us suppose that $J = 3\,I$. And the following form will be obtained:

$$E^{(s)} = A\,N \int_0^{\mu} (\,\varepsilon - 2\,\Lambda\,)\,\sqrt{\varepsilon}\,\mathrm{d}\varepsilon + 1/2\,A\,N \int_{\mu}^{\mu+I} [\,\varepsilon - (5\,I/2)\,]\,\sqrt{\varepsilon}\,\mathrm{d}\varepsilon\,.$$

As provided by the above equation, we can obtain:

$$E^{(s)} = N\,\varepsilon_F\,[\,(3/5) - 37\,I/(20\,\varepsilon_F) - 3\,I^2/(16\,\varepsilon_F^2) + \ldots\,]\,.$$

(3.7.41)

Now, we can find the difference in energy:

$$E^{(m)} - E^{(s)} = -\,(3/20)\,N\,I\,[\,1 - 5\,I/(4\,\varepsilon_F) + \ldots\,] < 0\,.$$

(3.7.42)

As we can see, the energy of state $E^{(m)}$ to be induced under the superposition of the superconductivity affect magnetic field occurs to be less than the superconductivity state energy $E^{(s)}$:

$$E^{(m)} < E^{(s)}.$$

(3.7.43)

This means that the superconductivity can vanish under magnetic field effect when its strength, as defined on a surface of a superconductor, is equal to a critical field. This condition meets Meissner – Ochsenfeld effect.

3.7.6. Meissner - Osnfeld Effect

Let us consider that parameter Λ is equal to I. In this case, the distribution pattern is displaced on its right. Now, we can find the critical magnetic-field strength, as $T = 0$, through the use of the formulas (3.7.27) and (3.7.33):

$$1/4\,\mu_B\,H_m(0)\,f(0) = I\,.$$

(3.7.44)

If temperature τ exceeds zero, the critical magnetic-field strength subject to the formula (3.7.25) is formulated as:

$$1/4\,\mu_B\,H_m(\tau)\,f(\tau) = \delta\varepsilon(\tau)\,,$$

(3.7.45)

where $\delta\varepsilon(\tau)$ is a width of the potential well (6.39) on the mean electron energy dependence of its kinetic energy.

With the graphical representation of the mean electron energy for $J = 3\ I$ dependence applied against its kinetic energy shown in Fig. (**3.4.9**), it is possible to plot the critical magnetic-field dependence of temperature. If to compare the plots demonstrated in Figs. (**3.1.3** and **3.4.9**) with the dependence of the critical-field strength considered against that of the theoretical one, it can be seen that they match with each other.

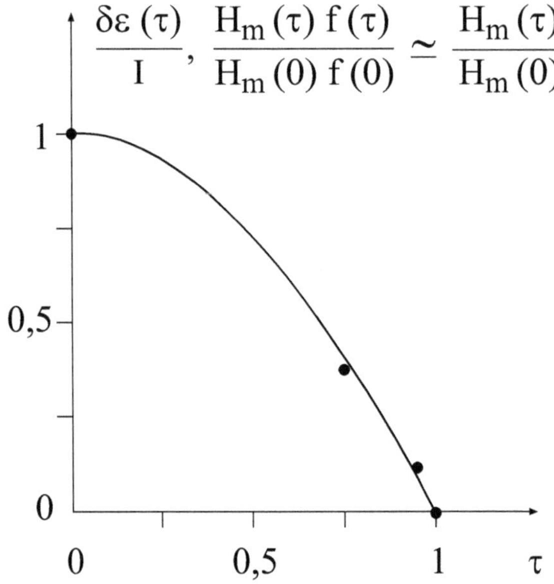

Fig. (3.7.2). Meissner – Ochsenfeld effect.

By what way does the width of the energy well behave at constant temperature (τ = const), provided that the external magnetic-field strength jumps from zero to value *H*? In the case of the absence of the magnetic field, the width of a well will be rated at $\delta\varepsilon(\tau)$. When the magnetic field of the strength *H* is produced on a superconductor surface, the kinetic energy (3.7.31) is displaced on its right by a certain value. In this case, energy wells are not subject to variation but displaced on their right. It is possible to assume that the electron energy will take up the least value when some portion of a well is displaced by value $\mu_B\ H\ f(\tau)/4$. Thus, the superconducting well width is represented by the formula:

$$\Delta\varepsilon(\tau) = \delta\varepsilon(\tau) - \mu_B H f(\tau)/4 .$$

(3.7.46)

Let us substitute $\delta\varepsilon(\tau)$ in the equation (3.7.47) by using the formula (3.7.46). Now, we will obtain:

$$\Delta\varepsilon(\tau) = 1/4\, \mu_B f(\tau) [H_m(\tau) - H].$$

(3.7.47)

When the magnetic-field strength H is equal to the critical value $H_m(\tau)$, the superconducting well width will be brought to zero.

3.7.7. Superconducting Current

If there is no external magnetic field, then $\Lambda = 0$ and the distribution function for $J = 3\,I$ for $T = 0$ takes the form (see Fig. **3.7.3a**).

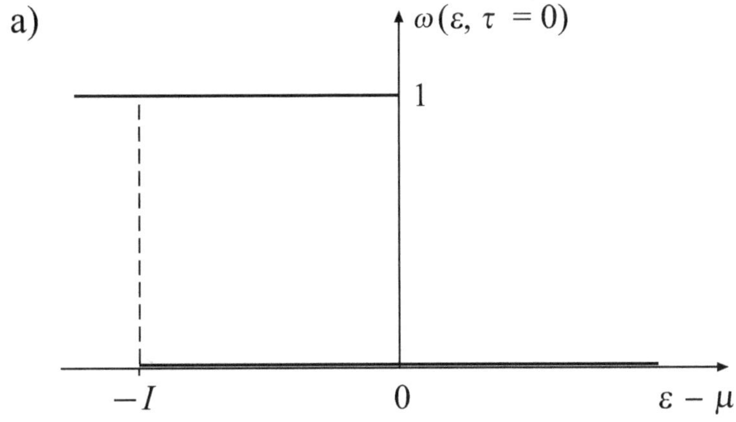

Fig. (3.7.3a). The real distribution of conduction electrons by energy at a temperature of $\tau = 0$. The magnetic field included. The distribution function shifts to the right by Λ. But the superconductivity hasn't disappeared yet.

If there is no magnetic field, the width of the pit will be $\delta\varepsilon(\tau)$. When the magnetic field of H is present on the surface of the superconductor, the electron distribution function $w(\varepsilon, \tau)$ shifts to the right by Λ. In this case, the energy well does not change, but only shifts to the right (see Fig. **3.7.3b**). The width of the superconducting band will be $I - \Lambda$.

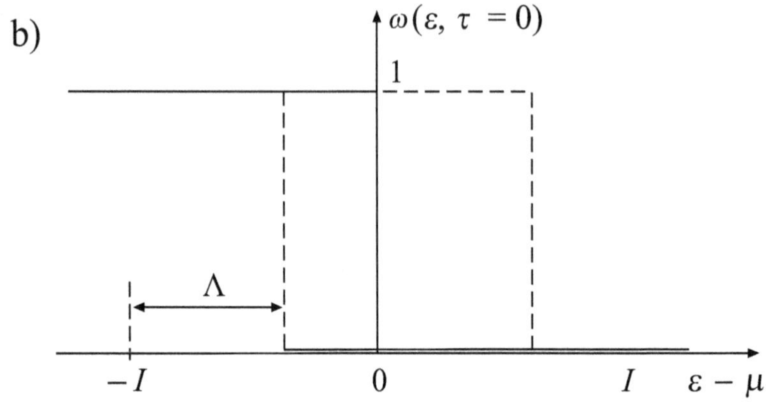

Fig. (3.7.3b). The real distribution of conduction electrons by energy at a temperature of $\tau = 0$. Magnetic field included. The distribution function shifts to the right by Λ. But the superconductivity hasn't disappeared yet.

It can be assumfed that according to (3.7.25) the electron energy will have the smallest value when the part of the pit is shifted by the value of

$$\Lambda = \mu_B H f(\tau)/4 .$$

Thus, the superconducting width of the pit, which provides superconductivity, will be equal to

$$\Delta\varepsilon(\tau) = I - \Lambda = \delta\varepsilon(\tau) - \mu_B H f(\tau)/4 .$$

Replace in equation (3.7.47) the value $\delta\varepsilon(\tau)$ by the formula (21.2). We will have

$$\Delta\varepsilon(\tau) = 1/4\, \mu_B f(\tau) [H_m(\tau) - H] .$$

When the magnetic field strength H is equal to the critical value $H_m(\tau)$, the superconducting width of the pit will be zero.

Any superconductor current flow is featured with its density matrix shown in Fig. (**3.4.7**). Mean electron velocity is:

$$\boldsymbol{v} = (\hbar G/m N) \sum_k \boldsymbol{k}\, w_k .$$

(3.7.48)

Now, we substitute the sum (3.7.49) for the following integral:

$$v = (\hbar\, G\, V / 8\, \pi^3\, m\, N) \int \boldsymbol{k}\, w_{\boldsymbol{k}}\, \mathrm{d}^3 k\,.$$

(3.7.49)

Using the above formula, we calculate the mean velocity for $T = 0$ with the formula (7.25) considered as true. This formula makes it possible to find radii $\boldsymbol{k}_1$ and $\boldsymbol{k}_2$ of the spheres shown in Fig. (**3.6.7**):

$$\hbar^2 k_1^2/(2\,m) = \mu\,, \qquad \hbar^2\, k_2^2/(2\,m) = \mu + I\,.$$

As a result, we will obtain as follows:

$$k_1 = \sqrt{2\,m\,\mu}/\hbar,\;\; k_2 = \sqrt{2\,m\,(\,\mu + I\,)}/\hbar.$$

The layer thickness, as specified in Fig. (**3.4.7**), will be approximately equal to:

$$\delta k = k_2 - \,k_1 \cong I/\hbar\sqrt{m/2\,\varepsilon_{\mathrm{F}}}\,.$$

(3.7.50)

In case of the absence of external magnetic fields, when an electron distribution pattern is described by the formula (7.39), the largest value of mean electron velocity produced in the flow of supercurrent will be formulated as:

$$v_{\max} = 3\,I/4\sqrt{2\,m\,\varepsilon_F} \;\;\text{or}\;\; v_{\max} = 3\,\delta\varepsilon(0)/4\sqrt{2\,m\,\varepsilon_F}\,.$$

(3.7.51)

As it is follows the respective electric current density vector can be expressed as:

$$j_{\max} = \mathrm{e}\, n\, v_{\max}\,.$$

(3.7.52)

With the external magnetic field applied while its strength is kept on the level of less than the critical value, the following mean electron velocity can be obtained:

$$v(\tau) = 3\,\Delta\varepsilon(\tau)/4\sqrt{2\,m\,\varepsilon_{\mathrm{F}}}\,.$$

If to substitute the formula (3.7.48), we will obtain the expression as shown below:

$$v(\tau) = 3\,\mu_{\rm B}\, f(\tau)/16\,\sqrt{2\,m\,\varepsilon_{\rm F}}\,\left[\,H_m(\tau) - H\,\right].$$

(3.7.53)

Now, the current flowing along the superconductor will have the density determined by the current velocity (3.7.54):

$$j(\tau) = {\rm e}\,n\,v(\tau)\,.$$

As a result, we obtain the formula expressed as:

$$j(\tau) = 3\,\mu_{\rm B}\,{\rm e}\,n\,f(\tau)/16\,\sqrt{2\,m\,\varepsilon_F}\,\left[\,H_m(\tau) - H\,\right],$$

(3.7.54)

where $H \leq H_m(\tau)$. When the external field strength exceeds its critical value, the superconductivity vanishes and the density vector will be determined by the substance conductivity ρ .

Now consider the case when the magnetic field destroys the superconductivity at $T = 0$. The energy of the valence electrons is now defined by the formula (3.7.30). Let

$$\Lambda = I\,.$$

(3.7.55)

The graph of the distribution function is shifted to the right on this value (see. Fig. **3.7.1**).

3.7.8. The Magnetic Field in a Flat Superconductor

We investigated the equilibrium state of the superconductor. Time independent functions are to be considered only in this state. These functions satisfy the stationary Maxwell equations expressed as:

$$\operatorname{rot} \boldsymbol{H} = 4\,\pi\,\boldsymbol{j}/c\,,$$

(3.7.56)

$$\operatorname{div} \boldsymbol{B} = 0\,.$$

(3.7.57)

Let us consider the magnetic field effect produced on a planar border of the superconductor, provided that the magnetic field vector ***H*** runs parallel to the border. We arrange axes x lengthwise the conductor border and parallel to vector ***H*** and axis y – perpendicularly to the border (see Fig. **3.7.4**).

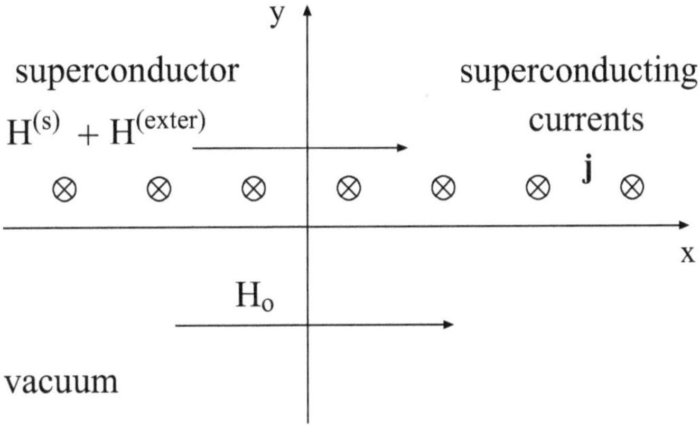

Fig. (3.7.4). Layout of the coordinate axes lengthwise the planar border of the superconductor. Currents produced by superconducting electrons are graphically represented here.

In this case, the equation (3.7.56) takes up the form as:

$$-\,\mathrm{d}H_x/\mathrm{d}y = 4\,\pi\,j_z/c\,.$$

(3.7.58)

Since any superconducting current is produced by a negative component j_z of vector ***j*** :

$$j_z = -\,j\,,$$

substituting the value (3.7.55) in this equation, we will obtain the formula expressed as:

$$\mathrm{d}H_x/\mathrm{dy} = g(\tau)\,[\,H_m(\tau) - H_x]\,,$$

(3.7.59)

where

$$g(\tau) = 3\,\pi\,\mathrm{e}\,n\,\mu_\mathrm{B}\,f(\tau)\,/(4\,c\,\sqrt{2\,m\,\varepsilon_\mathrm{F}}\,)\;.$$

(3.7.60)

In view of the superposition principle, the magnetic field vector $\boldsymbol{H}$ can be considered as equal to the sum of the external and magnetic fields to be created by supercurrent:

$$\boldsymbol{H} = \boldsymbol{H}^{(\mathrm{exter})} + \boldsymbol{H}^{(\mathrm{s})}\,.$$

(3.7.61)

Let us consider that any supercurrent flowing along axis z , as demonstrated in Fig. (**3.7.3**), is produced in a substance with the external magnetic field absent:

$$\boldsymbol{H}^{(\mathrm{exter})} = 0\,.$$

Such current may flow along the superconductor as long as it can. As regards electrons, they achieve their mean velocity. The supercurrent produces a self-magnetic field featured with the strength of that $H_x^{(s)}$ satisfies the equation as:

$$\mathrm{d}H_x^{(\mathrm{s})}/\mathrm{dy} = g(\tau)\,[\,H_m(\tau) - \;H_x^{(\mathrm{s})}]\,.$$

(3.7.62)

Any strength produced by the supercurrent will be equal to zero at the superconductor border:

$$H_x^{(s)}(0) = 0\,.$$

(3.7.63)

Since the strength satisfies the condition, it can be expressed as:

$$H_x^{(s)}(y) = H_m(\tau)\,[\,1 - e^{-g(\tau)\,y}\,].$$

(3.7.64)

For the graphical representation of this function, see Fig. (**3.7.5**).

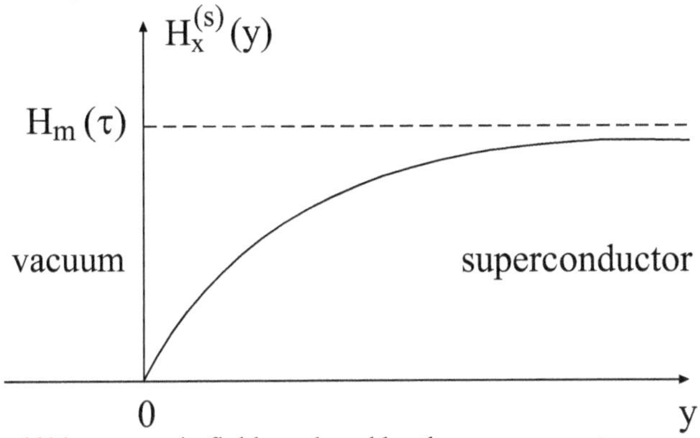

Fig. (3.7.5). The self\hh -magnetic field produced by the supercurrent.

Now, we create the external magnetic field running lengthwise axis x with its strength represented by $H_x^{(exter)}(y)$. It is evident that the above field strength may have any alternate mark:

$$H_x^{(exter)}(y) < 0 \quad \text{or} \quad H_x^{(exter)}(y) > 0\,.$$

The self-magnetic field of the superconductor is added to a just created field as specified by the principle of superposition. Therefore, the complete field will be formulated as:

$$H_x(y) = H_x^{(exter)}(y) + H_x^{(s)}(y)\,.$$

(3.7.65)

Let us substitute the complete field (3.7.65) in equation (3.7.60). We will obtain the formula shown below:

$$\mathrm{d}H_x^{(exter)}/\mathrm{d}y + \mathrm{d}H_x^{(s)}/\mathrm{d}y = g(\tau)\,[\,H_m(\tau) - H_x^{(exter)} - H_x^{(s)}].$$

(3.7.66)

We subtract the equation (3.7.66) from the equation (3.7.62). And the following external field equation is obtained:

$$dH_x^{(\text{exter})}/dy = -\, g(\tau)\, H_x^{(\text{exter})}\,.$$

(3.7.67)

The external field strength, as determined at the border of the superconductor, will be equal to the given value $\pm\, H_0$ to be subject to change throughout an experiment:

$$H_x^{(\text{exter})}(0) = \pm\, H_o\,.$$

(3.7.68)

Since the function $H_x^{(\text{exter})}(y)$ satisfies the starting condition (3.7.68) it can be formulated as follows:

$$H_x^{(\text{exter})}(y) = \pm\, H_o\, e^{-g(\tau)\,y}\,.$$

(3.7.69)

This function is graphically represented in Fig. (**3.7.6**). Here the value (3.7.69) is positive:

$$H_x^{(\text{exter})}(y) > 0\,.$$

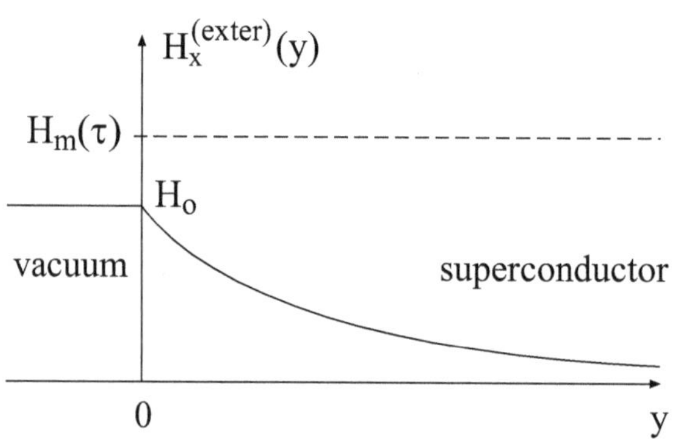

Fig. (3.7.6). External magnetic field penetrating inside the superconductor.

Thus, we could obtain the strength of the external magnetic field going down inside the superconductor. The complete field (3.7.65) is equal to the magnetic field produced by supercurrent and to the external field, provided that $H_x^{(\text{exter})}(y) > 0$ takes up the form as follows:

$$H_x(y) = H_o\, \mathrm{e}^{-g(\tau)\,y} + H_m(\tau)\,[1 - \mathrm{e}^{-g(\tau)\,y}]\,.$$

(3.7.70)

When the external field strength takes up the following form

$$H_o = H_m(\tau)\,,$$

(3.7.71)

the function (3.7.70) will be equal to the critical field:

$$H_x(y) = H_m(\tau)\,.$$

(3.7.72)

With H_o subsequently increased, the self-magnetic field vanishes and the complete field strength will be equal to that of the external field:

$$H_x(y) = H_o\,.$$

(3.7.73)

So, it is proved that when value H_o belongs to the interval

$$0 \leq H_o < H_m(\tau)\,,$$

(3.7.74)

the superconductivity will be applicable in the conductor under the Meissner – Ochsenfeld effect. But if value H_o exceeds the critical one:

$$H_o \geq H_m(\tau)\,,$$

(3.7.75)

the superconductivity vanishes.

We have discussed in detail the case when the strengths $\boldsymbol{H}^{(\text{exter})}$ and $\boldsymbol{H}^{(\text{s})}$ of the external and self-magnetic fields runs in the same direction. And this is time to consider the strengths being directed differently. So we can obtain the following formula:

$$H_x(y) = -H_o\, \mathrm{e}^{-g(\tau)\, y} + H_m(\tau)\, [\, 1 - \mathrm{e}^{-g(\tau)\, y}\,]\,. \tag{3.7.76}$$

At $H_o = H_m(\tau)$, this function will take up the form:

$$H_x(y) = H_m(\tau)\, [\, 1 - 2\, \mathrm{e}^{-g(\tau)\, y}\,]\,.$$

From the above, the Meissner – Ochsenfeld effect cannot be applicable. To remedy such a condition, it is necessary to specify the external field as that changing direction of supercurrent flow. Thus, the direction of the superconductor self-magnetic field vector is considered as that matching the direction of the external magnetic field strength. As provided by the given calculations, the external magnetic field goes down inside the superconductor thanks to presence of the magnetic field produced by supercurrent.

Using the formula (3.7.55), let us find the supercurrent density as dependent on coordinate y. We substitute the formula (3.7.55) in the formula (3.7.70). As a result, we will obtain the following formula:

$$j(\, y \mid \tau,\ H_o) = K(\tau)\, [\, H_m(\tau) - H_o]\, \mathrm{e}^{-g(\tau)\, y}\ , \tag{3.7.77}$$

where

$$K(\tau) = 3\, \mathrm{e}\, n\, \mu_{\mathrm{B}}\, f\, (\tau)/(\, 16\, \sqrt{2\, m\, \varepsilon_{\mathrm{F}}}\ )\,.$$

As follow from the above formula, the density of current goes down from the superconductor border to an exponent. In case when the external magnetic field is absent ($H_o = 0$), the density will be equal to

$$j(\, y \mid \tau,\, 0\,) = K(\tau)\, H_m(\tau)\; \mathrm{e}^{-g(\tau)\, y}\,. \tag{3.7.78}$$

At $H_o = H_m(\tau)$, the supercurrent density vanishes:

$$j\,[\, y \mid \tau,\; H_m(\tau)] = 0\,. \tag{3.7.79}$$

3.7.9. The Magnetic Field in the Superconducting Sphere

Let's discuss the behavior of the ***H*** magnetic field inside and outside a spherically shaped superconductor. The conductor with no superconducting properties induced at temperature $T > T_c$ is shown in Fig. (**3.7.6**). External magnetic field $\boldsymbol{H} = \boldsymbol{H}^{(\mathrm{exter})}$ penetrates inwards the conductor in such a way as described by the solution of the Maxwell equation.

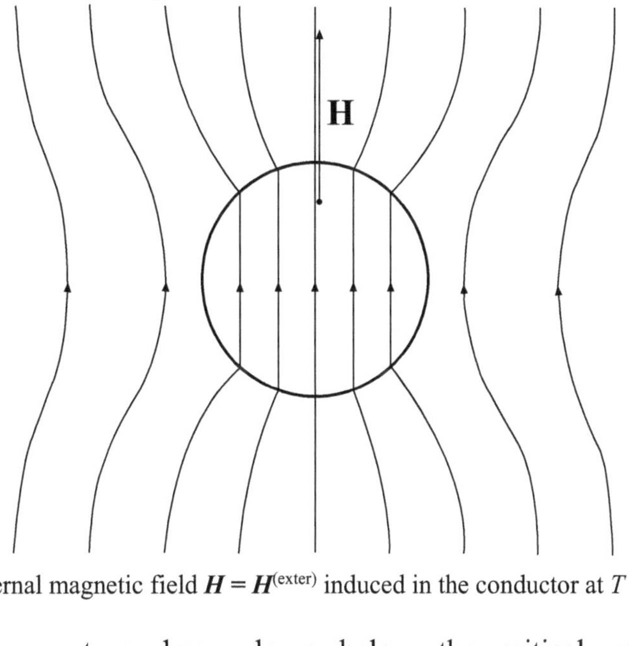

Fig. (3.7.7). External magnetic field $\boldsymbol{H} = \boldsymbol{H}^{(\mathrm{exter})}$ induced in the conductor at $T > T_c$.

When the temperature drops down below the critical value: $T < T_c$ superconductive function is induced in the conductor. This is the condition when the magnetic field is forced out of the conductor. This condition is demonstrated in Fig. (**3.7.7**). Such a pattern is produced under the Meissner – Ochsenfeld effect and exists until the strength of the external field modulus applicable to the superconductor surface remains less than that of the critical field: $H_0^{(\mathrm{exter})} <$

$H_m(T)$. Should the external field strength go to its critical value, the superconductivity disappears and the function is brought to the pattern shown in Fig. (**3.7.8**).

Let's implement spherical coordinates r and ϑ inside the sphere where r is a distance from the sphere center to any arbitrary point of space; ϑ is a latitude angle. The modulus of vector $\boldsymbol{j}$ to be produced by the superconductive current is equal to:

$$j(r, \vartheta \mid \tau, H_o) = K(\tau)\,[\,H_m(\tau) - H_o\,]\,\mathrm{e}^{-g(\tau)\,(R-r)} \cos \vartheta\,, \tag{3.7.80}$$

where R is a sphere radius. For current density, conditions $\tau < 1$ and $H_o < H_m(\tau)$ are to be followed to apply the superconductive function. As soon as $\tau = 1$ or $H_o = H_m(\tau)$, the current density (3.7.80) goes down to zero and the superconductive function disappears. Supercurrent $\boldsymbol{j}$ flows along the spherical surface (see Fig. **3.7.7**).

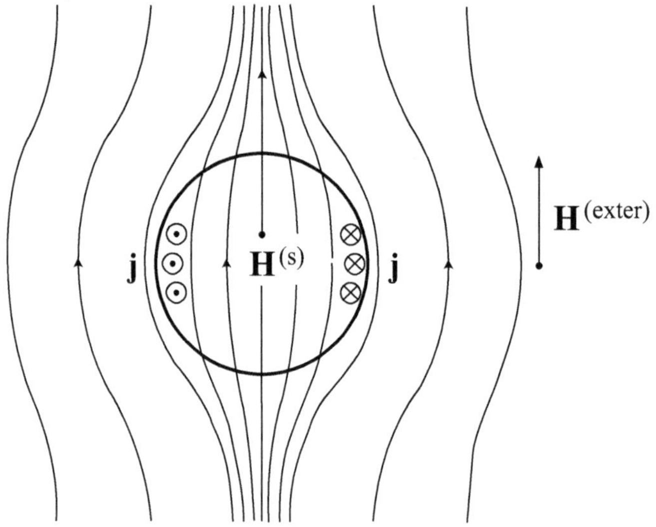

Fig. (3.7.8). An external magnetic field applied to the conductor at $T < T_c$.

External magnetic field $\boldsymbol{H}^{(\text{exter})}$ is forced out of the superconductor and its strength applicable to the conducting surface is less than that of the critical field: $H^{(\text{exter})} < H_m(T)$.

External magnetic field $\boldsymbol{H}^{(\text{exter})}$ is not capable to penetrate deep inwards the superconductor and its modulus is equal to

$$H^{(\text{exter})}(r,\vartheta \mid \tau, H_o) = H_o \, e^{-g(\tau)(R-r)} \cos\vartheta \quad \text{at } r \leq R .$$

(3.7.81)

The above formula demonstrates the influence of the external field on the supercurrent density. As soon as $H_o = H_m(\tau)$, the current density (3.7.80) goes down to zero. As a result, the superconductive function disappears. Now, we can use magnetic field $\boldsymbol{H}^{(s)}$ induced by the supercurrent inside the conductor. Its modulus is equal to:

$$H^{(s)}(r,\vartheta \mid \tau) = H_m(\tau)[\,1 - e^{-g(\tau)(R-r)} \cos\vartheta\,] \ \text{at}\, r \leq R .$$

(3.7.82)

These fields are shown in Fig. (**3.7.8**).

Now, we switch off the external magnetic field. In this case, its modulus applicable to the conductor surface is to be equal to zero ($H_o = 0$). When the temperature goes down below the critical value ($T < T_c$), the superconductive function does not disappear. The supercurrent being expressed by the following equation

$$j(r,\vartheta \mid \tau, 0) = K(\tau)\, H_m(\tau)\, e^{-g(\tau)(R-r)} \cos\vartheta$$

(3.7.83)

will be much the same as before. The pattern of self-magnetic field lines only will obtain a few changes (see Fig. **3.7.9**).

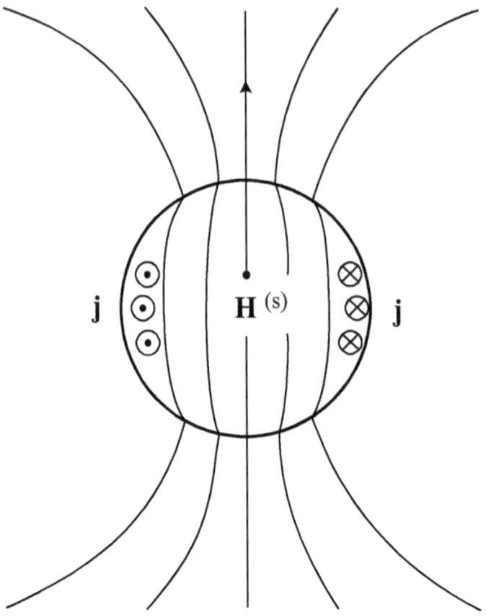

Fig. (3.7.9). Supercurrent-induced self-magnetic field in the conductor at $T < T_{c.}$. No external magnetic field is applied.

3.7.10. Magnetic Field in the Flat Disc of Superconductor

Let the superconductor be a flat disk. An external magnetic field is perpendicular to the plane of the disk (see Fig. **3.7.10**). The self-magnetic field will be directed in the same direction as the external field. If the temperature is less than critical: $T < T_c$, then the metal will be superconductive. But this external field strength should be less than the critical field:

$$H_x^{(\text{exter})} = H_0 < H_m(\tau) \,.$$

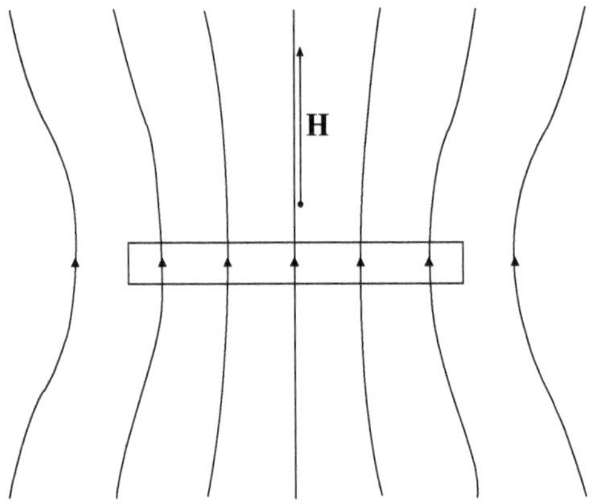

Fig. (3.7.10). External magnetic field $\boldsymbol{H} = \boldsymbol{H}^{(\text{exter})}$ produced in the flat disk of conductor at $T > T_{c.}$

The superconducting current density $\boldsymbol{j}$ will flow over the disk surface (see Fig. **3.7.10**). We write the approximate expression for the unit j of the superconducting current density $\boldsymbol{j}$ by analogy with the formula (3.6.17):

$$j(r \mid \tau,\, H_0) = K(\tau)\,[\,H_m(\tau) - H_0]\,\mathrm{e}^{-\,g(\tau)\,(R-r)}. \tag{3.7.84}$$

where r is the radial coordinate, R is the radius of the superconducting disk. The moduli of the tension of self-magnetic field produced by superconducting current and the external magnetic field inside the drive will be approximately equal, if these formulas are constructed like formulas:

$$H_x^{(s)}(\,r \mid \tau) = H_m(\tau)\left[\,1 - \mathrm{e}^{-\,g(\tau)\,(R-r)}\,\right], \tag{3.7.85}$$

$$H_x^{(\mathrm{exter})}(r \mid \tau, H_0) = H_0\,\mathrm{e}^{-\,g(\tau)\,(R-r)}\,. \tag{3.7.86}$$

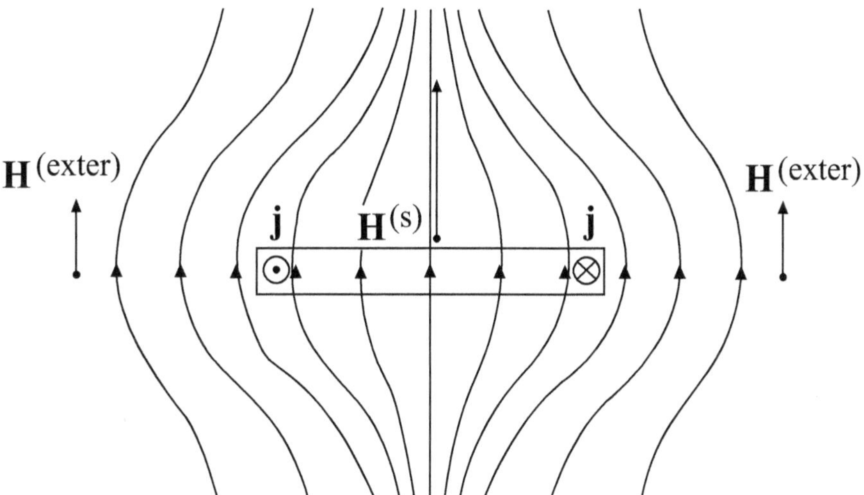

Fig. (3.7.11). The magnetic field in the flat disk of conductor at $T < T_c$. The external magnetic field is forced out of the superconductor and its strength gets less than that of the critical field: $H^{(\mathrm{exter})} < H_m$.

If you turn off the external magnetic field, a superconducting current will flow through the disk, the magnitude of which is equal to

$$j(r \mid \tau, 0) = K(\tau)\, H_m(\tau)\, \mathrm{e}^{-g(\tau)\,(R-r)}$$

(3.7.87)

and the magnitude of the tension of the self-magnetic field remains the same. Fig. (**3.7.12**) shows field lines.

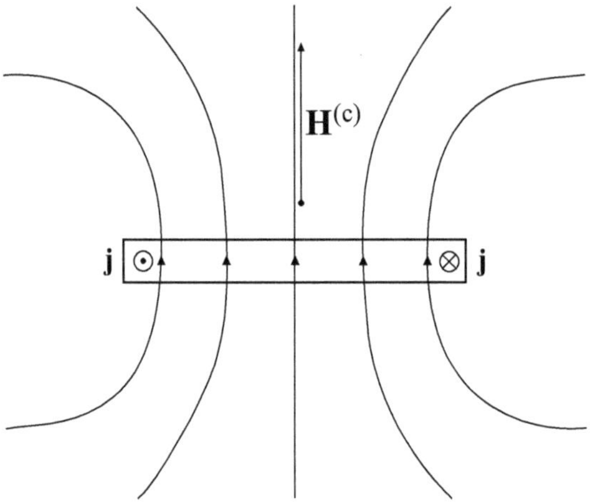

Fig. (3.7.12).: Self-magnetic field $H^{(s)}$ in the flat disk conductor produced by supercurrent at $T < T_c$. No external magnetic field $\boldsymbol{H}^{(\mathrm{exter})}$ is applied.

3.7.11. Supercurrent Flowing Through the Coil

Let us discuss a superconducting wire coil. Current flows through such conductor passing each circular loop. Much the same current flows over a thin-coat disk surface (see Fig. **3.7.12**). For making a circular coil a core of the disk may be cut out (see Fig. **3.7.13**). Both the self-magnetic field and current density shall save their characteristics:

$$H_x^{(s)}(r \mid \tau) = H_m(\tau)\,[1 - \mathrm{e}^{-g(\tau)\,(R-r)}]\,,$$

(3.7.88)

and coil plain:

$$j(r \mid \tau, 0) = K(\tau)\, H_m(\tau)\, \mathrm{e}^{-g\,(\tau)\,(R-r)}\,.$$

(3.7.89)

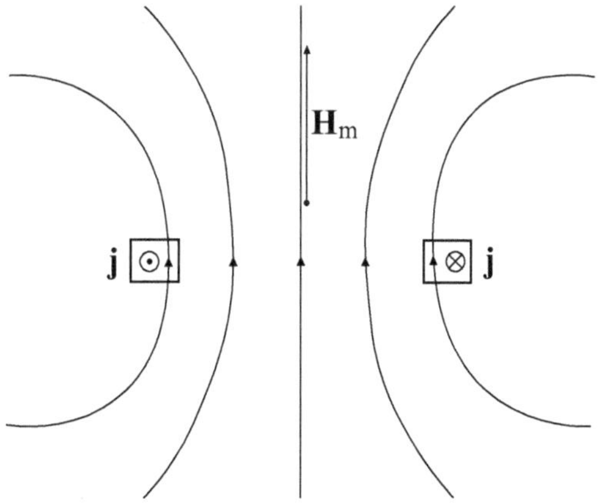

Fig. (3.7.13). Supercurrent-induced self-magnetic field in the wire coil.

3.7.12. Superconductivity Flowing Through the Solenoid

Now, let's discuss a superconducting solenoid. Current may flow passing through a circular loop similar to that described in the previous section. But the solenoid will have exceeded the self-magnetic field. If there are N coils in the solenoid, then the strength of the magnetic field within its core will go to $N\,\boldsymbol{H}_m$ – *i.e.* the strength will grow up proportionally (see Fig. **3.7.14**).

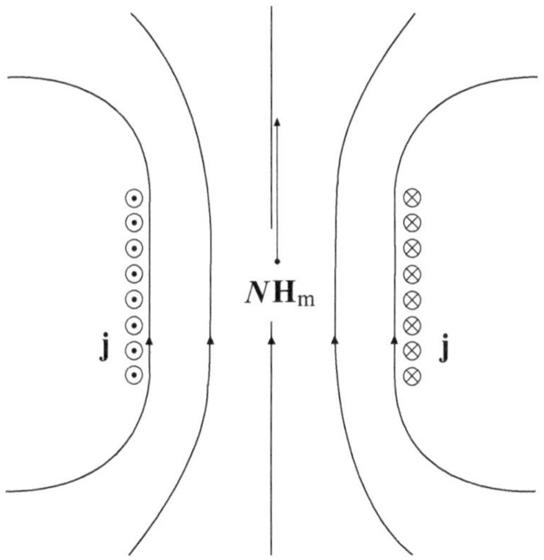

Fig. (3.7.14). Supercurrent-induced self-magnetic field in the solenoid.

3.7.13. Quantum Levitation and Quantum Trapping

Let's consider the quantum levitation state when a superconductor is located above a static permanent magnet less any suspension or support and it is capable to move up and down. When the distance between the superconductor and magnet is large, force $\boldsymbol{F}$ that acts on the superconductor will be directed downwards (see Fig. **3.7.15**).

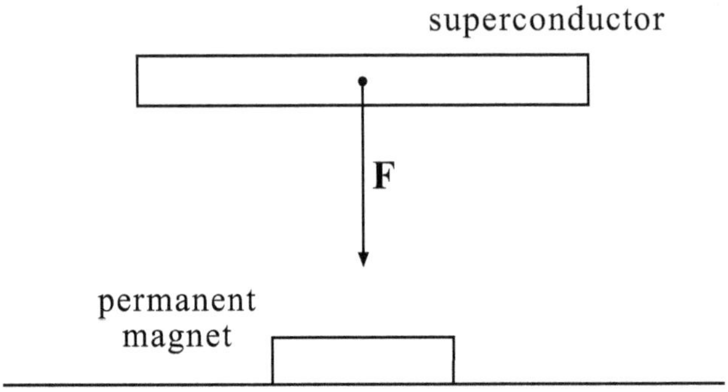

Fig. (3.7.15). Quantum levitation of a superconductor at $T > T_c$. When the distance between the superconductor and the permanent magnet is large, the current on the superconductor, the force is directed down.

But with the distance being rather small, such force will be directed upwards (see Fig. **3.7.16**). It means that the force will be equal to zero – *i.e.* $\boldsymbol{F} = 0$. This is a stable equilibrium position. There is a key quantum levitation question to be arisen: Why does the superconductor-acting force become up-directed?

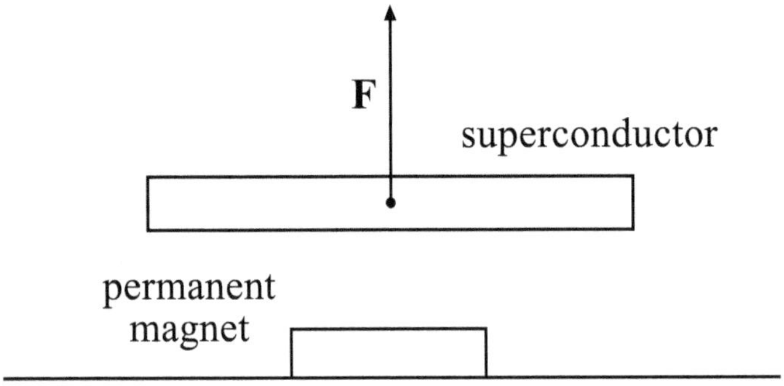

Fig. (3.7.16). When the distance between the superconductor and the permanent magnet is small, acting on the superconductor, the force is directed upwards.

Let's the superconductor is shaped as a disk. Let's assume that the external magnetic-field vectors $\boldsymbol{H}^{(\text{exter})}$, proper field vectors $\boldsymbol{H}^{(s)}$, and vector $\boldsymbol{p}_m$ of the magnetic superconductor, moment are unidirectional. Vector $\boldsymbol{H}^{(s)}$ and vector $\boldsymbol{p}_m$

are created by super-current which density is expressed by parameter ***j***. Now, we draw x-axis upwards off the permanent magnet as shown in Fig. (**3.7.17**).

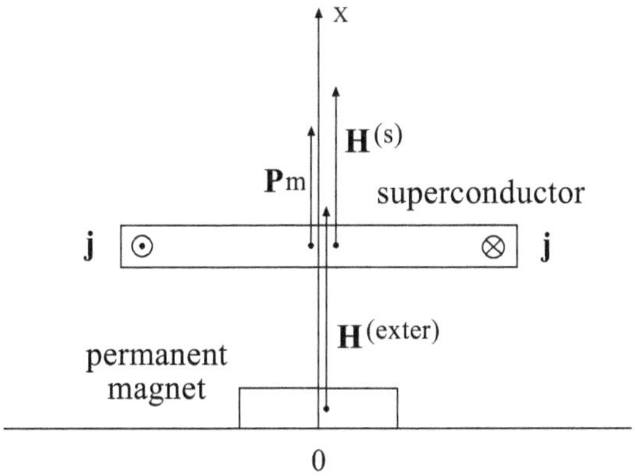

Fig. (3.7.17). Quantum levitation of a superconductor at $T < T_c$. Shows the vector of the external field $\boldsymbol{H}^{(\mathrm{exter})}$, the own field of the superconductor $\boldsymbol{H}^{(\mathrm{s})}$ and the vector $\boldsymbol{p}_m$ magnetic moment of the superconductor.

As it is provided by the theoretical physics, any potential energy of the superconductor produced in this case within the external field will be equal to

$$U(x) = -\boldsymbol{p}_m \boldsymbol{B}^{(\mathrm{exter})}(x) + M g x ,$$

(3.7.90)

where $\boldsymbol{B}^{(\mathrm{exter})}(\mathrm{x})$ is the magnetic induction of the external field, M is the superconductor mass, g is the fall free acceleration. In this formula, the first component is

$$U_m(x) = -p_m B^{(\mathrm{exter})}(x),$$

(3.7.91)

where is the energy of interacting external and proper magnetic fields. The second component

$$U_g = M g x$$

(3.7.92)

is the potential energy of the superconductor-to-terrestrial gravitational attraction. Such gravity F_g is directed downwards and is equal to Mg.

The external magnetic field is heterogeneous and its magnetic induction $B^{(\text{exter})}(x)$ grows up near to a permanent magnet. Consequently, energy $U_m(x)$ will be expressed by a negative value being increased as soon as it gets closer to a permanent magnet (see Fig. **3.7.18**).

Projection of the force applied to the superconductor from the external field is expressed by formula

$$F_x^{(m)}(x) = -\,\mathrm{d}U(x)/\mathrm{d}x\,.$$

(3.7.93)

Hence, force vector $\boldsymbol{F}_m$ will be directed downwards when $x > x_0$ (see Fig. **3.7.18**). Parameter x_0 is taken as a coordinate of the equilibrium point

$$F_x^{(m)}(x_0) - Mg = 0\,.$$

(3.7.94)

This equation demonstrates that the quantum levitation exists when the magnetic force exceeds that of the superconductor gravity

$$F_x^{(m)}(x) > Mg \ \text{ at } \ x < x_o\,.$$

(3.7.95)

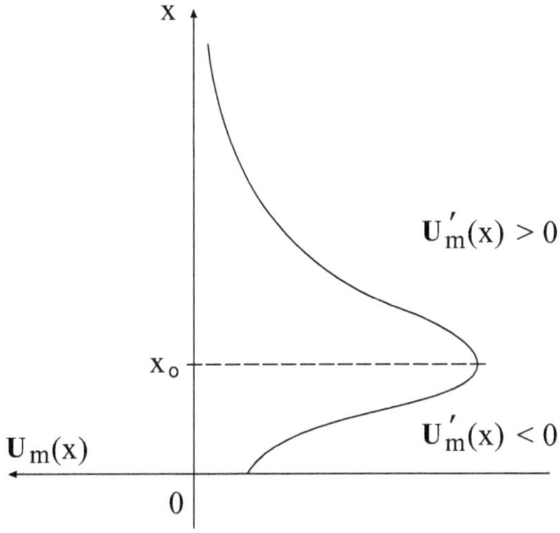

Fig. (3.7.18). Potential energy $U_m(x)$ of a superconductor under the action of external fields in quantum levitation.

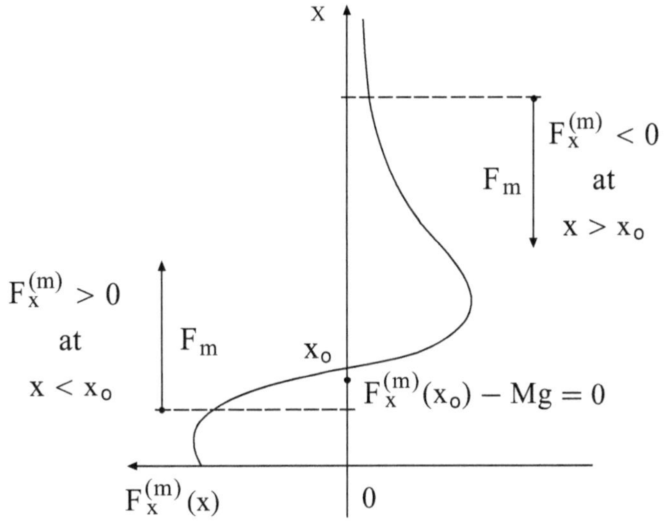

Fig. (3.7.19). Force $\mathbf{F}_m$ acting on the superconductor.

Now, let us take a closer look at behavior exhibited by the superconductor when $x < x_0$. The supercurrent density will be approximated to

$$j\,[\,r \mid \tau,\, H_0(x)\,] = K(\tau)\,[\,H_m(\tau) - H_0(x)]\,\mathrm{e}^{-g(\tau)\,(R-r)}\,, \tag{3.7.96}$$

where R is a distance from the superconductor center to its periphery; r is a radial coordinate; $H_0(x)$ is an external magnetic field strength distributed over the surface of the superconducting disk at its periphery.

The most interesting thing is that this formula exhibits the difference of the critical field and strength of the external field distributed over the superconductor surface :

$$H_m(\tau) - H_0(x) .$$

Strength $H_0(x)$ of the external field grows up closer to a permanent magnet and it may be even equal to critical field $H_m(\tau)$. Hence, the density will be equal to zero. In this case, the superconductivity disappears. The self-magnetic field distributed over the superconductor will also get equal to zero and the superconductivity will start dropping. But till that moment when the current remains different to zero, the magnetic moment of the superconductor to be proportional to the current density

$$p_m(x) \cong j\,[\, r \mid \tau,\, H_0(x)\,] . \tag{3.7.97}$$

It depends on coordinate x and goes down. Consequently, the potential energy gets expressed by a decay function. Therefore, the potential energy $U_m(x)$ of the superconductor taken as an x-function will be approximately expressed by such a form as shown in Fig. (**3.7.17**).

The distance dependence of force $F_x^{(m)}(x)$ curve is shown in Fig. (**3.7.18**). If $x < x_0$, the projection of force $F_x^{(m)}(x)$ will be expressed by a positive value

$$F_x^{(m)}(x) > 0 \text{ at } x < x_0 , \tag{7.98}$$

and force vector $\boldsymbol{F}_m$ will be directed upwards. If force $F_x^{(m)}(x)$ occurs to exceed gravity $M\,g$, the superconductor will find itself in the quantum levitation condition.

If we aim the external magnetic field in the opposite direction, it will also redirect supercurrent to another side and, as a result, the self-magnetic field will get inverted, as well. Hence, the external magnetic field, self-magnetic field, and vector p_m will exhibit the unilateral direction and the force will remain unchanged.

The position when the superconductor is exposed to a permanent magnet is known as the quantum capture (see Fig. **3.7.20**). This condition is explained similarly to that when the superconductor is subject to quantum levitation.

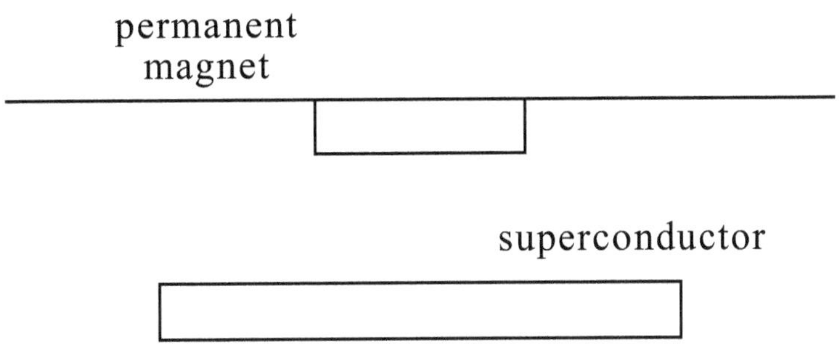

Fig. (3.7.20). Quantum capture superconductor.

REFERENCES

[1] H. Kamerlingh Onnes, *Comm. Phys. Lab. Univ. Leiden.*, vol. 122, pp.13, 1911.
[2] W. Meissner, R. Ochsenfeld, "Ein neuer Effekt bei Eintritt der Supraleitfähigkeit", *Naturwiss*, vol. 21, pp. 787, 1933.
[3] V.L. Ginszburg, L.D. Landau, "On the theory of superconductivity", *Journal of Experimental and Theoretical Physics*, vol. 20, pp. 1064, 1950.
[4] J. Bardeen, L.N. Cooper, J.R. Schrieffer, "Microscopic Theory of Superconductivity" *Phys. Rev.*, vol. 106, pp. 162, 1957.
[5] B.V. Bondarev, "On some features of the electron distribution function the Bloch states". *Vestnik MAI*, vol. 3, no. 2, pp. 56-65, 1996.
[6] B.V. Bondarev, "Method of equilibrium density matrix. Anisotropy and superconductivity. Energy gap. Springer proceeding in physics. Advanced materials, Manufacturing", *Physics, Mechanics and Applications*, vol. 175, pp. 157-178, 2016.

CHAPTER 4

New Theory of Superfluidity

4.1. NEW THEORY OF SUPER-FLUIDITY EQUILIBRIUM DENSITY MATRIX METHOD

4.1.1. Liquid Helium

Gaseous helium at atmospheric pressure becomes liquid when its temperature reaches a value of 4.44 K. Solid helium can only exist at a pressure of at least 25 atm. At lower pressures, helium remains liquid down to zero absolute temperature.

There are two isotopes of helium: He^3 and He^4. In the liquid He^4, a phase transition occurs at a temperature $T_\lambda = 2{,}18$ K, *i.e.* two helium phases are distinguished, which denote He I and He II. At temperature $T < T_\lambda$, helium He^4 is in a phase He II. In this case, helium behaves so, as if it is a mixture of two liquids, which are called normal and superfluid components. Distinctive features of the latter are 1) its zero entropy and 2) the lack of friction of this component with the normal component and the walls of the vessel.

The phenomenon of superfluidity of helium He II was discovered in 1937 by the Soviet physicist, Peter Leonidovich Kapitsa. In 1978, he received the Nobel prize for this discovery.

As the temperature decreases, the density ρ_N of the normal component decreases, and the density ρ_S of the superfluid increases. The density of He II is:

$$\rho = \rho_N + \rho_S.$$

Temperature dependence of the relationship ρ_N/ρ is set experimentally. The dependence presented in this graph is satisfactorily described by the empirical formula as,

$$\rho_N/\rho = \begin{cases} (T/T_\lambda)^{5,6} & \text{at} \quad T < T_\lambda, \\ 1 & \text{at} \quad T \geq T_\lambda. \end{cases}$$

Boris V. Bondarev

Experimental dependence of $C = C(T)$ helium heat capacity on temperature it resembles the letter λ. Therefore, this phase transition is called λ-transition. At the point $T = T_\lambda$ the function $C = C(T)$ becomes infinitely large.
Atoms He^3 are fermions, and atoms He^4 -- bosons. It is therefore natural to assume that the λ-transition in He^4 is somehow related to the possible Bose condensation of helium atoms.

Today, the question of Bose condensation has aroused quite a keen interest in the study of the nature of this phenomenon and the thermodynamic properties of multifrequency systems. The most general statistical formulation of a multifrequency system in quantum mechanics is obtained by applying a density matrix. To describe the density matrix of the boson equilibrium system, the variational principle [1] is used in this paper.

4.1.2. Uniform Distribution of Particles in Space

For analysis of equilibrium states of quantum gas consisting of angular momentum zero-spin particles, we apply a variational density matrix method. We assume that N of the above particles finds itself within a certain space area with the volume equal to V. If the particles have their mean uniform distribution, a single-particle density matrix, which in coordinate representation depends on radius vectors $\boldsymbol{r}$ and $\boldsymbol{r}'$, will obtain the form as follows:

$$\varrho_{rr'} = 1/V \sum_{k} p_{k} \, e^{i\boldsymbol{k}(\boldsymbol{r}-\boldsymbol{r}')} , \quad \mathbf{(4.1.1)}$$

where p_k is the probability that an arbitrarily accepted particle is in the state defined by wave vector $\boldsymbol{k}$. Particle momentum

$$\boldsymbol{p} = \hbar\, \boldsymbol{k}.$$

Probability p_k meets its normalizing condition,

$$\sum_{k} p_{k} = 1 . \quad \mathbf{(4.1.2)}$$

As formula (4.1.1) indicates, the momentum representation density matrix is diagonal, *i.e.*

$$\varrho_{kk'} = p_k \, \delta_{kk'} ,$$

and transition from momentum representation to the nodal one is affected by a unitary matrix

$$u_{rk} = 1/\sqrt{V}\, e^{\,i\boldsymbol{k}\boldsymbol{r}}\,, \qquad \textbf{(4.1.3)}$$

which is subject to the condition as follows:

$$\int u_{rk}\, u^*_{rk'}\, d\boldsymbol{r} = \delta_{kk'} \qquad \text{or} \qquad 1/V \int e^{\,i(\boldsymbol{k}-\boldsymbol{k}')\boldsymbol{r}}\, d\boldsymbol{r} = \delta_{kk'}\,.$$

4.1.3. Kinetic Energy of Particle

Particle kinetic energy operator $\widehat{H}^{(1)}$ takes the form as follows:

$$\widehat{H}^{(1)} = -\hbar^2\nabla^2/(2\,m)\,, \qquad \textbf{(4.1.4)}$$

where m – particle mass.

Applying formula:

$$H_{kk'} = \int u^*_{rk}\, \widehat{H}^{(1)} u_{rk'}\, d\boldsymbol{r}$$

we can find elements of operator $\widehat{H}^{(1)}$ in momentum representation:

$$H_{kk'} = \varepsilon_k\, \delta_{kk'}\,. \qquad \textbf{(4.1.5)}$$

where ε_k is the particle kinetic energy with pulse $\boldsymbol{p} = \hbar\,\boldsymbol{k}$:

$$\varepsilon_k = \hbar^2 \boldsymbol{k}^2/(2\,m)\,. \qquad \textbf{(4.1.6)}$$

4.1.4. Particle Interaction Energy

We assume that $U_{rr'} = U_{r-r'}$ is potential two particle interaction energy. In virtue of translation invariance, the potential energy depends on vectors difference $\boldsymbol{r}_1 - \boldsymbol{r}_2$ only and it is represented by its symmetrical function:

$$U_{rr'} = U_{r'r}\,.$$

In this case, interaction Hamiltonian matrix elements shall be formulated by the following method:

$$H_{r_1r_2\, r_1'r_2'} = 1/2\, U_{r_1r_2}\, (\, \delta_{r_1r_1'}\, \delta_{r_2r_2'} + \delta_{r_1r_2'}\, \delta_{r_2r_1'}\,), \qquad \textbf{(4.1.7)}$$

where $\delta_{r_1r_1'}$ is a delta-Dirac function.

We can find interaction Hamiltonian matrix elements in momentum representation using the following expression:

$$H_{k_1k_2\, k_1'k_2'} = \int u^*_{r_1k_1}\, u^*_{r_2k_2}\, H_{r_1r_2\, r_1'r_2'}\, u_{r_1'k_1'}\, u_{r_2'k_2'}\, \mathrm{d}\boldsymbol{r}_1\, \mathrm{d}\boldsymbol{r}_2\, \mathrm{d}\boldsymbol{r}_1'\, \mathrm{d}\boldsymbol{r}_2' \,.$$

Substitution of matrices (4.1.3) and expression (4.1.7) into the formula makes it possible to obtain the expression as follows:

$$H_{k_1k_2\, k_1'k_2'} = 1/(2\,V)\; \delta_{k_1+\,k_2-\,k_1'-\,k_2'} \int U_r\, [\, \mathrm{e}^{\mathrm{i}\,(k_1-\,k_1')\,r} + \mathrm{e}^{\mathrm{i}\,(k_2-\,k_2')\,r}\,]\, \mathrm{d}\boldsymbol{r}. \qquad \textbf{(4.1.8)}$$

In this case, two-particle and wave $\boldsymbol{k}_1 - \boldsymbol{k}_2$ vector interaction energy will be represented by the following function,

$$V_{k_1k_2} = (\,2 - \delta_{k_1k_2}\,)\, H_{k_1k_2\, k_1'k_2'} = 1/(2\,V)\,(\,2 - \delta_{k_1k_2}\,) \int U_r\, [\,1 + \mathrm{e}^{\mathrm{i}\,(k_1-\,k_2)\,r}\,]\, \mathrm{d}\boldsymbol{r}\,. \qquad \textbf{(4.1.9)}$$

This kind of expression can be represented by the following form:

$$V_{k_1k_2} = 1/V\, [\,(\,2 - \delta_{k_1k_2}\,)\, U + v_{k_1-\,k_2}\,], \qquad \textbf{(4.1.10)}$$

where

$$U = \int U_r\, \mathrm{d}\boldsymbol{r}\,, \qquad \textbf{(4.1.11)}$$

$$v_{k_1-\,k_2} = -\int U_r\, [\,1 - \mathrm{e}^{\mathrm{i}\,(k_1-\,k_2)\,r}\,]\, \mathrm{d}\boldsymbol{r}\,. \qquad \textbf{(4.1.12)}$$

Any real closely interspaced particle tends to be exposed to strong repulsion. Therefore, it can be assumed that all the values of function (4.1.12) are rather small against the values of function (11) which will be considered as a positive one. It is believed that gas properties could be kept out of their significant change, provided that particle interaction is represented by the following simplified formula:

$$V_{k_1k_2} = 1/V\,(\,2 - \delta_{k_1k_2}\,)\, U, \qquad \textbf{(4.1.13)}$$

where $U > 0$.

4.1.5. Gas Internal Energy

As stated in the approximation of statistically independent particles, the internal energy is defined by the following expression:

$$E = \sum_k \varepsilon_k \bar{n}_k + 1/2 \sum_{k_1 k_2} V_{k_1 k_2} \bar{n}_{k_1} \bar{n}_{k_2} , \qquad \textbf{(4.1.14)}$$

where

$$\bar{n}_k = N p_k$$

is a mean number of particles with wave vector ***k***,

$$\sum_k \bar{n}_k = N, \qquad \textbf{(4.1.15)}$$

Substitution of expression (4.1.10) into formula (4.1.14) makes it possible to modify it into:

$$E = \sum_k \varepsilon_k \bar{n}_k + 1/(2V) [2UN^2 - U \sum_k \bar{n}_k^2 + \sum_{k_1 k_2} v_{k_1 - k_2} \bar{n}_{k_1} \bar{n}_{k_2}] . \qquad \textbf{(4.1.16)}$$

4.1.6. Particle Pulse Distribution Function

As stated in the approximation of statistically independent particles, the equation of function $\bar{n}_k$ which describes the particle pulse distribution process takes up the following form:

$$\vartheta \ln \{ (1 + \bar{n}_k)/\bar{n}_k \} = \bar{\varepsilon}_k - \mu , \qquad \textbf{(4.1.17)}$$

where average particle energy is formed as:

$$\bar{\varepsilon}_k = \partial E / \partial \bar{n}_k = \varepsilon_k + 1/V [- U \bar{n}_k + \sum_{k'} v_{k - k'} \bar{n}_{k'}] . \qquad \textbf{(4.1.18)}$$

We regard values $v_{k - k'}$, as small perturbations: $v_{k - k'} \ll U$ and we can find the distribution function using the formula as follows:

$$\bar{n}_k = f[\beta (\varepsilon_k - \mu)] + \delta \bar{n}_k , \qquad \textbf{(4.1.19)}$$

where $f(\xi)$ – one-variable function

$$\xi = \beta (\varepsilon_k - \mu) . \qquad \textbf{(4.1.20)}$$

Upon some simple transformations, substitution of expression (4.1.19) into equation (4.1.17) makes it possible to equate function $f(\xi)$:

$$\ln \{ (1 + f)/f \} = \xi - b f, \qquad \textbf{(4.1.21)}$$

where

$$b = \beta U/V,$$

and to formulate the following formula:

$$\delta\bar{n}_k = -\beta (1 + f_k) f_k \Sigma_{k'} v_{k-k'} f_{k'} 1/[V - \beta U (1 + f_k) f_k] , \qquad \textbf{(4.1.22)}$$

where

$$f_k = f[(\varepsilon_k - \mu)] .$$

For the graph of function $f = f(\xi)$ being the solution of equation (4.1.21), as applied to one of the values of parameter b, (see Fig. **4.1.1**). The curve has two asymptotes: horizontal one:

$$f = 0$$

and oblique one:

$$f = \xi / b.$$

Function $f = f(\xi)$ can be suitable for the positive values of argument ξ only, like: $0 < \xi_0 \leq \xi$, where the least value of ξ_0 depends on parameter b. Value b within the thermodynamic limit ($V \to \infty$) becomes zero at any temperature range. In this case, $\lim_{b \to \infty} \xi_0 = 0$.

At any value of $\xi_0 > 0$, function $f = f(\xi)$ takes up two values with the minor one defined as $f_n(\xi)$ and with the greater one $f_c(\xi)$. As provided by equation (4.1.21), if specific macroscopic values of volume V are applied, these values are defined by the following expressions:

$$f_n(\xi) = 1/(e^{\xi} - 1), \quad \textbf{(4.1.23)}$$

$$f_c(\xi) = V\,\xi/(\beta\, U) = N\,\xi/(\beta\, c\, U), \quad \textbf{(4.1.24)}$$

where c = N/V – particle concentration. It should be noted that an average number of particles $f_c(\xi)$ is macroscopically great against number $f_n(\xi)$.

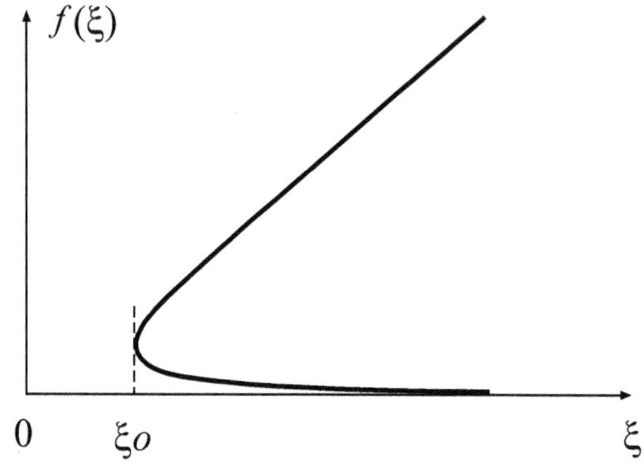

Fig. (4.1.1). Distribution function $f = f(\xi)$, as applied for solution of equation (6.5).

4.1.7. Chemical Potential

According to formula (4.1.19), the particle pulse distribution stated in its first approximation is described by the following function:

$$\bar{n}_{\boldsymbol{k}} = f[\,\beta\,(\,\varepsilon_{\boldsymbol{k}} - \mu\,),\ N/(\beta\, c\, U)\,], \quad \textbf{(4.1.25)}$$

which parametrically contains temperature, chemical potential and dimensionless quantity $N/(\beta\, c\, U)$. Substitution of function (4.1.25) into equation (4.1.16) results in the following equation:

$$\Sigma_{\boldsymbol{k}} f\,[\,\beta\,(\,\varepsilon_{\boldsymbol{k}} - \mu\,),\ N/(\beta\, c\, U)\,] = N, \quad \textbf{(4.1.26)}$$

which implicitly sets chemical potential dependence on temperature.

Should a certain function $\varphi_{\boldsymbol{k}}$ of argument $\boldsymbol{k}$ be of continuous nature, the summation by wave vectors $\boldsymbol{k}$ may be substituted for the following interaction:

$$\sum_k \varphi_k = V/(2\,\pi)^3 \int \varphi_k \,\mathrm{d}\boldsymbol{k}\,.$$

If function φ_k is a complex one for vector $\boldsymbol{k}$, where kinetic energy ε_k acts as an intervening variable, the foregoing equation integral may be transformed with formula (4.1.6) by the method as follows:

$$\int \varphi_k \,\mathrm{d}\boldsymbol{k} = 4\,\pi\, m \sqrt{2\,m}/\hbar^3 \int_0^\infty \phi\,(\varepsilon)\,\sqrt{\varepsilon}\,\mathrm{d}\varepsilon\,.$$

As a result, we obtain the known formula:

$$\sum_k \varphi_k = A\,N \int_0^\infty \phi\,(\varepsilon)\,\sqrt{\varepsilon}\,\mathrm{d}\varepsilon\,, \qquad \textbf{(4.1.27)}$$

where

$$A = m\sqrt{2\,m}\,V/(\,2\,\pi^2\,\hbar^3 N\,)\,. \qquad \textbf{(4.1.28)}$$

We assume that at high temperatures, the distribution of particles by their states is described by dependence (4.1.23). In this case, condition (4.1.26) may be formulated (4.1.27) by the method as follows:

$$A \int_0^\infty \sqrt{\varepsilon}/[\,\mathrm{e}^{\,\beta\,(\varepsilon\,-\,\mu)} - 1\,]\,\mathrm{d}\varepsilon = 1\,.$$

If a new integration variable is applied, in particular:

$$y = \beta\,\varepsilon,$$

this equation may be transformed into the following form:

$$A\,\vartheta^{3/2} J(\,-\,\mu/\vartheta\,) = 1\,, \qquad \textbf{(4.1.29)}$$

where

$$J(x) = \int_0^\infty \sqrt{y}/(\,\mathrm{e}^{\,x\,+y} - 1\,)\,\mathrm{d}y\,. \qquad \textbf{(4.1.30)}$$

As it is known, the temperature value ϑ_{B}, when chemical potential becomes zero, is called Bose condensation temperature:

$$\vartheta_{\mathrm{B}} = 1/[\,A\,J(0)]^{2/3} = \hbar^2/m\,\{\,\sqrt{2}\,\pi^2\,N/[\,V\,J(0)]\,\}^{\,2/3}\,. \qquad \textbf{(4.1.31)}$$

We apply dimensionless quantities:

$$\tau = \vartheta/\vartheta_{\mathrm{B}}\,, \qquad \tilde{\mu} = \mu/\vartheta_{\mathrm{B}}\,. \tag{4.1.32}$$

Now, equation (4.1.29) may be represented in the form suitable for numerical solution:

$$\tau^{3/2}\, J(-\tilde{\mu}/\tau) = J(0)\,, \tag{4.1.33}$$

This kind of equation defines dependence $\tilde{\mu} = \tilde{\mu}(\tau)$ which can be significant when $\tau \geq 1$ only. If $\vartheta < \vartheta_{\mathrm{B}}$, equation (4.1.29) cannot be solved and should be substituted for another one. We can obtain the above new equation by the method as follows. We assume that at temperatures below the specific critical value ϑ_{c}, the distribution function takes up the following form:

$$\bar{n}_{\boldsymbol{k}} = \begin{cases} f_n[\beta\,(\varepsilon_{\boldsymbol{k}} - \mu)]\,, & \text{if} \quad \boldsymbol{k} \neq 0\,, \\ f_c(-\beta\,\mu)]\,, & \text{if} \quad \boldsymbol{k} = 0\,. \end{cases} \tag{4.1.34}$$

With this function substituted in normalizing condition (4.1.26), the following equation can be obtained:

$$N_n + N_c = N\,, \tag{4.1.35}$$

where value

$$N_n = \sum\nolimits_{\boldsymbol{k} \neq 0} f_n\,[\beta\,(\varepsilon_{\boldsymbol{k}} - \mu)] \tag{4.1.36}$$

will be named as a normal state particle number and value

$$N_c = f_c(-\beta\,\mu)] \tag{4.1.37}$$

– as a condensed state particle number. The above particles produce zero pulse and so-called condensate. On substituting expressions (4.1.23) and (4.1.24) in formulas (4.1.36) and (4.1.37) and changing sum $\boldsymbol{k}$ for energy ε integral in (4.1.36), we transform equation (4.1.35) into the following form:

$$\tau^{3/2}/J(0\,J(-\tilde{\mu}/\tau) - \tilde{\mu}/I = 1\,, \tag{4.1.38}$$

where I – particle interaction parameter, $I = c\,U/\vartheta_{\mathrm{B}}$.

For graphs of functions $\tilde{\mu} = \tilde{\mu}(\tau)$ being the solution of equations (4.1.33) and (4.1.38), as applied to various values of parameter I, (see Fig. **4.1.2**).

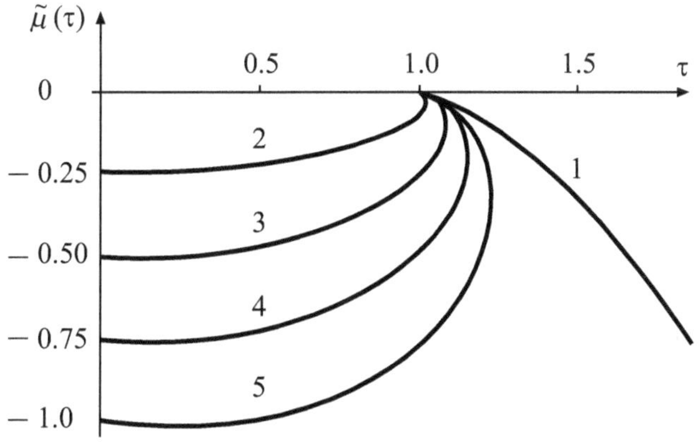

Fig. (4.1.2). Chemical potential $\tilde{\mu}$, as a τ-temperature function for various values of interaction parameter I: 1 – I = 0; 2 – I = 0,25; 3 – I = 0,5; 4 – I = 0,75; 5 – I = 1,0.

If $I = 0$, we produce the known chemical potential temperature dependence by applying the theory of boson interaction Bose condensation. If $I \neq 0$, the nature of chemical potential dependence on temperature is significantly changed. If $\tau = 0$, chemical potential $\tilde{\mu}(0)$ is equal to I. Function $\tilde{\mu}(\tau)$ is monotonically increased within the range of zero to critical value $\tau > 0$. If $\tau \in (1, \tau_c)$, function $\tilde{\mu}(\tau)$ starts zigzagging accepting three values from the interval against every value τ and joining to function $\tilde{\mu}(\tau)$ of respective $I = 0$ at point (1, 0). Value τ_c is a monotonically increased function of parameter I. If the concerned value is duly defined, the critical temperature $\vartheta_c = \tau_c\, \vartheta_B$ will be produced.

4.1.8. Order Parameter

If $\boldsymbol{k} = 0$, particle macroscopic number transition in its condensed state shall be referred to a specific change of phase. As provided by the phase transition theories, the level of particle-phase transformation is conventionally characterized by its η-order parameter. In the case analyzed, the order parameter shall be defined, as the ratio of condensed state particle number N_c to the full number of particles N:

$$\eta = N_c/N. \tag{4.1.39}$$

When formulae (4.1.36) and (4.1.37) are applied, it is not difficult to find the relationship between chemical potential and order parameter:

$$\mu = -c\,U\eta \quad \text{or} \quad \tilde{\mu} = -I\eta\,, \text{if } \vartheta \leq \vartheta_{\text{c}}\,. \qquad \textbf{(4.1.40)}$$

Substitution of the second expression (4.1.40) into formula (4.1.28) makes it possible to obtain the equation as follows:

$$\tau^{3/2}/J(0)\,J(\,I\,\eta/\tau) + \eta = 1\,, \qquad \textbf{(4.1.41)}$$

For graphs of function $\eta = \eta(\tau)$ being the solution of this equation, (see Fig. **4.1.3**). If $I = 0$, equation (4.1.41) produces the dependence as follows:

$$\eta = 1 - \tau^{3/2}\,, \qquad \textbf{(4.1.42)}$$

which describes particle interaction Bose condensation.

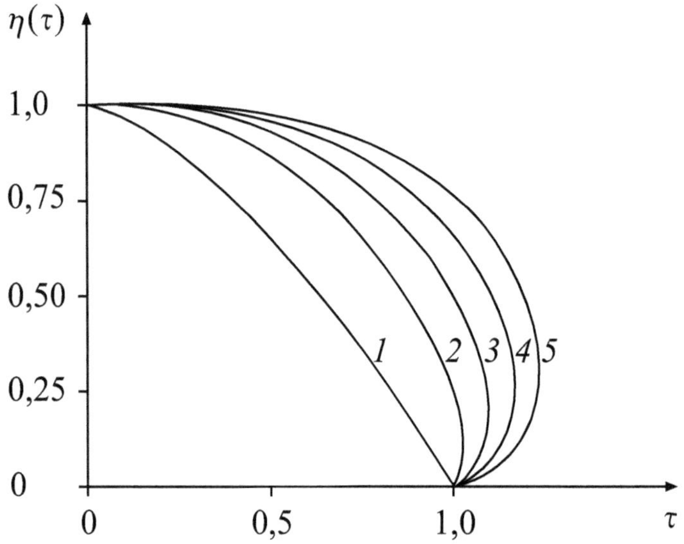

Fig. (4.1.3). Order parameter η, as a τ-temperature function for various values of interaction parameter I: 1 – I = 0; 2 – I = 0,25; 3 – I = 0,5; 4 – I = 0,75; 5 – I = 1,0.

4.1.9. Gas Internal Energy Dependence on Temperature

If we base on the particle pulse distribution function, we can find thermal dependence of gas internal energy by applying formula (4.1.16). As duly specified by the first approximation, *i.e.* less correction $\delta\bar{n}_{\boldsymbol{k}}$ and members containing $v_{\boldsymbol{k}-\boldsymbol{k}'}$, we will apply the form:

$$E = \textstyle\sum_{\boldsymbol{k}} \varepsilon_{\boldsymbol{k}}\,\bar{n}_{\boldsymbol{k}} - U/(2\,V)\sum_{\boldsymbol{k}} \bar{n}_{\boldsymbol{k}}^2\,. \qquad \textbf{(4.1.43)}$$

where constant summand $U\,N^2/V$ is rejected for simplicity and function $\bar{n}_k$ is set by formula (4.1.34) subject to an application for the purpose of transformation of expression (4.1.43) into the form as follows:

$$E = E_n + E_c\,, \tag{4.1.44}$$

where the first summand

$$E_n = \sum_{\boldsymbol{k}} \varepsilon_{\boldsymbol{k}}\, f_n[\,\beta\,(\,\varepsilon_{\boldsymbol{k}} - \,\mu\,)] \tag{4.1.45}$$

is particle energy in the normal state. If we substitute $\boldsymbol{k}$-summation for energy integration ε, the above summand may be described as follows:

$$E_n = N\,\vartheta_{\mathrm{B}}\,\tau^{5/2}/J(0)\,K(\,I\,\eta/\tau)\,, \tag{4.1.46}$$

where

$$K(x) = \int_0^\infty y\,\sqrt{y}/(\,\mathrm{e}^{\,x+y} - 1\,)\,\mathrm{d}y \tag{4.1.47}$$

Formula (4.1.45) is applicable for the temperatures of below critical point, *i.e.* when $\tau < \tau_c$. When temperature values exceed the one of Bose condensation, *i.e.* when $\tau > 1$, formula (4.1.45) will be adequate as given below :

$$E_n = N\,\vartheta_{\mathrm{B}}\,\tau^{5/2}/J(0)\,K(\,-\,\tilde{\mu}/\tau\,)\,, \tag{4.1.48}$$

where function $\tilde{\mu}(\tau)$ is the solution of equation (4.1.40).

The second summand in formula (4.1.44)

$$E_c = -\,c\,U\,N\,\eta^2/2 \tag{4.1.49}$$

is the condensed state particle energy.

Now, we apply dimensionless quantity

$$\tilde{\varepsilon} = E/(N\,\vartheta_{\mathrm{B}})\,, \tag{4.1.50}$$

which is rated as per one gas energy particle and measured in ϑ_{B}. Dependence of specific energy $\tilde{\varepsilon}$ on temperature is formulated as follows:

$$\tilde{\varepsilon}(\tau)=\tau^{5/2}/J(0)\,K(I\,\eta/\tau)-I\,\eta^2/2\,, \qquad (4.1.51)$$

at $\tau \leq \tau_c$.

If $I = 0$, we produce the following formula:

$$\tilde{\varepsilon}(\tau)=K(0)/J(0)\,\tau^{5/2} \qquad (4.1.52)$$

at $\tau \leq 1$.

Besides, if $\tau \geq 1$, the following formula is produced from (4.1.48)

$$\tilde{\varepsilon}(\tau)=\tau^{5/2}/J(0)\,K(-\tilde{\mu}/\tau) \qquad (4.1.53)$$

for particle specific energy in the normal state.

For the graph of function $\tilde{\varepsilon} = \tilde{\varepsilon}(\tau)$, as applied to interaction parameter value $I = 1$, (see Fig. **4.1.4**).

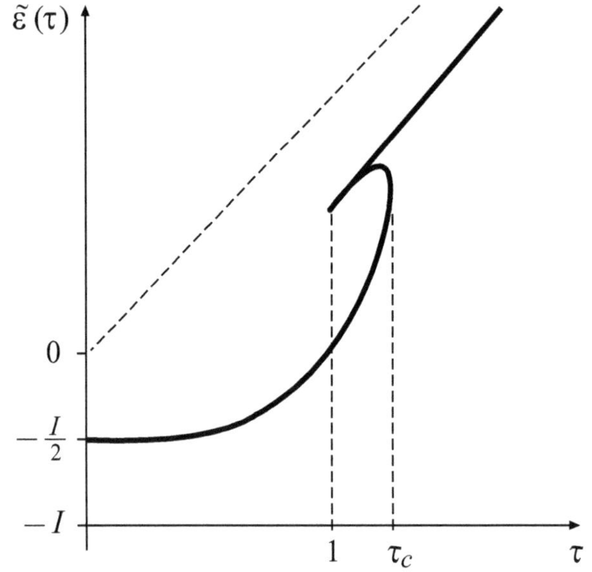

Fig. (4.1.4). Bose gas internal energy $\tilde{\varepsilon}$, as τ-temperature function for interaction parameter value I = 1.

For $I = 0$ dependence is represented, as monotonically increasing function to be reduced to zero at $\tau = 0$. If $\tau \in (1,\ \tau_c)$, the graph of function $\tilde{\varepsilon} = \tilde{\varepsilon}(\tau)$ starts zigzagging at any $I \neq 0$. Function $\tilde{\varepsilon} = \tilde{\varepsilon}(\tau)$ accepts three values for every value τ within this interval. If $\tau = 1$, the graph of function $\tilde{\varepsilon} = \tilde{\varepsilon}(\tau)$ is joined to that of the similar function of respective $I = 0$. The least energy state is stable against other several possible macrosystem states. Therefore, real dependence $\tilde{\varepsilon} = \tilde{\varepsilon}(\tau)$ at point $\tau = \tau_c$ will take up specific discontinuity. This means that the discontinuous decrease of gas internal energy shall occur at the temperature being dropped down to value T_c. As specified in the graph of order parameter dependence on temperature (see Fig. **4.1.3**), macroscopic number of particles shall pass to the condensed state at $T = T_c$. The rest normal-state particles shall condense, as far as the rate of temperature goes down. As a matter of fact, Bose gas condensation may be specified as heat release-accompanied phase transition. If $\tau = \tau_c$, the specific heat of phase transition is defined by the rate of discontinuity of function $\tilde{\varepsilon} = \tilde{\varepsilon}(\tau)$.

4.1.10. Heat Capacity of Gas

As it is finally stated by the data above, we can use the following Bose gas specific energy formula:

$$\tilde{\varepsilon}(\tau) = \begin{cases} \tau^{5/2}/J(0)\, K(I\,\eta/\tau) - I\,\eta^2/2 & \text{at} \quad \tau \leq \tau_c\,, \\ \tau^{5/2}/J(0)\, K(-\,\tilde{\mu}/\tau) & \text{at} \quad \tau \geq 1\,. \end{cases} \qquad \textbf{(4.1.54)}$$

where dependence $\eta = \eta(\tau)$ is defined by equation (4.1.41) and dependence $\mu = \mu(\tau)$ – by equation (4.1.40). Applying this formula, we can find gas heat capacity dependence

$$C_V = \mathrm{d}E/\mathrm{d}T$$

on temperature. As provided by ratio (4.1.50), *i.e.* $E = N\,\vartheta_{\mathrm{B}}\,\tilde{\varepsilon}(\tau)$, the rate of heat capacity may be calculated by the following formula:

$$C_V = N\,k_{\mathrm{B}}\,\mathrm{d}\tilde{\varepsilon}/\mathrm{d}\tau\,. \qquad \textbf{(4.1.55)}$$

For graphs of dependence of derivative $\mathrm{d}\tilde{\varepsilon}/\mathrm{d}\tau$ on temperature τ, as applied to parameter $I = 1$, see (Fig. **4.1.5**). If any value of $I > 0$ is applied, the rate of Bose gas heat capacity for $\tau \in (0, \tau_c)$ is monotonically increased from zero to infinity. If $\tau > \tau_c$, particle interaction in the approximation involved does not affect temperature dependences of internal energy and gas heat capacity. Therefore, if $\tau > \tau_c$, the above dependencies are similar to those produced by ideal Bose gas.

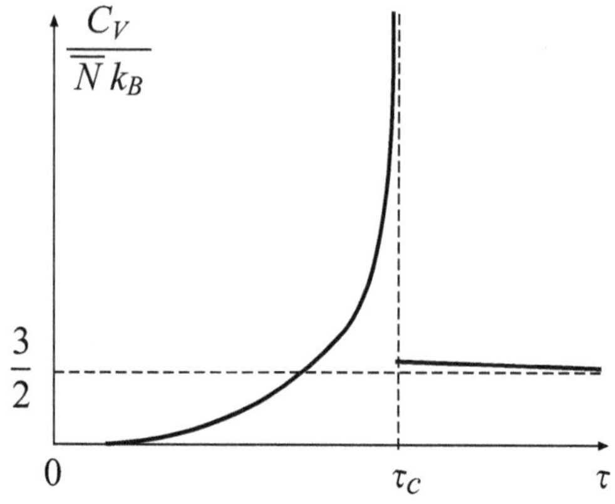

Fig. (4.1.5). Bose gas heat capacity as a function of temperature τ for interaction parameter value $I = 1$.

4.1.11. Energy Spectrum of Particles

Now, let us find $\varepsilon_{\boldsymbol{k}}$-energy dependence of a particle on wave vector $\boldsymbol{k}$. Interaction of one particle with other gas particles makes energy obtain functions of the particle pulse distribution . If stated in the approximation of statistically independent particles, this function is defined by formula (4.1.18). As provided by formulae (4.1.20) and (4.1.34), the equilibrium distribution function may be defined by the following expression:

$$\bar{n}_{\boldsymbol{k}} = N_c\,\delta_{\boldsymbol{k}} + f_n[\,\beta\,(\,\varepsilon_{\boldsymbol{k}} - \mu\,)\,] + \delta\bar{n}_{\boldsymbol{k}}\,. \qquad \textbf{(4.1.56)}$$

Substitution of this expression into formula (4.1.18) gives the following expression:

$$\bar{\varepsilon}_{\boldsymbol{k}} = \varepsilon_{\boldsymbol{k}} + c\,\eta\,(\,-\,U\,\delta_{\boldsymbol{k}} + v_{\boldsymbol{k}}\,)\,. \qquad \textbf{(4.1.57)}$$

While deriving this formula, it is taken into account that due to availability of a very small multiplier $1/V$ in formula (4.1.18) the macroscopic large bracketed summands only, *i.e.* those containing number N_c, may be specified as the substantial ones.

4.1.12. Superfluidity

Now we consider Bose gas at the temperature of $T = 0$ to be in the state described by a distribution function:

$$\bar{n}_{\boldsymbol{k}} = N_1 \, \delta_{\boldsymbol{k}\boldsymbol{k}_1} + N_2 \, \delta_{\boldsymbol{k}\boldsymbol{k}_2} \,, \tag{4.1.58}$$

where

$$N_1 + N_2 = N, \qquad \boldsymbol{k}_1 \neq \boldsymbol{k}_2 \,. \tag{4.1.59}$$

Formula (4.1.58) means that all the gas particles are distributed into two groups. Each particle of the first group has a pulse stated as $\boldsymbol{p}_1 = \hbar \, \boldsymbol{k}_1$. The particle number is equal to N_1. As for each particle in the second group, its pulse is stated as $\boldsymbol{p}_2 = \hbar \, \boldsymbol{k}_2$ and the number is equal to N_2.

Now, we find gas energy in this state by applying the following formula (4.1.43):

$$E = \varepsilon_{\boldsymbol{k}_1} N_1 + \varepsilon_{\boldsymbol{k}_2} N_2 - U \, (N_1^2 + N_2^2)/(2 \, V) \,. \tag{4.1.60}$$

We can prove that gas being in the state when all particles migrate at similar velocity remains stable, *i.e.* gas may be reconditioned under external effects only, but not spontaneously. This is possible when other gas states close to that under consideration have exceeded energy. Let us assume that number $N_1 = N$ at a certain time t_1, *i.e.* all gas particles will have pulse $\boldsymbol{p}_1$. In this case, $N_2 = 0$, and gas energy is expressed by the following formula:

$$E = \varepsilon_{\boldsymbol{k}_1} N_1 - U \, N/(2 \, V) \,. \tag{4.1.61}$$

Let us assume that at some time $t_2 > t_1$ one or several particles ($N_2 \geq 1$) gain pulse $\boldsymbol{p}_2$. As a result, gas energy takes on value E_2 defined by formula (4.1.60), *i.e.* it increments.

$$\Delta E = E_2 - E_1 = (\varepsilon_{\boldsymbol{k}_2} - \varepsilon_{\boldsymbol{k}_1}) \, N_2 + \varepsilon_{\boldsymbol{k}_2} N_2 + U \, (N - N_2) \, N_2/V \,. \tag{4.1.62}$$

If $N_2 = N_0$, function (4.1.62) has a maximum, where

$$N_0 = \left[c + \left(\varepsilon_{\boldsymbol{k}_2} - \varepsilon_{\boldsymbol{k}_1} \right)/U \right] V/2 \,. \tag{4.1.63}$$

As $U > 0$, function ΔE of N_2 will be positive for $N_2 \in (1, 2 \, N_0)$ provided that $N_0 > 0$, *i.e.* when

$$\varepsilon_{\boldsymbol{k}_1} < \varepsilon_{\boldsymbol{k}_2} + c \, U \,. \tag{4.1.64}$$

If $\boldsymbol{k}_2 = 0$, value ΔE takes on the least value like the function of $\boldsymbol{k}_2$. Thus, to stop one of the particles within the flux, it shall be duly energized

$$\Delta E_1 = - \varepsilon_{\boldsymbol{k}_1} + (N - 1) U/V .$$

This value will be positive, if

$$\varepsilon_{\boldsymbol{k}_1} < (N - 1) U/V , \qquad \textbf{(4.1.65)}$$

i.e. provided that $\boldsymbol{k}_1 < \boldsymbol{k}_c$, where the critical value is represented as follows:

$$k_c = \sqrt{2\, m\, c\, U}/\hbar .$$

The respective critical value of speed will take the following form:

$$v_c = \sqrt{2\, c\, U/m} .$$

Condition (4.1.65) is stronger than condition (4.1.64), *i.e.* condition (4.1.64) is also executed for all $\mathbf{k}_1$ satisfying (4.1.65).

So, to take one or several $(1 \leq N_2 \leq 2N_0)$ particles out of the flux with the particles running at velocity $\upsilon < \upsilon_c$, it is necessary to additionally energize them. The simultaneous and spontaneous transition of a macroscopic number of particles from state $\boldsymbol{k}_1$ into particular state $\boldsymbol{k}_2$ is hardly probable. Since this kind of spontaneous transition is hardly probable, state of gas, where all particles migrate at similar velocity v, will be stable, provided that $\upsilon < \upsilon_c$. This means that Bose gas exhibits its superfluidity property.

If required, the aforesaid arguments and calculations may be repeated, providing that gas temperature fits the condition,

$$0 < T < T_c.$$

In this case, the least energy, when one of numbers N_1 or N_2 is equal to zero, provided that vectors $\boldsymbol{k}_1$ and $\boldsymbol{k}_2 \neq \boldsymbol{k}_1$ satisfy inequation (4.1.65), will also correspond to the state described by the following distribution function

$$\bar{n}_{\boldsymbol{k}} = N_1\, \delta_{\boldsymbol{k}\boldsymbol{k}_1} + N_2\, \delta_{\boldsymbol{k}\boldsymbol{k}_2} + f_n[\, \beta\, (\varepsilon_{\boldsymbol{k}} - \mu)\,],$$

where $N_1 + N_2 = N$. Thus, we can summarize that critical velocity does not depend on temperature and the particles producing or having produced condensate tend to gain similar velocity. A "team" of particles prevents individual particles and large groups of particles from leaving off. So, if $\upsilon < \upsilon_c$, the condensate particles involved inside the directed movement and left to their own will migrate at their constant velocity as long as possible. But ganged particles exhibit joint directed motion to find out inside the space areas which dimensions do not exceed the coherence length. In this areas, particles move according to a quantum mechanics law. The particles spaced by a distance exceeding the coherence length may have different velocities distributed in space, as described by microscopic hydrodynamics laws. Since the directed motion of particles of the superfluid gas component saves its stability against changes of individual particle velocities, any gas convection current produced by temperature gradients will be free of attenuation over time. Such currents tend to cause condensate exhaustion. Zero-velocity particles may have very low fraction even at low temperatures produced at macroscopic space areas.

4.1.13. Conclusion

Now, having written the last formula, the matter of theory applications is arisen. What kind of real multifrequency system could be described with the help of the obtained relations? It may be assumed that this system may be represented by liquid He^4. The likeness of liquid helium and quantum Bose gas thermal capacity temperature dependences argues in favor of this assumption. Herewith, the η-order parameter may be interpreted as the relation of superfluid component densities ϱ_c/ϱ to the full liquid density. The temperature dependence of an order parameter, as specified in this paper, resembles experimental dependence of ratio ϱ_c/ϱ on temperature. But the literature about liquid helium provides no information of discontinuity of internal energy being transmitted into its superfluid state. It is known that medium field theories refer to the first approximation only accounting for the interacted particle effect produced on thermodynamic particle properties. Perhaps, the internal energy jump at $T = T_c$ is caused by irregularity of the accepted approximation and any correlations calculated would result in clarifying the gas internal energy temperature dependence.

This work is chiefly targeted to logical formulation of quantum Bose gas thermodynamic properties under the theory of density matrix. The author suggests analyzing compliance of this model with any real multifrequency system and matching theoretical relations with experimental results in an individual paper.

REFERENCE

[1] B.V. Bondarev, "Application of the variational density matrix method to describe the thermodynamic properties of quantum Bose gas". *Vestnik Moscow Aviation Institute*, vol. 5, no. 2, pp. 53-60, 1998.

CHAPTER 5

New Theory of Arbitrary Atom

5.1. METHOD OF DENSITY MATRIX NEW CALCULATION OF ENERGY LEVEL OF ELECTRONS IN ATOM

In this chapter, we apply a stationary Shrödinger equation, the solution of which allows us to find the energy levels of electrons in an arbitrary atom. This approach is based on the density matrix method. The density matrix was used to find the average electron energy in the atom. **The main idea of this paper is that all two-electron matrices must be antisymmetric.** To ensure that the two-electron Hamiltonian is anti-symmetric we use the Slater two-particle wave function.

5.1.1. Introduction

The density matrix is the most general description of the system in quantum mechanics. The state of the system, which is described by the density matrix, is called mixed. In a particular case, the density matrix can be proportional to the product of two wave functions. In this case, the state of the system is called pure.

At first, the concept of density matrix was defined [1, 2]. The method of calculation of energies in the spectrum of an arbitrary atom, which is called the **Hartree – Fock** method, is used. Based on the variational principle, this method applies different combinations of single-electron wave functions that take into account their anti-symmetry. This makes electrons to be fermions.

Here, the density matrix method is used to derive the average energy of electrons in an atom. The Slater wave function is used to obtain an anti-symmetric two-electron Hamiltonian. The stationary Schrödinger equation for the wave function is written. The solution of this equation makes it possible to find the energy levels of electrons in an atom.

5.1.2. Statistical Operator

The system of identical particles in quantum mechanics is characterized by a hierarchical sequence of statistical operators as given below:

$$\hat{\varrho}^{(1)}, \hat{\varrho}^{(2)}, \hat{\varrho}^{(3)}, \ldots$$

Boris V. Bondarev

We need only the first two operators from this sequence: $\hat{\varrho}^{(1)}$ and $\hat{\varrho}^{(2)}$. Let q_1 and q_2 be quantum coordinates that determine the States of two particles of the system. The first of these operators $\hat{\varrho}^{(1)}$ depends only on one of these numbers:

$$\hat{\varrho}^{(1)} = \hat{\varrho}^{(1)}(q)\,. \tag{5.1.2.1}$$

The statistical operator $\hat{\varrho}^{(2)}$ depends on these two numbers:

$$\hat{\varrho}^{(2)} = \hat{\varrho}^{(2)}(q_1, q_2)\,. \tag{5.1.2.2}$$

Electrons are identical particles. The consequence of the indistinguishability of two electrons is the symmetry of the operator $\hat{\varrho}^{(2)}$, *i.e.*

$$\hat{\varrho}^{(2)}(q_1, q_2) = \hat{\varrho}^{(2)}(q_2, q_1)\,. \tag{5.1.2.3}$$

By definition, we will use equality

$$\mathrm{Tr}_q\, \hat{\varrho}^{(1)}(q) = N\,, \tag{5.1.2.4}$$

where N is the number of particles in the system;

$$\mathrm{Tr}_{q_1 q_2}\, \hat{\varrho}^{(2)}(q_1, q_2) = N(N-1)\,, \ldots \tag{5.1.2.5}$$

5.1.3. Density Matrix

The density matrix of any system in quantum mechanics is of the form $\varrho_{\alpha\alpha'}$, where α and α' are quantum numbers that determine the States of one particle. The density matrix corresponds to the statistical operator $\hat{\varrho}^{(1)}$. This correspondence is determined by the following formula:

$$\varrho_{\alpha\alpha'} = \int \varphi^*_\alpha(q)\, \hat{\varrho}^{(1)}(q)\, \varphi_{\alpha'}(q)\, \mathrm{d}q\,, \tag{5.1.3.1}$$

where $\varphi_\alpha(q)$ is some wave function satisfying the ortho-normalization condition:

$$\int \varphi^*_\alpha(q)\, \varphi_{\alpha'}(q)\, \mathrm{d}q = \delta_{\alpha\alpha'}\,. \tag{5.1.3.2}$$

where $\delta_{\alpha\alpha'}$ – Kronecker symbol.

The meaning of the density matrix is that the diagonal elements $\varrho_{\alpha\alpha'}$ of this matrix are equal to the probability

$$w_\alpha = \varrho_{\alpha\alpha}$$

detecting the system in the state α. Probability w_α, according to (5.1.2.4), satisfies the normalization condition as follows:

$$\sum_\alpha w_\alpha = N, \qquad \textbf{(5.1.3.3)}$$

The two-part operator $\hat{\varrho}^{(2)}$ corresponds to the density matrix $\varrho^{(2)}$:

$$\varrho^{(2)} = \varrho_{\alpha_1\alpha_2,\alpha_1'\alpha_2'} . \qquad \textbf{(5.1.3.4)}$$

Density matrix in abbreviated form can be written in any coordinate as follows:

$$\varrho^{(1)} = \varrho_{11'} , \qquad \varrho^{(2)} = \varrho_{12,1'2'} .$$

Since electrons are fermions, their two-electron density matrix must be anti-symmetric. This means that the ratios are fair:

$$\varrho_{12,1'2'} = - \varrho_{21,1'2'} = - \varrho_{12,2'1'} = \varrho_{21,2'1'} . \qquad \textbf{(5.1.3.5)}$$

The anti-symmetry ratio can be approximately satisfied if the two-electron density matrix is approximately equal to

$$\varrho_{12,1'2'} \simeq \varrho_{11'} \varrho_{22'} - \varrho_{12'} \varrho_{21'} . \qquad \textbf{(5.1.3.6)}$$

Exactly two-electron density matrix will satisfy the ratio (5.1.3.5), if it has the form

$$\varrho_{12,1'2'} = \varrho_{11'} \varrho_{22'} - \varrho_{12'} \varrho_{21'} + \xi_{12,1'2'} .$$

Here, the correlation function $\xi_{12,1'2'}$ is symmetric:

$$\xi_{12,1'2'} = \xi_{21,1'2'} = \xi_{12,2'1'} = \xi_{21,2'1'} .$$

One-electron density matrix is related to the two-electron matrix by the ratio

$$\sum_{\alpha_2} \varrho_{12,1'2} = (N-1)\, \varrho_{11'} . \qquad \textbf{(5.1.3.7)}$$

Let's sum is the ratio of α_1. Condition (3.3) gives

$$\sum_{\alpha_1} \sum_{\alpha_2} \varrho_{12,12} = (N-1)N. \quad \textbf{(5.1.3.8)}$$

5.1.4. Hamiltonian of the Electron System

Consider a system that consists of N electrons. The system of identical electrons will be characterized by a hierarchical sequence of Hamiltonians

$$\widehat{H}^{(1)}, \widehat{H}^{(2)}, \widehat{H}^{(3)}, \ldots$$

Only the first two of this sequence will be needed:

$$\widehat{H}^{(1)} = \widehat{H}^{(1)}(q), \quad \widehat{H}^{(2)} = \widehat{H}^{(2)}(q_1, q_2). \quad \textbf{(5.1.4.1)}$$

Electrons are identical particles. Therefore,

$$\widehat{H}^{(2)}(q_1, q_2) = \widehat{H}^{(2)}(q_2, q_1). \quad \textbf{(5.1.4.2)}$$

Now let us define the physical meaning of the Hamiltonian. The average energy E of the system of N electrons will be equal to:

$$E = \int \widehat{H}^{(1)}\, \hat{\varrho}\, dq + 1/2 \int \widehat{H}^{(2)}\, \hat{\varrho}^{(2)}\, dq_1\, dq_2, \quad \textbf{(5.1.4.3)}$$

According to this formula, we can say that the Hamiltonian $\widehat{H}^{(1)}$ gives the energy of electrons that do not take into account the energy of their interaction, and the Hamiltonian $\widehat{H}^{(2)}$ describes the energy of the interaction of two electrons.

5.1.5. Matrix Elements of the Hamiltonian

The Hamiltonian of the system of non-interacting electrons in the α-representation is determined by the formula similar to (5.1.3.1):

$$H_{\alpha\alpha'} = \int \varphi^*_{\alpha}(q)\, \widehat{H}^{(1)}\, \varphi_{\alpha'}(q)\, dq, \quad \textbf{(5.1.5.1)}$$

The Hamiltonian of the system of interacting electrons in the α-representation is determined accordingly.

$$H_{\alpha_1\alpha_2,\alpha'_1\alpha'_2} = \int \Phi^*_{\alpha_1\alpha_2}\, \widehat{H}^{(2)}\, \Phi_{\alpha'_1\alpha'_2}\, dq_1\, dq_2, \quad \textbf{(5.1.5.2)}$$

where $\Phi_{\alpha_1\alpha_2} = \Phi_{\alpha_1\alpha_2}(q_1, q_2)$ – some function of two electrons.

Electrons are fermions. Therefore, two-electron matrix elements must be anti-symmetric, *i.e.*

$$H_{12,1'2'} = -H_{21,1'2'} = -H_{12,2'1'} = H_{21,2'1'} \ . \qquad \textbf{(5.1.5.3)}$$

where marked

$$H_{\alpha_1\alpha_2,\,\alpha_1'\alpha_2'} = H_{12,1'2'} \ .$$

Hence, the function

$$\Phi_{\alpha_1\alpha_2} = \Phi_{\alpha_1\alpha_2}(q_1, q_2)$$

must be a Slater two-particle wave function:

$$\Phi_{\alpha_1\alpha_2}(q_1, q_2) = 1/\sqrt{2}\,[\,\varphi_{\alpha_1}(q_1)\,\varphi_{\alpha_2}(q_2) - \varphi_{\alpha_1}(q_2)\,\varphi_{\alpha_2}(q_1)\,] \, . \qquad \textbf{(5.1.5.4)}$$

Function (5.1.5.4) satisfies the normalization condition:

$$\begin{aligned} &\int \Phi^*_{\alpha_1\alpha_2}(q_1, q_2)\ \Phi_{\alpha_1'\alpha_2'}(q_1, q_2)\, dq_1\, dq_2 = \\ &= 1/2 \int [\,\varphi^*_{\alpha_1}(q_1)\,\varphi^*_{\alpha_2}(q_2) - \varphi^*_{\alpha_1}(q_2)\,\varphi^*_{\alpha_2}(q_1)\,] \cdot \\ &\cdot [\,\varphi_{\alpha_1'}(q_1)\,\varphi_{\alpha_2'}(q_2) - \varphi_{\alpha_1'}(q_2)\,\varphi_{\alpha_2'}(q_1)\,]\, dq_1\, dq_2 = \\ &= \delta_{\alpha_1\alpha_1'}\,\delta_{\alpha_2\alpha_2'} - \delta_{\alpha_2\alpha_1'}\,\delta_{\alpha_1\alpha_2'} \ . \end{aligned}$$

(5.1.5.5)

If $\alpha_1' = \alpha_1$ and $\alpha_2' = \alpha_2$, this formula gives

$$\int \Phi^*_{\alpha_1\alpha_2}(q_1, q_2)\ \Phi_{\alpha_1\alpha_2}(q_1, q_2)\, dq_1\, dq_2 = 1 \, . \qquad \textbf{(5.1.5.6)}$$

If $\alpha_1' = \alpha_2$ and $\alpha_2' = \alpha_1$, the formula (5.1.5.5) has the form

$$\int \Phi^*_{\alpha_1\alpha_2}(q_1, q_2)\ \Phi_{\alpha_2\alpha_1}(q_1, q_2)\, dq_1\, dq_2 = -1 \, . \qquad \textbf{(5.1.5.7)}$$

Substituting the function (5.1.5.4) in the formula (5.1.5.2), we obtain an anti-symmetric matrix

$$H_{12,1'2'} = 1/2\,(V_{12,1'2'} - V_{21,1'2'} - V_{12,2'1'} + V_{21,2'1'}), \qquad \textbf{(5.1.5.8)}$$

where

$$V_{12,1'2'} = \int \varphi^*_{\alpha_1}(q_1)\,\varphi^*_{\alpha_2}(q_2)\,\hat{H}^{(2)}\,\varphi_{\alpha'_1}(q_1)\,\varphi_{\alpha'_2}(q_2)\,\mathrm{d}q_1\,\mathrm{d}q_2\,. \qquad \textbf{(5.1.5.9)}$$

5.1.6. Wave Function of One Electron Moving Around an Arbitrary Nucleus

In this case, the electron has only one Hamiltonian:

$$\hat{H}^{(1)} = -\,\hbar^2/(2\,m)\,\nabla^2 - Z\,e^2/r\,, \qquad \textbf{(5.1.6.1)}$$

where m и e – mass and charge of an electron, Z – number of electrons in an atom.

As the function $\varphi_\alpha(q)$ we will use the eigenfunctions of an electron in an atom that consists of a nucleus and a single electron, *i.e.* this function is the solution of the stationary Schrödinger equation:

$$\hat{H}^{(1)}\,\varphi_\alpha = \varepsilon^{(\mathrm{o})}_\alpha\,\varphi_\alpha\,. \qquad \textbf{(5.1.6.2)}$$

Here, $\varepsilon^{(\mathrm{o})}_\alpha$ is the electron's energy. Taking into account Hamiltonian (5.1.6.1), we write equation (6.2) in more detail as follows:

$$[-\,\hbar^2/(2\,m)\,\nabla^2 - Z\,e^2/r\,]\,\varphi_\alpha = \varepsilon^{(\mathrm{o})}_\alpha\,\varphi_\alpha. \qquad \textbf{(5.1.6.3)}$$

We introduce an orthonormal system of functions that describes only one electron moving around an arbitrary nucleus:

$$\varphi_\alpha(q) \equiv \phi_s(\boldsymbol{r})\,\chi_\zeta(\sigma)\,, \qquad \textbf{(5.1.6.4)}$$

where $q = \{\boldsymbol{r}, \sigma\}$, $\phi_s(\boldsymbol{r})$, $\alpha = \{s, \zeta\}$ there is a set of quantum numbers that determine the state of the electron; $\phi_s(\boldsymbol{r})$ is a wave function of the electron, $\chi_\zeta(\sigma)$ – spin function. The function (6.3) satisfies the normalization conditions (5.1.3.2). Given that $q = \{\boldsymbol{r}, \sigma\}$, the normalization condition can be written as:

$$\int \phi^*_s(\boldsymbol{r})\,\phi_{s'}(\boldsymbol{r})\,\mathrm{d}\boldsymbol{r} = \delta_{ss'}\,, \qquad \textbf{(5.1.6.5)}$$

$$\textstyle\sum_{\sigma=-1/2}^{1/2} \chi^*_\zeta(\sigma)\,\chi_{\zeta'}(\sigma) = \delta_{\zeta\zeta'}. \qquad \textbf{(5.1.6.6)}$$

Function $\varphi_\alpha(q)$, depending on the polar coordinates r, ϑ and φ, has the form:

$$\varphi_\alpha(q) \equiv \phi_s(\boldsymbol{r})\, \chi_\zeta(\sigma) = R_{nl}(r)\, Y_{lm}(\vartheta, \varphi)\, \chi_\zeta(\sigma)\,, \tag{5.1.6.7}$$

Parameters

$$\alpha = \{n, l, m, \zeta\}\,, \qquad s = \{n, l, m\}\,, \tag{5.1.6.8}$$

where

$$n = 1, 2, \ldots$$

– main quantum number,

$$l = 1, 2, \ldots, \ n - 1$$

there is an orbital number,

$$m = 0, \pm 1, \pm 2 \ldots, \pm (\, l - 1\,), \pm\, l$$

there is the magnetic quantum number, $R_{nl}(r)$ is the radial function of the module r is the radius vector $\boldsymbol{r}$, $Y_{lm}(\vartheta, \varphi)$ – own function of the operator of the angular momentum $\boldsymbol{M}^2$ ($\widehat{\boldsymbol{M}} = [\,\widehat{\boldsymbol{r}}\,\widehat{\boldsymbol{p}}\,]$), which depends on the polar coordinates ϑ and φ. The number α for a fixed number n takes $2\,n^2$ different values. The eigenvalue $\varepsilon_\alpha^{(0)}$ is described by the following formula:

$$\varepsilon_\alpha^{(0)} = \varepsilon_n^{(0)} = -\, Z^2 m\, e^4 \,/(2\, \hbar^2 n^2) = -\, Z^2 R\, \hbar \,/n^2, \tag{5.1.6.9}$$

where R – Rydberg constant.

Now we will find the matrix elements $H_{\alpha\alpha'}$:

$$H_{\alpha\alpha'} = \int \varphi_\alpha^*(q)\, \widehat{H}^{(1)}\, \varphi_{\alpha'}(q)\, \mathrm{d}q = \varepsilon_n^{(0)} \int \varphi_\alpha^*(q)\, \varphi_{\alpha'}(q)\, \mathrm{d}q = \varepsilon_n^{(0)}\, \delta_{\alpha\alpha'}\,. \tag{5.1.6.10}$$

5.1.7. Matrix Elements of the Electron Interaction Hamiltonian

In a real atom, electrons repel each other. This action is characterized by potential energy

$$\hat{H}^{(2)} = e^2/r_{12} \ , \qquad \textbf{(5.1.7.1)}$$

where r_{12} is the distance between two electrons from the total number Z. The value (5.1.5.9) is equal to

$$V_{12,1'2'} = \int \varphi^*_{\alpha_1}(q_1)\, \varphi^*_{\alpha_2}(q_2)\, e^2/r_{12}\, \varphi_{\alpha'_1}(q_1)\, \varphi_{\alpha'_2}(q_2)\, \mathrm{d}q_1\, \mathrm{d}q_2 \, . \qquad \textbf{(5.1.7.2)}$$

Substituting here the function (5.1.6.4), we will have:

$$\begin{aligned} V_{12,1'2'} &= e^2 \textstyle\sum_{\zeta_1\zeta_2=-1/2}^{1/2} \chi^*_{\sigma_1}(\zeta_1)\, \chi^*_{\sigma_2}(\zeta_2)\, \chi_{\sigma'_1}(\zeta_1)\, \chi_{\sigma'_2}(\zeta_2)\, \cdot \\ &\cdot \int \phi^*_{s_1}(\boldsymbol{r}_1)\, \phi^*_{s_2}(\boldsymbol{r}_2)\, 1/|\,\boldsymbol{r}_1 - \boldsymbol{r}_2\,|\, \phi_{s'_1}(\boldsymbol{r}_1)\, \phi_{s'_2}(\boldsymbol{r}_2)\, \mathrm{d}\boldsymbol{r}_1\, \mathrm{d}\boldsymbol{r}_2 = \\ &= e^2 \textstyle\sum_{\zeta_1=-1/2}^{1/2} \chi^*_{\sigma_1}(\zeta_1)\, \chi_{\sigma'_1}(\zeta_1) \sum_{\zeta_2=-1/2}^{1/2} \chi^*_{\sigma_2}(\zeta_2)\, \chi_{\sigma'_2}(\zeta_2) \cdot \\ &\cdot \int \phi^*_{s_1}(\boldsymbol{r}_1)\, \phi^*_{s_2}(\boldsymbol{r}_2)\, 1/|\,\boldsymbol{r}_1 - \boldsymbol{r}_2\,|\, \phi_{s'_1}(\boldsymbol{r}_1)\, \phi_{s'_2}(\boldsymbol{r}_2)\, \mathrm{d}\boldsymbol{r}_1\, \mathrm{d}\boldsymbol{r}_2 \, . \end{aligned}$$

Taking into account the conditions of normalization (5.1.6.6) of spin functions, we obtain

$$V_{12,1'2'} = e^2\, \delta_{\sigma_1\sigma'_1}\, \delta_{\sigma_2\sigma'_2}\, \Phi(s_1, s_2;\ s'_1, s'_2) \, , \qquad \textbf{(5.1.7.3)}$$

where the function is indicated

$$\Phi(s_1, s_2;\ s'_1, s'_2) = \int \phi^*_{s_1}(\boldsymbol{r}_1)\, \phi^*_{s_2}(\boldsymbol{r}_2)\, 1/|\,\boldsymbol{r}_1 - \boldsymbol{r}_2\,|\, \phi_{s'_1}(\boldsymbol{r}_1)\, \phi_{s'_2}(\boldsymbol{r}_2)\, \mathrm{d}\boldsymbol{r}_1\, \mathrm{d}\boldsymbol{r}_2 \, . \qquad \textbf{(5.1.7.4)}$$

Let us substitute matrix (5.1.7.3) in expression (5.1.5.6) for matrix elements of the Hamiltonian of electron interaction, we will have:

$$\begin{aligned} H_{12,1'2'} &= e^2/2\, [\, \delta_{\sigma_1\sigma'_1}\, \delta_{\sigma_2\sigma'_2}\, \Phi(s_1, s_2;\ s'_1, s'_2) - \delta_{\sigma_2\sigma'_1}\, \delta_{\sigma_1\sigma'_2}\, \Phi(s_2, s_1;\ s'_1, s'_2) - \\ &- \delta_{\sigma_1\sigma'_2}\, \delta_{\sigma_2\sigma'_1}\, \Phi(s_1, s_2;\ s'_2, s'_1) + \delta_{\sigma_2\sigma'_2}\, \delta_{\sigma_1\sigma'_1}\, \Phi(s_2, s_1;\ s'_2, s'_1)\,] = \\ &= e^2/2 \{\, \delta_{\sigma_1\sigma'_1}\, \delta_{\sigma_2\sigma'_2}\, [\, \Phi(s_1, s_2;\ s'_1, s'_2) + \Phi(s_2, s_1;\ s'_2, s'_1)\,] - \\ &- \delta_{\sigma_2\sigma'_1}\, \delta_{\sigma_1\sigma'_2}\, [\, \Phi(s_2, s_1;\ s'_1, s'_2) + \Phi(s_1, s_2;\ s'_2, s'_1)\,]\, \} \, . \end{aligned}$$

(5.1.7.5)

Compare the two functions $\Phi(s_2, s_1;\ s'_2, s'_1)$ and $\Phi(s_1, s_2;\ s'_1, s'_2)$:

$$\Phi(s_2, s_1;\ s'_2, s'_1) = \int \phi^*_{s_2}(\boldsymbol{r}_1)\, \phi^*_{s_1}(\boldsymbol{r}_2)\, 1/|\,\boldsymbol{r}_1 - \boldsymbol{r}_2\,|\, \phi_{s'_2}(\boldsymbol{r}_1)\, \phi_{s'_1}(\boldsymbol{r}_2)\, \mathrm{d}\boldsymbol{r}_1\, \mathrm{d}\boldsymbol{r}_2 = (s_1, s_2;\ s'_1, s'_2) \, . \qquad \textbf{(5.1.7.6)}$$

These two functions were equal to each other.

Now let's compare functions $\Phi(s_2, s_1;\ s_1', s_2')$ and $\Phi(s_1, s_2;\ s_2', s_1')$:

$$\Phi(s_2, s_1;\ s_1', s_2') = \int \phi^*_{s_2}(\boldsymbol{r}_1)\, \phi^*_{s_1}(\boldsymbol{r}_2)\, 1/|\,\boldsymbol{r}_1 - \boldsymbol{r}_2\,|\, \phi_{s_1'}(\boldsymbol{r}_1)\, \phi_{s_2'}(\boldsymbol{r}_2)\, \mathrm{d}\boldsymbol{r}_1\, \mathrm{d}\boldsymbol{r}_2\,. \qquad \textbf{(5.1.7.7)}$$

$$\Phi(s_1, s_2;\ s_2', s_1') = \int \phi^*_{s_1}(\boldsymbol{r}_1)\, \phi^*_{s_2}(\boldsymbol{r}_2)\, 1/|\,\boldsymbol{r}_1 - \boldsymbol{r}_2\,|\, \phi_{s_2'}(\boldsymbol{r}_1)\, \phi_{s_1'}(\boldsymbol{r}_2)\, \mathrm{d}\boldsymbol{r}_1\, \mathrm{d}\boldsymbol{r}_2 =$$
$$= \int \phi^*_{s_1}(\boldsymbol{r}_2)\, \phi^*_{s_2}(\boldsymbol{r}_1)\, 1/|\,\boldsymbol{r}_1 - \boldsymbol{r}_2\,|\, \phi_{s_2'}(\boldsymbol{r}_2)\, \phi_{s_1'}(\boldsymbol{r}_1)\, \mathrm{d}\boldsymbol{r}_1\, \mathrm{d}\boldsymbol{r}_2 = (s_2, s_1;\ s_1', s_2')\,. \qquad \textbf{(5.1.7.8)}$$

Functions (5.1.7.7) and (5.1.7.8) are equal; therefore, we have in equation (5.1.7.5)

$$H_{12,1'2'} = e^2\,[\,\delta_{\sigma_1\sigma_1'}\,\delta_{\sigma_2\sigma_2'}\,\Phi(s_1, s_2;\ s_1', s_2') - \delta_{\sigma_2\sigma_2'}\,\delta_{\sigma_1\sigma_1'}\,\Phi(s_2, s_1;\ s_2', s_1')\,]. \qquad \textbf{(5.1.7.9)}$$

5.1.8. The Energy of the Electrons in Core is Recorded Using Density Matrix

Unknown density matrix $\varrho_{\alpha\alpha'}$ and $\varrho_{12,1'2'}$, written in the α-representation, allow to transform the expression (5.1.4.3) which can be written as:

$$E = \textstyle\sum_{\alpha\alpha'} H_{\alpha\alpha'}\,\varrho_{\alpha'\alpha} + 1/2 \sum_{\{\alpha\}} H_{12,1'2'}\,\varrho_{1'2',12}\,, \qquad \textbf{(5.1.8.1)}$$

where $\{\alpha\} = \alpha_1,\ \alpha_2,\ \alpha_1',\ \alpha_2'$. We express a two-electron matrix through single-electron matrices (3.6). Let's substitute here an approximate expression (5.1.3.6). Get

$$E = \textstyle\sum_{\alpha\alpha'} H_{\alpha\alpha'}\,\varrho_{\alpha'\alpha} + \sum_{\{\alpha\}} H_{12,1'2'}\,\varrho_{1'1}\,\varrho_{2'2}\,, \qquad \textbf{(5.1.8.2)}$$

If the density matrix were diagonal, *i.e.*

$$\varrho_{\alpha\alpha'} = w_\alpha\,\delta_{\alpha\alpha'},$$

then it would be possible to write the entropy

$$S = -\,k_B \textstyle\sum_\alpha\,[\,w_\alpha \ln w_\alpha + (\,1 - w_\alpha\,) \ln (\,1 - w_\alpha\,)\,]$$

and to express the thermodynamic potential through the function w_α of electron distribution:

$$\Omega = E - S\,T - \mu\,Z\,. \qquad \textbf{(5.1.8.3)}$$

where T is temperature, and μ is the chemical potential. By differentiating this function by w_α and by equating the resulting expression to zero:

$$\partial\Omega/\partial w_\alpha = 0\,.$$

we would get an equation for the electron distribution function. Unfortunately, we have no reason to consider the density matrix diagonal.

5.1.9. The Pure State of Electrons in Atom

When finding the energy levels of electrons in an atom, their state can be considered pure. The pure state by definition is such that the density matrix is equal to the product of the wave functions $\psi_\alpha(q)$ of the system. Let the statistical operator be:

$$\hat{\varrho}^{(1)} = Z\ \delta(q - q_o)\,, \qquad \textbf{(5.1.9.1)}$$

where $\delta(q - q_o)$ – delta function. According to the formula (5.1.3.1) we will have

$$\varrho_{\alpha\alpha'} = Z\int \psi^*_\alpha(q)\,\delta(q - q_o)\,\psi_{\alpha'}(q)\,\mathrm{d}q_o = Z\psi^*_\alpha(q)\,\psi_{\alpha'}(q)\,. \qquad \textbf{(5.1.9.2)}$$

The number $\psi_\alpha(q)$ satisfy the normalization condition for the

$$\textstyle\sum_\alpha \psi^*_\alpha\,\psi_\alpha = 1. \qquad \textbf{(5.1.9.3)}$$

In the equilibrium net state, the numbers ψ_α must satisfy the stationary Schrödinger equation.

We record the average energy of electrons in the pure state. To do this, the density matrix is substituted

$$\varrho_{\alpha\alpha'} = Z\psi^*_\alpha\,\psi_\alpha \qquad \textbf{(5.1.9.4)}$$

in formula (5.1.8.2) to obtain:

$$E = Z\textstyle\sum_{\alpha\alpha'} H_{\alpha\alpha'}\,\psi^*_{\alpha'}\,\psi_\alpha + Z(Z-1)\textstyle\sum_{\{\alpha\}} H_{12,1'2'}\,\psi^*_{1'}\,\psi_1\psi^*_{2'}\,\psi_2\,. \qquad \textbf{(5.1.9.5)}$$

Here, we substituted the exact number $Z - 1$, which is in the formula (3.8).

If we use the formula (6.7), the matrix $H_{\alpha\alpha'}$ will be diagonal (5.1.6.10) in the α-representation, and the matrix $H_{12,1'2'}$ will be anti-symmetric (5.1.7.9):

$$E = Z\sum_{\alpha} \varepsilon_{\alpha}^{(0)} \psi_{\alpha}^{*} \psi_{\alpha} + Z(Z-1)\sum_{\{\alpha\}} H_{12,1'2'} \psi_{1'}^{*} \psi_{1}\psi_{2'}^{*} \psi_{2}, \qquad \textbf{(5.1.9.6)}$$

Now we write the stationary Schrödinger equation for the numbers ψ_{α}. This equation has the form:

$$\varepsilon_{n}^{(0)} \psi_{\alpha} + (Z-1)\sum_{\alpha_1\alpha_2\alpha_1'} H_{12,1'\alpha} \psi_{1}\psi_{2} \psi_{1'}^{*} = \varepsilon_{\alpha} \psi_{\alpha}, \qquad \textbf{(5.1.9.7)}$$

where $\alpha = \{n, l, m, \sigma\}$ is the eigenvalue, and ε_{α} is the eigenvalue of the electron energy. If this equation is multiplied by $Z\psi_{\alpha}^{*}$ and summed by α, we obtain:

$$E = Z\sum_{\alpha} \varepsilon_{\alpha} \psi_{\alpha}^{*} \psi_{\alpha}. \qquad \textbf{(5.1.9.8)}$$

If $Z = 1$, then for the hydrogen atom we will have:

$$\varepsilon_{\alpha} = \varepsilon_{n}^{(0)}. \qquad \textbf{(5.1.9.9)}$$

When $Z = 2, 3,...$ the equation (5.1.9.7) can be be solved.

5.1.10. Conclusion

The solution of the equation (5.1.9.7) gives the electron energy spectrum in an arbitrary equilibrium atom as follows:

$$\varepsilon_{\alpha} = \varepsilon_{n\,m\,l\,\sigma}.$$

In order to solve this problem, a computer is required.

5.1.11. Comment

Given that the Slater function does not have quite normal normalization conditions (5.1.5.6) and (5.1.5.7), it is possible to introduce a constant coefficient η in the formula (5.1.9.6) and equation (5.1.9.7) to obtain the following equations:

$$E = Z\sum_{\alpha} \varepsilon_{\alpha}^{(0)} \psi_{\alpha}^{*} \psi_{\alpha} + Z(Z-1)\,\eta\sum_{\{\alpha\}} H_{12,1'2'} \psi_{1'}^{*} \psi_{1}\psi_{2'}^{*} \psi_{2}, \qquad \textbf{(5.1.11.1)}$$

$$\varepsilon_{\mathrm{n}}^{(\mathrm{o})}\,\psi_{\alpha} + (\,Z - 1\,)\,\eta \sum_{\alpha_1\alpha_2\alpha_1'} H_{12,1'\alpha}\,\psi_1\psi_2\,\psi_{1'}^{*} = \varepsilon_{\alpha}\,\psi_{\alpha}\,. \qquad \textbf{(5.1.11.2)}$$

REFERENCES

[1] B.V. Bondarev, "Density Matrix Method in Theory of Atom. Advanced Materials: Manufacturing", *Physics, Mechanics and Applications*; ed. Ivan A. Parinov. New York, London, Springer Proceedings in Physics, vol. 207, pp. 145-159, 2018.

[2] B.V. Bondarev, "Density matrix method. New calculation of electron energy levels in atom", *Scientific Discussion*, vol. 1, no. 18, pp. 32-36, 2018.

CHAPTER 6

New Theory of Laser

6.1. DENSITY MATRIX METHOD IN TWO-LEVEL LASER THEORY

In his earlier work, the author obtained the quantum kinetic equation for the density matrix. The equation contains two summands in the right part. The first one is the same as in Liouville – von Neumann equation. The second one describes the dissipative members. This quantum equation can be written in any representation. We show that in perturbation theory, this equation has the order associated with the order parameter. The order parameter expansion leads to two equations, which determine the zero and the first approximation of the density matrix. Because of the presence of time-dependent perturbations in the Hamiltonian, even the zero representation of the density matrix suggests that the states of quantum systems are mixed, *i.e.*, the density matrix will not be equal to the product of the wave functions.

In this chapter, we use the method of density matrix in the theory of lasers with two energy levels [1]. Written for the atom, the equation of the density matrix of zero approximation is very simple and widely known. In the second approximation, the quantum kinetic equation for the density matrix becomes the classical kinetic equation, if the density matrix has a diagonal form. The resulting Hamiltonian also has a diagonal form. In quantum mechanics, this form can be obtained using a unitary matrix. The elements of the diagonal Hamiltonian in this representation are the eigenvalues of the atom energy. We found the density matrix in this representation. We obtained the dissipative matrices, which characterize the operations of pumping and damping. In the representation where the Hamiltonian is diagonal, we wrote the equations that describe the work of the laser.

6.1.1. Introduction

In 1964, N. G. Basov, A. M. Prokhorov and C. H. Townes won the Nobel Prize. They were awarded this prize for their fundamental research in quantum electronics. These studies have led to the development of masers and lasers [2, 3].

Let us consider the principle of operation of the quantum generator, laser. The basic element of the quantum generator is the active environment, *i.e.* the substance that creates the inverse population of levels. The active medium typically has the shape of a long cylinder (see Fig. **1**). At the ends of this cylinder, there are two plane-

Boris V. Bondarev

parallel mirrors, one at each end, perpendicular to its axis. The purpose of these mirrors is to increase the length of the path where increased radiation occurs, by means of the multiple passages of the beam through the active medium. The mirrors form a so-called resonator. Between them appears a standing electro-magnetic wave. One of the mirrors is half-translucent. Through this mirror, the electromagnetic radiation comes out of the resonator in the form of a narrow, almost non-divergent beam. The process, in which the energy is transferred in some way to the active medium and the inversion of the population of levels is created, is called pumping (see Fig. **6.1.1**).

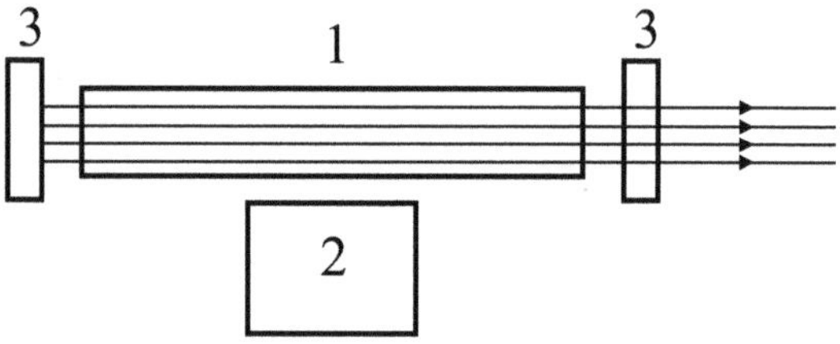

Fig. (6.1.1). The scheme of the quantum generator: 1 – active substance; 2 – pumping; 3 - mirrors.

The process in which energy is transferred to the active medium in some way creating a population inversion of the levels is called pumping. Pump energy may be in the form of light, electric current, energy, chemical or nuclear reactions, thermal or mechanical energy.

Various methods have been proposed to create a population inversion of energy levels. The method of two levels proposed by Basov and Prokhorov in 1955 is the most convenient and common. The atoms or molecules of the active substance are greatly influenced in some way, so that the electrons in them move from the ground state $\varphi_1(\boldsymbol{r})$ with energy ε_1, into the excited state $\varphi_2(\boldsymbol{r})$ with energy ε_2 (see Fig. **6.1.2**).

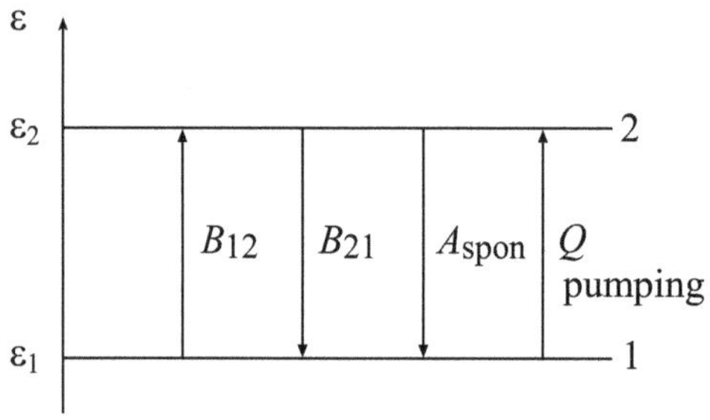

Fig. (6.1.2). The two-level scheme of the interaction of atom and radiation.

Due to the intense pumping, the saturation is reached, in which the number of electrons N_2 in the state $\varphi_2(\boldsymbol{r})$ becomes equal to the number N_1 of electrons in the ground state $\varphi_1(\boldsymbol{r})$. At the same time for a pair of levels ε_1 and ε_2, there is population inversion ($N_2 > N_1$). Only a small part of the energy given to the active medium during the pumping is converted into the energy of the generated radiation. Most of this energy is converted into heat. The active medium becomes very hot and sometimes it requires intense cooling.

6.1.2. Kinetics of Quantum Transitions

Consider a system of non-interacting atoms in which valence electron can make the quantum jump from one stationary state to another. The community of atoms in the state φ_1 with energy ε_1 can pass into the state φ_2 with bigger energy ε_2 during the absorption of photons by atoms. The number $\mathrm{d}N_{12}^{(\mathrm{absor})}$ of transitions made by electrons from the state φ_1 into the state φ_2 during the time from t to $t + \mathrm{d}t$ of the absorption of a photon by the atom, is proportional to the number $N_1(t)$ of atoms in the state φ_1, time dt and the number of photons $W(\omega)$ with frequency ω, flying into atoms:

$$\mathrm{d}N_{12}^{(\mathrm{absor})} = B_{12} W(\omega)\, N_1(t)\, \mathrm{d}t, \tag{6.1.2.1}$$

where B_{12} is the coefficient of proportionality, the frequency ω of these photons is determined by the formula,

$$\omega = (\,\varepsilon_2 - \varepsilon_1)/\hbar\,, \tag{6.1.2.2}$$

Where the spectral energy density $W(\omega)$ is equal to the energy of electromagnetic radiation, which falls per unit volume $\mathrm{d}V = 1$ and the unit of frequency $\mathrm{d}\omega = 1$:

$$W(\omega)\, \mathrm{d}V\, \mathrm{d}\omega. \tag{6.1.2.3}$$

Let at time t this system have the number $N_2(t)$ of atoms in which the electron is in the state φ_2 with energy ε_2. The number $\mathrm{d}N_{21}^{(\mathrm{spon})}$ of spontaneous transitions made by electrons from the state φ_2 to the state φ_1 with smaller energy ε_1 with the emission of a photon during the time from the moment t to the moment $t + \mathrm{d}t$, is proportional to the number $N_2(t)$ of atoms and the time interval $\mathrm{d}t$:

$$\mathrm{d}N_{21}^{(\mathrm{spon})} = A^{(\mathrm{spon})} N_2(t)\, \mathrm{d}t, \tag{6.1.2.4}$$

where $A^{(spon)}$ is a positive coefficient. The value A is called the probability of spontaneous transition of the atom.

In fact, in addition to spontaneous transitions from the state φ_2 to the state φ_1, the atom can make the forced transitions between these states. The number $dN_{21}^{(ind)}$ of atoms, making forced transitions from φ_2 to φ_1 in the time from t to $t + dt$, is proportional to their number $N_2(t)$ in the state φ_2, the time interval dt and the number of photons forcing atoms to make such transitions. So, for the numbers of forced (induced) transitions, we can write the following formula:

$$dN_{21}^{(ind)} = B_{21} W(\omega)\, N_2(t)\, dt, \tag{6.1.2.5}$$

where B_{21} is the coefficient of proportionality. The values B_{12}, B_{21} and A are called the Einstein coefficients.

Finally, electrons can be forced to move from state 1 to state 2 as a result of the process called pumping. Here, the pumping process is not discussed, but in this process, the number of electrons in state 1 will decrease by the amount

$$dN_{12}^{(pump)} = Q\, N_1(t)\, dt. \tag{6.1.2.6}$$

But in this process, the transition of electrons is not accompanied by emission or absorption of photons.

Adding these increments, we will have a kinetic equation

$$dN_1/dt \;=\; -\left[\, Q + B_{12}\, W(\omega)\right] N_1 + \left[\, A^{(spon)} + B_{21} W(\omega)\right] N_2\,, \tag{6.1.2.7}$$

where

$$N_1 + N_2 = N \tag{6.1.2.8}$$

is the total number of atoms.

If the numbers N_1 and N_2 were constant, then the number of photons espousing or absorbed by atoms is equal to

$$B_{12} W(\omega) N_1 = \left[A^{(spon)} + B_{21} W(\omega)\right] N_2.$$

When atoms are in equilibrium, *i.e.* $N_1 \sim e^{-\beta\,\varepsilon_1}$ and $N_2 \sim e^{-\beta\,\varepsilon_2}$, then

$$e^{\beta(\varepsilon_2 - \varepsilon_1)} = [A^{(\mathrm{spon})} + B_{21}W(\omega)]\,/[B_{12}W(\omega)].$$

Now we obtain a formula for the spectral energy density of radiation:

$$W(\omega) = (A^{(\mathrm{spon})}/B_{12})/[\,e^{\beta\,\hbar\,\omega} - B_{21}/B_{12}]\,. \qquad \textbf{(6.1.2.9)}$$

In quantum optics, the dependence $W(\omega)$ must be described by Planck's formula:

$$W^{(\mathrm{Planck})}\,(\omega) = \hbar\,\omega^3/(\pi^2c^3)\,(e^{\beta\,\hbar\,\omega} - 1)\,. \qquad \textbf{(6.1.2.10)}$$

Comparing formulas (6.1.2.8) and (6.1.2.9) we find that,

$$B_{12} = B_{21}\,,\ A^{(\mathrm{spon})}/B_{12} = \hbar\,\omega^3/(\pi^2c^3)\,. \qquad \textbf{(6.1.2.11)}$$

Taking these equalities into consideration, we write the equation (6.1.2.6) as,

$$\mathrm{d}N_1/\mathrm{d}t \;=\; -\,[\,Q + B\,W(\omega)\,]\,N_1 + \,[\,A + B\,W(\omega)\,]\,\boldsymbol{N_2}\,, \qquad \textbf{(6.1.2.12)}$$

This is a phenomenological equation, which was derived only based on considerations of truth, but without the quantum laws.

6.1.3. Density Matrix

The density matrix $\varrho_{nn'}$ is the most general description of the system in quantum mechanics. Here n and n' is the quantum parameters that define the system state.

This description is called **mixed**, if only the density matrix is not equal to the product of the wave functions. Then this description is called **pure**.

The meaning of the density matrix is that its diagonal elements are the probability w_n that the system is in quantum state n:

$$\varrho_{nn} = w_n\,. \qquad \textbf{(6.1.3.1)}$$

6.1.4. Equation for the Density Matrix

In the quantum theory of non-equilibrium processes, the kinetic equation for the density matrix $\varrho_{nn'}(t)$ is derived from the equation Liouville - von Neumann. This equation has the form,

$$\mathrm{i}\,\hbar\,\dot{\varrho}_{nn'} = \sum_m \left(H_{nm}\varrho_{mn'} - \varrho_{nm}H_{mn'}\right) + \mathrm{i}\,\hbar\{\sum_{m\,m'} \gamma_{nm,m'n'}\,\varrho_{mm'} - 1/2\sum_m(\gamma_{nm}\varrho_{mn'} + \varrho_{nm}\,\gamma_{mn'})\}\,, \quad \textbf{(6.1.4.1)}$$

where H_{nm} – matrix elements of the Hamiltonian of the system, $\gamma_{nm,m'n'}$ – a dissipative matrix,

$$\gamma_{nn'} = \sum_m \gamma_{mn',nm}\,. \quad \textbf{(6.1.4.2)}$$

Equation (4.1) can be written as,

$$\mathrm{i}\,\hbar\,\dot{\varrho}_{nn'} = \sum_m (H_{nm}\varrho_{mn'} - \varrho_{nm}H_{mn'}) + \mathrm{i}\,\hbar\,D_{nn'}\,,$$

where $D_{nn'}$ – dissipative matrix.

The diagonal element $\varrho_{nn'}$ is the probability w_n that the system is in state *n*. This value satisfies the condition of normalization as shown below:

$$\sum_n \varrho_{nn} = 1. \quad \textbf{(6.1.4.3)}$$

In addition, the density matrix ϱ_{mn} satisfies the condition,

$$\varrho_{nm}^* = \varrho_{mn}\,. \quad \textbf{(6.1.4.4)}$$

The Hamiltonian is subject to the same condition:

$$H_{nm}^* = H_{mn}. \quad \textbf{(6.1.4.5)}$$

Let us consider the case when the density matrix ϱ_{nm} at a random time *t* is diagonal:

$$\varrho_{nm} = w_n\,\delta_{nm}\,, \quad \textbf{(6.1.4.6)}$$

where δ_{nm} – the Kronecker symbol. Then from equation (6.1.4.1), we get

$$\dot{w}_n = \sum_m (p_{nm} w_m - p_{mn} w_n), \qquad \textbf{(6.1.4.7)}$$

where the probability w_n satisfies the normalization condition

$$\sum_n w_n = 1, \qquad \textbf{(6.1.4.8)}$$

$$p_{nm} = \gamma_{nm,mn} = (2\,\pi/\hbar) \sum_N \sum_M |v_{nN,mM}|^2 W_M \delta(\varepsilon_n - \varepsilon_m + E_N - E_M) \qquad \textbf{(6.1.4.9)}$$

is the probability of transition per unit time of a system from state *m* to state *n*,

$$W_N = \nu \exp(-\beta E_N)$$

is the probability that the system environment is in the equilibrium state with quantum numbers *N*, and E_N is the energy in this state, ν is the normalization factor, $\beta = 1/(k_B T)$ – inverse temperature; $v_{nN,mM}$ – matrix elements of the interaction of the system with its environment. Formula (6.1.4.9) is Fermi Golden rule.

Judging by Fermi rule, the transition probability p_{nm} can always be represented in the form

$$p_{nm} = p_{nm}^{(o)} \, \mathrm{e}^{-\beta(\varepsilon_n - \varepsilon_m)/2}, \qquad \textbf{(6.1.4.10)}$$

$$p_{nm}^{(o)} = p_{mn}^{(o)} . \qquad \textbf{(6.1.4.11)}$$

Equation (6.1.4.1) can be written in the operator form. This equation is called the Lindblad equation and it has the form,

$$\mathrm{i}\,\hbar\,\dot{\hat{\varrho}} = [\,\hat{H}\,\hat{\varrho}\,] + \mathrm{i}\,\hbar \sum_{jk} C_{jk} [\,\hat{a}_j \hat{\varrho}\,\hat{a}_k^+ - 1/2\,(\,\hat{a}_k^+ \hat{a}_j\,\hat{\varrho} + \hat{\varrho}\,\hat{a}_k^+ \hat{a}_j\,)\,],$$

$$\textbf{(6.1.4.12)}$$

where $\hat{H}$ – the Hamiltonian, C_{jk} – some numbers, $\hat{a}_j$ – arbitrary operators. From the comparison of equations (6.1.4.1) and (6.1.4.12), we obtain

$$\gamma_{nm,m'n'} = \sum_{jk} C_{jk}\, a_{nm,j}\, a^+_{m'n',k}, \qquad \textbf{(6.1.4.13)}$$

where $a_{nm,j}$ – matrix elements of operators $\hat{a}_j$. Equation (6.1.4.12) has the ability to use the operators $\hat{a}_j$ for the solution of this equation.

6.1.5. Equation for Density Matrix of the First-order Approximation

It was shown that equation (6.1.4.1) for the statistical operator $\hat{\varrho}$ was obtained in the first approximation by the Hamiltonian of the interaction of the system and the thermostat. Therefore, this equation can be written as:

$$\mathrm{i}\,\hbar\,\dot{\hat{\varrho}} = [\,\hat{H}\,\hat{\varrho}\,] + \lambda\,\mathrm{i}\,\hbar\,\hat{D}(\hat{\varrho}) + \dots\,, \tag{6.1.5.1}$$

here λ is the order parameter, $\hat{D}(\hat{\varrho})$ is a dissipative operator. We write the statistical operator $\hat{\varrho}$ as follows,

$$\hat{\varrho} = \hat{\varrho}^{(o)} + \lambda\,\hat{\varrho}^{(1)} + \dots, \tag{6.1.5.2}$$

where $\hat{\varrho}^{(o)}$, $\hat{\varrho}^{(1)}$, ... is the approximation of the density matrix.

Let us substitute the operator (6.1.5.2) in equation (6.1.5.1). We have

$$\mathrm{i}\,\hbar\,(\,\dot{\hat{\varrho}}^{(o)} + \lambda\,\dot{\hat{\varrho}}^{(1)} + \dots\,) = [\,\hat{H},\,\hat{\varrho}^{(o)} + \lambda\,\hat{\varrho}^{(1)} + \dots\,] + \lambda\,\mathrm{i}\,\hbar\,\hat{D}(\,\hat{\varrho}^{(o)} + \lambda\,\hat{\varrho}^{(1)} + \dots)\,. \tag{6.1.5.3}$$

If the order parameter $\lambda = 0$, we get the equation for the unperturbed statistical operator $\hat{\varrho}^{(o)}$:

$$\mathrm{i}\,\hbar\,\dot{\hat{\varrho}}^{(o)} = [\,\hat{H}\,\hat{\varrho}^{(o)}]\,. \tag{6.1.5.4}$$

The first parameter of $\lambda = 1$, gives

$$\mathrm{i}\,\hbar\,\dot{\hat{\varrho}}^{(1)} = [\,\hat{H}\,\hat{\varrho}^{(1)}] + \mathrm{i}\,\hbar\,\hat{D}(\,\hat{\varrho}^{(o)})\,, \tag{6.1.5.5}$$

Thus, the operator $\hat{\varrho}$ satisfies this system of equations.

6.1.6. Equation for the Density Matrix of Zero-order

For the two-level scheme on the first level, there are g_1 wave functions $\varphi_n(\boldsymbol{r})$, and on the second level, there are g_2 wave functions $\varphi_m(\boldsymbol{r})$. We now consider all functions $\varphi_n(\boldsymbol{r})$ to be approximately equal to one functions $\varphi_1(\boldsymbol{r})$ and all functions $\varphi_m(\boldsymbol{r})$ equal to one function $\varphi_2(\boldsymbol{r})$:

$$\varphi_n(\boldsymbol{r}) \approx \varphi_1(\boldsymbol{r}), \qquad \varphi_m(\boldsymbol{r}) \approx \varphi_2(\boldsymbol{r}).$$

Let us consider equation (6.1.4.1) of the density matrix $\varrho_{\alpha\alpha'}$ for the two-level scheme of interaction of an atom with radiation. Here α and α' are the quantum number of states. In this case, the density matrix will be a square matrix with the side equal to two: ϱ_{11}, ϱ_{12}, ϱ_{21} and ϱ_{22}. Here quantum numbers α, β, α' and β' are equal to only two values 1 and 2.

The normalization condition for the density matrix (6.1.4.3) in our case will look like this:

$$\sum_n \varrho_{nn} + \sum_m \varrho_{mm} = 1. \qquad \textbf{(6.1.6.1)}$$

Now the equation (6.1) will be as follows:

$$g_1\, \varrho_{11} + g_2\, \varrho_{22} = 1. \qquad \textbf{(6.1.6.2)}$$

Let us find the Hamiltonian of the system in a diagonal representation. The operator $\widehat{H}$ in the simplest representation might look like this:

$$\widehat{H} = \widehat{H}_0 + \widehat{V}, \qquad \textbf{(6.1.6.3)}$$

where $\widehat{H}_0$ is the Hamiltonian, which has the matrix elements

$H_{\alpha\beta} = \varepsilon_\alpha^{(0)} \delta_{\alpha\beta}$, $\varepsilon_\alpha^{(0)}$ is the energy of the atom; the operator $\widehat{V}$ of dipole interaction of the stationary atom and radiation is equal to,

$$\widehat{V} = -\,\widehat{\boldsymbol{d}} \cdot \boldsymbol{E}, \qquad \textbf{(6.1.6.4)}$$

$\widehat{\boldsymbol{d}} = \widehat{\boldsymbol{d}}(\boldsymbol{r})$ – the operator of the electric dipole moment of the atom, $\boldsymbol{E} = \boldsymbol{E}(t)$ is the vector of the electric field strength. Thus, the interaction operator depends only on the coordinates $\boldsymbol{r}$ and time t:

$\widehat{V} = \widehat{V}(\boldsymbol{r}, t)$. Since the Hamiltonian satisfies the condition $H^*_{\alpha\beta} = H_{\beta\alpha}$, the operator $\widehat{V}$ is also subject to this condition:

$$V^*_{\alpha\beta} = V_{\beta\alpha},$$

where $V_{\alpha\beta}$ are matrix elements of the operator $\widehat{V}$.

Consider the zero-order approximation of equation (6.1.5.4) for the statistical operator:

$$\mathrm{i}\,\hbar\,\dot{\hat{\varrho}}^{(\mathrm{o})} = [\,\hat{H}\,\hat{\varrho}^{(\mathrm{o})}]\,.$$

Writing the equation for the density matrix

$$\varrho^{(\mathrm{o})}_{\alpha\beta} = \int \varphi^{*}_{\alpha}\,\hat{\varrho}^{(\mathrm{o})}(\boldsymbol{r})\,\varphi_{\alpha'}\,\mathrm{d}\boldsymbol{r}\,. \qquad \textbf{(6.1.6.5)}$$

We have

$$\mathrm{i}\,\hbar\,\dot{\varrho}^{(\mathrm{o})}_{\alpha\alpha'} = \textstyle\sum_{\beta}\,(H_{\alpha\beta}\,\varrho^{(\mathrm{o})}_{\beta\alpha'} - \varrho^{(\mathrm{o})}_{\alpha\beta}\,H_{\beta\alpha'})\,, \qquad \textbf{(6.1.6.6)}$$

where the matrix elements of the Hamiltonian

$$H_{\alpha\beta} = \varepsilon^{(\mathrm{o})}_{\alpha}\,\delta_{\alpha\beta} + V_{\alpha\beta}\,, \qquad \textbf{(6.1.6.7)}$$

$\varepsilon^{(\mathrm{o})}_{1}$ and $\varepsilon^{(\mathrm{o})}_{2}$ – the energy at first and second levels. Substituting the Hamiltonian (6.1.6.7) in equation (6.6), we have ,

$$\mathrm{i}\,\hbar\,\dot{\varrho}^{(\mathrm{o})}_{\alpha\alpha'} = (\,\varepsilon^{(\mathrm{o})}_{\alpha} - \varepsilon^{(\mathrm{o})}_{\alpha'})\,\varrho^{(\mathrm{o})}_{\alpha\alpha'} + \textstyle\sum_{\beta}\,(\,V_{\alpha\beta}\,\varrho^{(\mathrm{o})}_{\beta\alpha'} - \varrho^{(\mathrm{o})}_{\alpha\beta}\,V_{\beta\alpha'}\,)\,. \qquad \textbf{(6.1.6.8)}$$

where

$$\hbar\,\omega_{\mathrm{o}} = \varepsilon^{(\mathrm{o})}_{2} - \varepsilon^{(\mathrm{o})}_{1} \qquad \textbf{(6.1.6.9)}$$

ω_{o} is the frequency of transition between levels, when the electric field is absent.

The density matrix will be as follows :

$$\begin{aligned}
\mathrm{i}\,\hbar\,\dot{\varrho}^{(\mathrm{o})}_{11} &= V_{12}\,\varrho^{(\mathrm{o})}_{21} - \varrho^{(\mathrm{o})}_{12}\,V_{21}\,,\\
\mathrm{i}\,\hbar\,\dot{\varrho}^{(\mathrm{o})}_{12} &= -\,\hbar\,\omega_{\mathrm{o}}\,\varrho^{(\mathrm{o})}_{12} + V_{12}\,(\,\varrho^{(\mathrm{o})}_{22} - \varrho^{(\mathrm{o})}_{11}\,)\,,\\
\mathrm{i}\,\hbar\,\dot{\varrho}^{(\mathrm{o})}_{21} &= \hbar\,\omega_{\mathrm{o}}\,\varrho^{(\mathrm{o})}_{21} + V_{21}\,(\,\varrho^{(\mathrm{o})}_{11} - \varrho^{(\mathrm{o})}_{22}\,)\,,\\
\mathrm{i}\,\hbar\,\dot{\varrho}^{(\mathrm{o})}_{22} &= V_{21}\,\varrho^{(\mathrm{o})}_{12} - \varrho^{(\mathrm{o})}_{21}\,V_{12}\,,
\end{aligned} \qquad \textbf{(6.1.6.10)}$$

Here we put,

$$V_{11} = 0\,, \qquad V_{22} = 0\,. \tag{6.1.6.11}$$

If the electric field is equal to,

$$\boldsymbol{E} = \boldsymbol{E}_{\text{o}} \cos \omega\, t\,, \tag{6.1.6.12}$$

where ω is the frequency of the electric field, then the matrix elements of the operator of dipole interaction are equal as shown below,

$$V_{\alpha\beta} = V_{\alpha\beta}^{(\text{o})} \cos \omega\, t\,, \quad V_{\alpha\beta}^{(\text{o})} = -\,(\widehat{\boldsymbol{d}} \cdot \boldsymbol{E}_o)_{\alpha\beta} = -\,\boldsymbol{d}_{\alpha\beta} \cdot \boldsymbol{E}_o\,, \tag{6.1.6.13}$$

where

$$\boldsymbol{d}_{\alpha\beta} = \int \varphi_\alpha^*(\boldsymbol{r})\, \widehat{\boldsymbol{d}}\, \varphi_\beta(\boldsymbol{r})\, \mathrm{d}\boldsymbol{r}. \tag{6.1.6.14}$$

Taking into account (6.1.6.11), we put the matrix

$$\begin{gathered} V_{\alpha\beta}^{(\text{o})} = -\,\boldsymbol{d} \cdot \boldsymbol{E}_{\text{o}}\,, \\ \boldsymbol{d}_{12} = \int \varphi_1^*(\boldsymbol{r})\, \widehat{\boldsymbol{d}}\, \varphi_2(\boldsymbol{r})\, \mathrm{d}\boldsymbol{r} = \boldsymbol{d}\,. \end{gathered} \tag{6.1.6.15}$$

Then

$$V_{12} = V_{21}^* = -\,\boldsymbol{d} \cdot \boldsymbol{E}_{\text{o}} \cos \omega\, t = -\,\hbar\, \Omega \cos \omega\, t\,, \tag{6.1.6.16}$$

where

$$\Omega = \boldsymbol{d} \cdot \boldsymbol{E}_o / \hbar\,, \tag{6.1.6.17}$$

here Ω is the Rabi frequency. Substituting formula (6.15) in equation (6.9), we have,

$$\mathrm{i}\,\dot{\varrho}_{11}^{(o)} = -\,\Omega\,(\,\varrho_{21}^{(o)} - \varrho_{12}^{(o)})\cos\omega\,t\,,$$
$$\mathrm{i}\,\dot{\varrho}_{12}^{(o)} = -\,\omega_0\,\varrho_{12}^{(o)} + \Omega\,(\,\varrho_{11}^{(o)} - \varrho_{22}^{(o)}\,)\cos\omega\,t\,,$$
$$\mathrm{i}\,\dot{\varrho}_{21}^{(o)} = \omega_0\,\varrho_{21}^{(o)} - \Omega\,(\,\varrho_{11}^{(o)} - \varrho_{22}^{(o)}\,)\cos\omega\,t\,,$$
$$\mathrm{i}\,\dot{\varrho}_{22}^{(o)} = \Omega\,(\,\varrho_{2}^{(o)} - \varrho_{12}^{(o)})\cos\omega\,t\,.$$

(6.1.6.18)

Recalling that the density matrix satisfies the condition

$$\varrho_{\alpha\alpha}^{*} = \varrho_{\alpha\alpha}.$$

This means that the diagonal elements of the density matrix are real. The sum of the first and the last equations is as follows:

$$\dot{\varrho}_{11}^{(o)} + \dot{\varrho}_{22}^{(o)} = 0.$$

(6.1.6.19)

This equation has the solution

$$\varrho_{11}^{(o)} + \varrho_{22}^{(o)} = 1.$$

(6.1.6.20)

The non-diagonal elements of the density matrix are such that,

$$\varrho_{12}^{*} = \varrho_{21}\,.$$

Therefore, the equation for the function $\dot{\varrho}_{21}^{(o)}$ can be discarded because knowing ϱ_{12}, we immediately find the first function.

If the electric field is absent, then

$$\dot{\varrho}_{11}^{(o)} = 0\,, \quad \mathrm{i}\,\dot{\varrho}_{12}^{(o)} = -\,\omega_0\,\varrho_{12}^{(o)}\,, \quad \mathrm{i}\,\dot{\varrho}_{21}^{(o)} = \omega_0\,\varrho_{21}^{(o)}\,, \quad \dot{\varrho}_{22}^{(o)} = 0\,.$$

(6.1.6.21)

Integrating these equations, we obtain

$$\varrho_{11}^{(o)} = C_{11}\,, \quad \varrho_{12}^{(o)} = C_{12}\,\mathrm{e}^{\mathrm{i}\,\omega_0\,t}\,, \quad \varrho_{21}^{(o)} = C_{12}^{*}\,\mathrm{e}^{-\mathrm{i}\,\omega_0\,t}\,, \quad \varrho_{22}^{(o)} = C_{22}\,.$$

(6.1.6.22)

The normalization condition (6.1.6.2) gives

$$C_{11} + C_{22} = 1 . \qquad \textbf{(6.1.6.23)}$$

6.1.7. Diagonal Hamiltonian

The atom energy eigenvalue can be simply found if we know the Hamiltonian $\widetilde{H}_{\kappa\kappa'}$, which is in the diagonal form:

$$\widetilde{H}_{\kappa\kappa'} = \varepsilon_\kappa \, \delta_{\kappa\kappa'} \, , \qquad \textbf{(6.1.7.1)}$$

where ε_κ is the sought eigenvalue of the energy of the atom.
To go from the Hamiltonian (6.1.6.7)

$$H_{\alpha\beta} = \varepsilon_\alpha^{(o)} \delta_{\alpha\beta} + V_{\alpha\beta} \, , \qquad \textbf{(6.1.7.2)}$$

to the diagonal Hamiltonian $\widetilde{H}_{\kappa\kappa'}$, it is necessary to find a unitary matrix $U_{\alpha\kappa}$, which satisfies the condition,

$$\sum_\alpha U_{\alpha\kappa} \, U^*_{\alpha\kappa'} = \delta_{\kappa\kappa'} \, . \qquad \textbf{(6.1.7.3)}$$

This transition from α-representations to κ-representation is carried out according to the formula

$$\widetilde{H}_{\kappa\kappa'} = \sum_{\alpha\beta} U_{\alpha\kappa} \, H_{\alpha\beta} U^*_{\beta\kappa'} \, . \qquad \textbf{(6.1.7.4)}$$

To find a unitary matrix

$$U_{\alpha\kappa} = \begin{Vmatrix} U_{11} & U_{12} \\ U_{21} & U_{22} \end{Vmatrix} ,$$

we write condition (6.1.7.3) in more detail as,

$$\begin{Vmatrix} U_{11} & U_{12} \\ U_{21} & U_{22} \end{Vmatrix} \begin{Vmatrix} U^*_{11} & U^*_{21} \\ U^*_{12} & U^*_{22} \end{Vmatrix} = \begin{Vmatrix} 1 & 0 \\ 0 & 1 \end{Vmatrix} .$$

The solution of this equation has the form:

$$U_{\alpha\kappa} = \begin{Vmatrix} a & -b^* \\ b & a^* \end{Vmatrix}, \quad U^*_{\alpha\kappa} = \begin{Vmatrix} a^* & b^* \\ -b & a \end{Vmatrix},$$

where

$$a = x + \mathrm{i}\, y, \quad b = z + \mathrm{i}\, u, \quad x^2 + y^2 + z^2 + u^2 = 1. \tag{6.1.7.5}$$

These numbers are called Cayley – Klein parameters.

When the electric field is absent, the Hamiltonian $\hat{H}$ is equal to,

$$H_{\alpha\beta} = \varepsilon_{\alpha}^{(o)} \delta_{\alpha\beta}.$$

In this case,

$$\widetilde{H}_{\kappa\kappa'} = \sum_{\alpha\beta} \delta_{\alpha\kappa}\, \varepsilon_{\alpha}^{(o)} \delta_{\alpha\beta}\, \delta^*_{\beta\kappa'} = \varepsilon_{\kappa}^{(o)} \delta_{\kappa\kappa'}. \tag{6.1.7.6}$$

Therefore, a unitary matrix is equal to

$$U^{(o)}_{\alpha\kappa} = \begin{Vmatrix} 1 & 0 \\ 0 & 1 \end{Vmatrix} = \delta_{\alpha\kappa},$$

where

$$a = 1, \quad b = 0. \tag{6.1.7.7}$$

Matrix elements of the Hamiltonian $\hat{H}$ are equal to,

$$H_{11} = \varepsilon_1^{(o)}, \quad H_{12} = -\hbar\,\Omega\, \cos\omega\, t, \quad H_{21} = -\hbar\,\Omega\, \cos\omega\, t, \quad H_{22} = \varepsilon_2^{(o)}. \tag{6.1.7.8}$$

Computing the diagonal Hamiltonian $\widetilde{H}_{\kappa\kappa'}$:

$$\widetilde{H}_{\kappa\kappa'} = \begin{Vmatrix} a & -b^* \\ b & a^* \end{Vmatrix} \begin{Vmatrix} \varepsilon_1^{(o)} & -\hbar\,\Omega\, \cos\omega\, t \\ -\hbar\,\Omega\, \cos\omega\, t & \varepsilon_2^{(o)} \end{Vmatrix} \begin{Vmatrix} a^* & b^* \\ -b & a \end{Vmatrix}.$$

Matrix elements of the Hamiltonian $\widetilde{H}_{\kappa\kappa'}$ will be:

$$\widetilde{H}_{11} = (\,\varepsilon_1^{(0)} a + \Omega\, b^* \cos\omega\, t)\, a^* + (\,\hbar\,\Omega\, a\, \cos\omega\, t + \varepsilon_2^{(0)} b^*) = \varepsilon_1\,, \qquad (6.1.7.9)$$

$$\widetilde{H}_{12} = (\,\varepsilon_1^{(0)} a + \hbar\,\Omega\, b^* \cos\omega\, t)\, b^* - (\,\hbar\,\Omega\, a\, \cos\omega\, t + \varepsilon_2^{(0)}\, b^*)\, a = 0\,, \qquad (6.1.7.10)$$

$$\widetilde{H}_{21} = (\,\varepsilon_1^{(0)} b - \,\hbar\,\Omega\, a^* \cos\omega\, t)\, a^* + (\,\hbar\,\Omega\, b\, \cos\omega\, t - \varepsilon_2^{(0)} a^*)\, b = 0\,, \qquad (6.1.7.11)$$

$$\widetilde{H}_{22} = (\,\varepsilon_1^{(0)} b - \,\hbar\,\Omega\, a^* \cos\omega\, \mathrm{t})\, b^* - \,(\,\hbar\,\Omega\, b\, \cos\omega\, \mathrm{t} - \varepsilon_2^{(0)} a^*)\, a = \varepsilon_2\,. \qquad (6.1.7.12)$$

Here we have two equations (6.1.7.10) and (6.1.7.11). These equations are identical:

$$\omega_o\, a\, b^* + \Omega\,(\,a^2 - {b^*}^2)\,\cos\omega\, t = 0\,. \qquad (6.1.7.13)$$

Let us add Cayley – Klein parameters

$$|a|^2 + |b|^2 = 1\,. \qquad (6.1.7.14)$$

We have two equations (6.1.7.13) and (6.1.7.14). When the electric field is absent, the number $a = 1$, and the number $b = 0$. When the electric field is present let $y = 0$ and $u = 0$. In this case

$$a = x \qquad \text{and} \qquad b = z\,. \qquad (6.1.7.15)$$

Then equations (6.1.7.13) and (6.1.7.14) take the form,

$$\omega_o\, x\, z + \Omega\,(\,x^2 - \,z^2)\,\cos\omega\, t = 0\,, \qquad (6.1.7.16)$$

$$x^2 + z^2 = 1\,. \qquad (6.1.7.17)$$

We introduce new variables:

$$x = \cos\vartheta, \qquad z = \sin\vartheta\,. \qquad (6.1.7.18)$$

Then we get equation (6.1.7.15) with one unknown ϑ:

$$\omega_o \cos\vartheta\, \sin\vartheta + \Omega\,(\,\cos^2\vartheta - \,\sin^2\vartheta\,)\,\cos\omega\, t = 0\,. \qquad (6.1.7.19)$$

To solve this equation:

$$\epsilon\, \mathrm{tg}^2\vartheta - 2\,\mathrm{tg}\,\vartheta - \epsilon = 0, \qquad (6.1.7.20)$$

where

$$\epsilon(t) = 2\,\Omega \cos \omega\, t/\omega_o \tag{6.1.7.21}$$

the solution of this equation is,

$$\mathrm{tg}\,\vartheta = (1 - \sqrt{1+\epsilon^2}\,)/\epsilon. \tag{6.1.7.22}$$

The parameter ϵ is extremely small. Therefore we expand the function in a row:

$$\sqrt{1+\epsilon^2} = 1 + \epsilon^2/2 + \ldots$$

In this case,

$$\mathrm{tg}\,\vartheta = -\,\epsilon/2 + \ldots = -\,\Omega \cos \omega\, t/\omega_o + \ldots \tag{6.1.7.23}$$

Formulas (7.18) correspond to the unitary matrix

$$U_{\alpha\kappa} = \begin{Vmatrix} \cos\vartheta & -\sin\vartheta \\ \sin\vartheta & \cos\vartheta \end{Vmatrix}. \tag{6.1.7.24}$$

Now taking into account formulas (6.1.7.15), we write the eigenvalues of the atom energy:

$$\begin{aligned} \varepsilon_1 &= \varepsilon_1^{(0)} x^2 + \varepsilon_2^{(0)} z^2 + \hbar\,\Omega\, x\, z\, \cos\omega\, t, \\ \varepsilon_2 &= \varepsilon_2^{(0)} x^2 + \varepsilon_1^{(0)} z^2 - \hbar\,\Omega\, x\, z\, \cos\omega\, t. \end{aligned}$$

Here x and z are expressed using the angle ϑ by the formula (6.1.7.18):

$$\begin{aligned} \varepsilon_1 &= \varepsilon_1^{(0)} \cos^2\vartheta + \varepsilon_2^{(0)} \sin^2\vartheta + \hbar\,\Omega \cos\vartheta \sin\vartheta\, \cos\omega\, t, \\ \varepsilon_2 &= \varepsilon_2^{(0)} \cos^2\vartheta + \varepsilon_1^{(0)} \sin^2\vartheta - \hbar\,\Omega \cos\vartheta \sin\vartheta\, \cos\omega\, t. \end{aligned} \tag{6.1.7.25}$$

Thus, we obtain the diagonal Hamiltonian

$$\widetilde{H}_{\kappa\kappa'} = \varepsilon_\kappa(t)\, \delta_{\kappa\kappa'}. \tag{6.1.7.26}$$

6.1.8. Density Matrix in κ-Representation

Zero density matrix corresponds to the equation

$$\mathrm{i}\,\hbar\,\dot{\varrho}^{(o)} = [\,\hat{H}\;\hat{\varrho}^{(o)}\,]\,.$$

In κ-representation, this equation has the form,

$$\mathrm{i}\,\hbar\,\dot{\tilde{\varrho}}_{\kappa\kappa'} = \sum\nolimits_{\kappa''} (\,\tilde{H}_{\kappa\kappa''}\,\tilde{\varrho}_{\kappa''\kappa'} - \tilde{\varrho}_{\kappa\kappa''}\,\tilde{H}_{\kappa''\kappa'})\,, \tag{6.1.8.1}$$

Substituting here the Hamiltonian $\tilde{H}_{\kappa\kappa'}$ in κ-representation, we get

$$\mathrm{i}\,\hbar\,\dot{\tilde{\varrho}}_{\kappa\kappa'} = [\,\varepsilon_\kappa(t) - \varepsilon_{\kappa'}(t)\,]\,\tilde{\varrho}_{\kappa\kappa'}\,. \tag{6.1.8.2}$$

Or in more detail,

$$\begin{gathered} \dot{\tilde{\varrho}}_{11} = 0, \quad \mathrm{i}\,\hbar\,\dot{\tilde{\varrho}}_{12} = -\,[\,\varepsilon_2(t) - \varepsilon_1(t)\,]\,\tilde{\varrho}_{12}\,, \\ \mathrm{i}\,\hbar\,\dot{\tilde{\varrho}}_{21} = [\,\varepsilon_2(t) - \varepsilon_1(t)\,]\,\tilde{\varrho}_{21}\,, \quad \dot{\tilde{\varrho}}_{22} = 0\,. \end{gathered} \tag{6.1.8.3}$$

The second equation can be written as,

$$\mathrm{d}\tilde{\varrho}_{12}/\tilde{\varrho}_{12} = \mathrm{i}/\hbar\,[\,\varepsilon_2(t) - \varepsilon_1(t)\,]\,\mathrm{d}t. \tag{6.1.8.4}$$

Integrating, we get,

$$\ln\tilde{\varrho}_{12} = \mathrm{i}/\hbar \int\,[\,\varepsilon_2(t) - \varepsilon_1(t)\,]\,\mathrm{d}t\,. \tag{6.1.8.5}$$

Using formula (6.1.7.25), we find the difference in the eigenvalues of the energy as

$$\varepsilon_2(t) - \varepsilon_1(t) \;=\; \hbar\,(\,\omega_o\cos 2\,\vartheta - \Omega\sin 2\,\vartheta\cos\omega\,t\,)\,. \tag{6.1.8.6}$$

Express $\cos 2\,\vartheta$ and $\sin 2\,\vartheta$ via $\mathrm{tg}\,\vartheta$:

$$\cos 2\,\vartheta = \;(\,1 - \;\mathrm{tg}^2\vartheta\,)/(\,1\;+\;\mathrm{tg}^2\vartheta\,)\,,$$

$$\sin 2\,\vartheta = 2\,\mathrm{tg}\,\vartheta/(\,1\;+\;\mathrm{tg}^2\vartheta\,)\,.$$

Substituting here the formula (6.1.7.23), we have

$$\cos 2\vartheta = [1 - (\epsilon/2)^2]/[1 + (\epsilon/2)^2],$$
$$\sin 2\vartheta = -\epsilon/[1 + (\epsilon/2)^2].$$

Expanding these functions in powers of ϵ:

$$\cos 2\vartheta = 1 - \epsilon^2/2 + \dots, \qquad \sin 2\vartheta = -\epsilon + \dots \tag{6.1.8.7}$$

Substitution of these formulas into (6.1.8.6) gives

$$[\varepsilon_2(t) - \varepsilon_1(t)]/(\hbar\,\omega_o) = 1 + 2\,(\Omega\cos\omega\,t/\omega_o)^2. \tag{6.1.8.8}$$

Using this function with the integral (6.1.8.5), we have

$$\ln\tilde{\varrho}_{12} = \mathrm{i}\int [\omega_o + 2\,(\Omega\cos\omega\,t)^2/\omega_o]\,\mathrm{d}t.$$

Finally,

$$\tilde{\varrho}_{12} = C_{12}\mathrm{e}^{\,\mathrm{i}\psi(t)}, \tag{6.1.8.9}$$

where

$$\psi(t) = (\omega_o + \Omega^2/\omega_o)\,t + \Omega^2/(2\,\omega_0\,\omega)\sin 2\,\omega\,t. \tag{6.1.8.10}$$

Other solutions of the system (6.1.8.3) will have the form,

$$\tilde{\varrho}_{11} = C_{11}, \qquad \tilde{\varrho}_{21} = C_{12}^{*}\mathrm{e}^{-\mathrm{i}\psi(t)}, \qquad \tilde{\varrho}_{22} = C_{22}. \tag{6.1.8.11}$$

We found energy eigenvalues ε_1 and ε_2 exactly. They are given by the formulas (6.1.7.25). To express them through time, we write:

$$\cos^2\vartheta = 1/(1 + \mathrm{tg}^2\vartheta) \sim 1 - \mathrm{tg}^2\vartheta \sim 1 - \epsilon^2/4,$$
$$\sin^2\vartheta = \mathrm{tg}^2\vartheta/(1 + \mathrm{tg}^2\vartheta) \sim \epsilon^2/4. \tag{6.1.8.12}$$

The formula (6.1.8.7) shows that sin ϑ is a negative value:

$$\cos\vartheta\,\sin\vartheta = -\,\epsilon/2\,. \tag{6.1.8.13}$$

Putting formulas (6.1.8.12) and (6.1.8.13) into equations (6.1.7.25); taking into account equations (6.1.6.9) and (6.1.7.21), we obtain

$$\varepsilon_1 = \varepsilon_1^{(o)} - \hbar\,\omega_o\,(\,\Omega\cos\omega\,t/\omega_o)^2, \tag{6.1.8.14}$$

$$\varepsilon_2 = \varepsilon_2^{(o)} + \hbar\,\omega_o\,(\,\Omega\cos\omega\,t/\omega_o)^2, \tag{6.1.8.15}$$

The difference of the eigenvalues (6.1.8.14) and (6.1.8.15) is proportional to the frequency $\widetilde{\omega}$:

$$\varepsilon_2 - \varepsilon_1 = \hbar\,\widetilde{\omega}\,, \tag{6.1.8.16}$$

which is equal to

$$\widetilde{\omega}(t) = \omega_o + 2\,(\,\Omega\cos\omega\,t)^2/\omega_o\,. \tag{6.1.8.17}$$

Finding the average energy of the atom when the electric field acts on it:

$$E = \sum_{\kappa\kappa'}\widetilde{H}_{\kappa\kappa'}\,\tilde{\varrho}_{\kappa'\kappa} = \sum_{\kappa\kappa'}\,\varepsilon_\kappa\,\delta_{\kappa\kappa'}\,\tilde{\varrho}_{\kappa'\kappa} = \sum_\kappa\varepsilon_\kappa\,\tilde{\varrho}_{\kappa\kappa} = \sum_\kappa\varepsilon_\kappa\,\widetilde{w}_\kappa = g_1\varepsilon_1\,\widetilde{w}_1 + g_2\varepsilon_2\widetilde{w}_2. \tag{6.1.8.18}$$

We substitute formulas (6.1.8.14) and (6.1.8.15) in this expression. We have

$$E = g_1\,\varepsilon_1^{(o)}\widetilde{w}_1 + g_2\,\varepsilon_2^{(o)}\widetilde{w}_2 + \hbar\,\omega_o\,(\,\Omega\cos\omega\,t/\omega_o)^2\,(\,g_2\,\widetilde{w}_2 - \,g_1\,\widetilde{w}_1)\,. \tag{6.1.8.19}$$

We see that the probability (6.1.8.11) remains constant, as they do not illustrate the operation of the pumping and damping. These phenomena are present in the second term

$$\mathrm{i}\,\hbar\,\widehat{D}(\hat{\varrho})$$

of the equation for the density matrix (6.1.4.1) and (6.1.5.1), which shows the dissipative effects.

6.1.9. The Transition Density Matrix from κ-Representation in Initial α-Representation

The transition density matrix from a κ-representation in α-representation is performed by means of unitary trans-formations:

$$\varrho_{\alpha\beta}^{(o)} = \sum_{\kappa\kappa'} U_{\alpha\kappa}\, \tilde{\varrho}_{\kappa\kappa'} U_{\beta\kappa'}^{*} \,. \tag{6.1.9.1}$$

where we have already found the unitary matrix $U_{\alpha\kappa}$. It has the form (6.1.7.24):

$$U_{\alpha\kappa} = \begin{Vmatrix} \cos\vartheta & -\sin\vartheta \\ \sin\vartheta & \cos\vartheta \end{Vmatrix} .$$

The density matrix $\tilde{\varrho}_{\kappa\kappa'}$ is

$$\tilde{\varrho}_{\kappa\kappa'} = \begin{Vmatrix} \tilde{\varrho}_{11} & \tilde{\varrho}_{12} \\ \tilde{\varrho}_{21} & \tilde{\varrho}_{22} \end{Vmatrix} . \tag{6.1.9.2}$$

We calculate the density matrix of

$$\varrho_{\alpha\beta}^{(o)} = \begin{Vmatrix} \cos\vartheta & -\sin\vartheta \\ \sin\vartheta & \cos\vartheta \end{Vmatrix} \begin{Vmatrix} \tilde{\varrho}_{11} & \tilde{\varrho}_{12} \\ \tilde{\varrho}_{21} & \tilde{\varrho}_{22} \end{Vmatrix} \begin{Vmatrix} \cos\vartheta & \sin\vartheta \\ -\sin\vartheta & \cos\vartheta \end{Vmatrix} . \tag{6.1.9.3}$$

Computing this product:

$$\varrho_{\alpha\beta}^{(o)} = \begin{Vmatrix} \cos\vartheta & -\sin\vartheta \\ \sin\vartheta & \cos\vartheta \end{Vmatrix} \begin{Vmatrix} a_{11} & a_{11} \\ a_{21} & a_{22} \end{Vmatrix} , \tag{6.1.9.4}$$

where

$$\begin{aligned} a_{11} &= \tilde{\varrho}_{11}\cos\vartheta - \tilde{\varrho}_{12}\sin\vartheta, \\ a_{12} &= \tilde{\varrho}_{11}\sin\vartheta + \tilde{\varrho}_{12}\cos\vartheta, \\ a_{21} &= \tilde{\varrho}_{21}\cos\vartheta - \tilde{\varrho}_{22}\sin\vartheta, \\ a_{22} &= \tilde{\varrho}_{21}\sin\vartheta + \tilde{\varrho}_{22}\cos\vartheta. \end{aligned}$$

Finally we have,

$$\varrho_{11}^{(o)} = a_{11}\cos\vartheta - a_{21}\sin\vartheta = \tilde{\varrho}_{11}\cos^2\vartheta + \tilde{\varrho}_{22}\sin^2\vartheta - (\tilde{\varrho}_{12} + \tilde{\varrho}_{21})\cos\vartheta \sin\vartheta\,,\ \varrho_{12}^{(o)} = a_{12}\cos\vartheta - a_{22}\sin\vartheta = (\tilde{\varrho}_{11} - \tilde{\varrho}_{22})\cos\vartheta\sin\vartheta + \tilde{\varrho}_{12}\cos^2\vartheta - \tilde{\varrho}_{21}\sin^2\vartheta\,,\ \varrho_{21}^{(o)} = a_{11}\sin\vartheta + a_{21}\cos\vartheta = (\tilde{\varrho}_{11} - \tilde{\varrho}_{22})\cos\vartheta\sin\vartheta - \tilde{\varrho}_{12}\sin^2\vartheta + \tilde{\varrho}_{21}\cos^2\vartheta\,,\ \varrho_{22}^{(o)} = a_{12}\sin\vartheta + a_{22}\cos\vartheta = \tilde{\varrho}_{11}\sin^2\vartheta + \tilde{\varrho}_{22}\cos^2\vartheta + (\tilde{\varrho}_{12} + \tilde{\varrho}_{21})\cos\vartheta\sin\vartheta\,. \tag{6.1.9.5}$$

The amount $\tilde{\varrho}_{12} + \tilde{\varrho}_{21}$ is equal to twice the real part of the values of $\tilde{\varrho}_{12}$:

$$\tilde{\varrho}_{12} + \tilde{\varrho}_{21} = \tilde{\varrho}_{12} + \tilde{\varrho}_{12}^{*} = 2\,\mathrm{Re}\,\tilde{\varrho}_{12}\,. \tag{6.1.9.6}$$

Substitute in formula (6.1.9.5) equation (6.1.9.6) and the approximate relation (6.1.8.13) and (8.14). Given formula (6.1.7.21) for the parameter ϵ, we have

$$\begin{aligned}
\varrho_{11}^{(o)} &= \tilde{\varrho}_{11} + (\tilde{\varrho}_{22} - \tilde{\varrho}_{11})\,\xi^2(\mathrm{t}) + 2\,\mathrm{Re}\,\tilde{\varrho}_{12}\,\xi(\mathrm{t})\,,\\
\varrho_{12}^{(o)} &= \tilde{\varrho}_{12} + (\tilde{\varrho}_{22} - \tilde{\varrho}_{11})\,\xi(\mathrm{t}) - 2\,\mathrm{Re}\,\tilde{\varrho}_{12}\,\xi^2(\mathrm{t})\,,\\
\varrho_{21}^{(o)} &= \tilde{\varrho}_{21} + (\tilde{\varrho}_{22} - \tilde{\varrho}_{11})\,\xi(\mathrm{t}) - 2\,\mathrm{Re}\,\tilde{\varrho}_{12}\,\xi^2(\mathrm{t})\,,\\
\varrho_{22}^{(o)} &= \tilde{\varrho}_{22} - (\tilde{\varrho}_{22} - \tilde{\varrho}_{11})\,\xi^2(\mathrm{t}) - 2\,\mathrm{Re}\,\tilde{\varrho}_{12}\,\xi(\mathrm{t})\,,
\end{aligned} \tag{6.1.9.7}$$

where

$$\xi(t) = \epsilon/2 = \Omega/\omega_0 \cos\omega\, t\,. \tag{6.1.9.8}$$

Matrix (6.1.9.7) can never be diagonal, if the external electric field acts on the atom $[\,\xi(t) \neq 0]$.

6.1.10. Dissipative Matrix

Now we need to figure out how the matrix $\gamma_{nm,m'n'}$ appears . The easiest way is to use formula (6.1.4.13). Suppose in this formula there is an operator $\hat{a}_j$ and this matrix looks like this

$$\gamma_{nm,m'n'} = \textstyle\sum_j C_j\, a_{nm,j}\, a^{+}_{m'n',j}\,, \tag{6.1.10.1}$$

where $a_{nm,j}$ are the matrix elements of the operator $\hat{a}_j$ and C_j is constant. By using this operator, you can write the dissipative operator, using the Lindblad equation (6.1.4.12):

$$\widehat{D} = 1/2 \sum_j C_j \left(2\, \hat{a}_j \hat{\varrho}\, a_j^+ - \hat{a}_j^+ \hat{a}_j\, \hat{\varrho} - \hat{\varrho}\, \hat{a}_j^+ \hat{a}_j \right), \qquad \textbf{(6.1.10.2)}$$

a dissipative matrix will be as follows,

$$D_{nn'} = \sum_{mm'} \gamma_{nm,m'n'}\, \varrho_{mm'} - 1/2 \sum_m \left(\gamma_{nm}\, \varrho_{mn'} + \varrho_{nm}\, \gamma_{mn'} \right), \qquad \textbf{(6.1.10.3)}$$

Now we need to give physical meaning to operators $\hat{a}_j$.

Let the two-level scheme, the operator $\hat{a}_1$ is equal to,

$$\hat{a}_1 = \begin{Vmatrix} 0 & a_{12} \\ a_{21} & 0 \end{Vmatrix}. \qquad \textbf{(6.1.10.4)}$$

Substituting this operator in formula (6.1.10.3), we can write the dissipative matrix in numbers:

$$\begin{aligned} D_{11}^{(1)} &= 1/2\, C_1 \sum_{mm'} \left(2\, a_{1m}\, \varrho_{mm'}\, a_{m'1}^{+} - a_{1m}^{+}\, a_{mm'}\, \varrho_{m'1} - \varrho_{1m}\, a_{mm'}^{+}\, a_{m'1} \right) = \\ &= 1/2\, C_1 \left(2\, a_{12}\, \varrho_{22}\, a_{21}^{+} - a_{12}^{+}\, a_{21}\, \varrho_{11} - \varrho_{11}\, a_{12}^{+}\, a_{21} \right) = \\ &= C_1 \left(a_{12}^{2}\, \varrho_{22} - a_{21}^{2}\, \varrho_{11} \right), \end{aligned} \qquad \textbf{(6.1.10.5)}$$

$$\begin{aligned} D_{12}^{(1)} &= 1/2\, C_1 \sum_{mm'} \left(2\, a_{1m}\, \varrho_{mm'}\, a_{m'2}^{+} - a_{1m}^{+}\, a_{mm'}\, \varrho_{m'2} - \varrho_{1m}\, a_{mm'}^{+}\, a_{m'2} \right) = \\ &= 1/2\, C_1 \left(2\, a_{12}\, \varrho_{21}\, a_{12}^{+} - a_{12}^{+}\, a_{21}\, \varrho_{12} - \varrho_{12}\, a_{21}^{+}\, a_{12} \right) = \\ &= 1/2\, C_1 \left[2\, a_{12}\, a_{21}\, \varrho_{21} - \left(a_{12}^{2} + a_{21}^{2} \right) \varrho_{12} \right], \end{aligned} \qquad \textbf{(6.1.10.6)}$$

where

$$a_{12}^{2} = p_{12}, \quad a_{21}^{2} = p_{21}. \qquad \textbf{(6.1.10.7)}$$

According toformulas (6.1.4.10) and (6.1.4.11)

$$p_{12} = p_{12}^{(0)} e^{-\beta(\varepsilon_1 - \varepsilon_2)/2}, \qquad p_{21} = p_{21}^{(0)} e^{-\beta(\varepsilon_2 - \varepsilon_1)/2}, \tag{6.1.10.8}$$

$$p_{12}^{(0)} = p_{21}^{(0)} = p\,. \tag{6.1.10.9}$$

Supposing the matrix $\hat{a}_2$ has the form

$$\hat{a}_2 = \begin{Vmatrix} 0 & a_1 \\ a_2 & 0 \end{Vmatrix}. \tag{6.1.10.10}$$

Dissipative matrix will now be,

$$D_{11}^{(2)} = C_2\,(a_1^2\,\varrho_{22} - a_2^2\,\varrho_{11})\,. \tag{6.1.10.11}$$

$$D_{12}^{(2)} = 1/2\,C_2\,[\,2\,a_1\,a_2\,\varrho_{21} - (a_1^2 + a_2^2)\,\varrho_{12}\,]\,. \tag{6.1.10.12}$$

We denote

$$a_1^2 = \gamma_1\,, \qquad a_2^2 = \gamma_2\,. \tag{6.1.10.13}$$

If the matrix $\hat{a}_3$ is such that

$$\hat{a}_3 = \begin{Vmatrix} 0 & a \\ a & 0 \end{Vmatrix}. \tag{6.1.10.14}$$

The third dissipative matrix by the formulas (6.1.10.11) and (6.1.10.12) will be,

$$D_{11}^{(3)} = C_3\,a^2\,(\varrho_{22} - \varrho_{11})\,, \tag{6.1.10.15}$$

$$D_{12}^{(3)} = C_3\,a^2\,(\varrho_{21} - \varrho_{12})\,. \tag{6.1.10.16}$$

If the matrix $\hat{a}_4$ is diagonal, *i.e.*

$$a_{nm} = b_n\,\delta_{nm}\,, \tag{6.1.10.17}$$

the dissipative matrix will be,

$$D^{(4)}_{nn'} = C_4\,(\,2\,b_n b_{n'} - b_n^2 - b_{n'}^2)\,\varrho_{nn'}\ . \qquad \textbf{(6.1.10.18)}$$

In particular,

$$D^{(4)}_{nn} = 0,\ D^{(4)}_{12} = \ 1/2\,C_4\,(\,2\,b_1 b_2 - b_1^2 - b_2^2)\,\varrho_{12}\ ,\ D^{(4)}_{21} = 1/2\,C_4\,(\,2\,b_2 b_1 - b_2^2 - b_1^2)\,\varrho_{21}\ . \qquad \textbf{(6.1.10.19)}$$

We can put,

$$D^{(4)}_{12} = -\,\Gamma\,\varrho_{12}\,, \quad D^{(4)}_{21} = -\,\Gamma\,\varrho_{21}\ . \qquad \textbf{(6.1.10.20)}$$

Where

$$\Gamma = \ 1/2\,C_4\,(\,b_1^2 + b_2^2 - 2\,b_1 b_2)\ .$$

Collecting all four options for dissipative matrices together and putting $C_1 = 1$, $C_2 = -\,2$ and $C_3 = 1$, we obtain

$$D_{11} = (\,p_{12} + \gamma_1 + a^2)\,\varrho_{22} - (\,p_{21} + \gamma_2 + a^2)\,\varrho_{11}\,, \qquad \textbf{(6.1.10.21)}$$

$$D_{12} = 1/2\,[\,2\,\sqrt{p_{12}\,p_{21}}\,\varrho_{21} - (\,p_{12} + p_{21})\,\varrho_{12}] + (\,\gamma_1 + \gamma_2\,)\,\varrho_{12} - \\ - 2\,\sqrt{\gamma_1\,\gamma_2}\,\varrho_{21} + a^2(\,\varrho_{21} - \varrho_{12}) - \Gamma\,\varrho_{12}\ . \qquad \textbf{(6.1.10.22)}$$

6.1.11. The Equation for the Density Matrix in κ-Representation

Write the equation for the density matrix in κ-representation, where the Hamiltonian is diagonal, *i.e.* has the form

$$\widetilde{H}_{\kappa\kappa'} = \varepsilon_\kappa(t)\,\delta_{\kappa\kappa'}\,, \qquad \textbf{(6.1.11.1)}$$

Energy eigenvalues $\varepsilon_\kappa(t)$ are determined by formulas (6.1.8.14) and (6.1.8.15).

The desired equation is:

$$\mathrm{i}\,\hbar\,\dot{\varrho}_{\kappa\kappa'} = \sum_{\kappa''} \left(\widetilde{H}_{\kappa\kappa''}\,\varrho_{\kappa''\kappa'} - \varrho_{\kappa\kappa'}\,\widetilde{H}_{\kappa''\kappa'} \right) + \mathrm{i}\,\hbar\,D_{\kappa\kappa'}\,. \tag{6.1.11.2}$$

Substituting here formulas (6.1.8.14) and (6.1.8.15), we obtain

$$\mathrm{i}\,\hbar\,\dot{\varrho}_{\kappa\kappa'} = [\,\varepsilon_{\kappa}(t) - \varepsilon_{\kappa'}(t)]\,\varrho_{\kappa\kappa'} + \mathrm{i}\,\hbar\,D_{\kappa\kappa'}\,. \tag{6.1.11.3}$$

Write equations for the density matrix ϱ_{11} and ϱ_{12}:

$$\dot{\varrho}_{11} = (\,p_{12} + \gamma_1 + a^2)\,\varrho_{22} - \,(\,p_{21} + \gamma_2 + a^2)\,\varrho_{11}\,, \tag{6.1.11.4}$$

$$\dot{\varrho}_{12} = \mathrm{i}\,\widetilde{\omega}\,\varrho_{12} + 1/2\,[\,2\,\sqrt{p_{12}\,p_{21}}\,\varrho_{21} - (\,p_{12} + p_{21})\,\varrho_{12}] + \\ + (\,\gamma_1 + \gamma_2)\,\varrho_{12} - 2\,\sqrt{\gamma_1\,\gamma_2}\,\varrho_{21} + a^2(\,\varrho_{21} - \varrho_{12}) - \Gamma\,\varrho_{12}\,. \tag{6.1.11.5}$$

These equations should be supplemented by the following equations

$$\varrho_{21} = \varrho_{12}^{*}\,, \tag{6.1.11.6}$$

$$\dot{\varrho}_{11} + \dot{\varrho}_{22} = 0\,. \tag{6.1.11.7}$$

Now you need to leave in equations (6.1.11.4) and (6.1.11.5) the density matrix that are in these equations, or to substitute here the density matrix in representation (6.1.9.7).

6.1.12. Kinetics of Laser

Now let us write equation (6.1.2.11) with probability w_1 and w_2. The number of atoms of N_1 and N_2 are connected with this probability by ratios

$$N_1 = N\,g_1\,w_1\,, \quad N_2 = N\,g_2\,w_2\,. \tag{6.1.12.1}$$

Equation (6.1.2.11) will look like this:

$$g_1\dot{w}_1 = -\,B\,W\,g_1 w_1 + (\,A + B\,W\,)\;g_2\,w_2\,. \tag{6.1.12.2}$$

Substituting the density matrix in this equation instead of probabilitics,

$$W_K \leftrightarrow \varrho_{KK},$$

We have,

$$\dot{\varrho}_{11} = -B\,W\,\varrho_{11} + (A + B\,W)\,\varrho_{22}\,. \qquad \textbf{(6.1.12.3)}$$

Now considering equation (6.1.11.4), obtained from the equation for the density matrix; equations (6.1.11.4) and (6.1.12.3) are remarkably similar to one another. First of all, we need to put

$$\gamma_2 = Q\,, \qquad \textbf{(6.1.12.4)}$$

And the values γ_1 and a^2 must be considered

$$\gamma_1 = A\,, \qquad a^2 = B\,W. \qquad \textbf{(6.1.12.5)}$$

Substituting these values into equation (6.1.11.4). Finally, we have

$$\dot{\varrho}_{11} = p_{12}\,\varrho_{22} - p_{21}\,\varrho_{11} + (A + B\,W)\,\varrho_{22} - (Q + B\,W)\,\varrho_{11}\,, \qquad \textbf{(6.1.12.6)}$$

We have proved that the phenomenological equation (6.1.2.11) coincides with the equation obtained by the method of quantum-mechanics. However, it is supplemented by two more terms.

We write the equation (6.1.11.6) in more detail

$$\dot{\varrho}_{11} = p\,(\mathrm{e}^{\beta\hbar\tilde{\omega}/2}\varrho_{22} - \mathrm{e}^{-\beta\hbar\tilde{\omega}/2}\varrho_{11}) + \\ + [A + B\,W(t,\tilde{\omega})]\,\varrho_{22} - [Q + B\,W(t,\tilde{\omega})]\,\varrho_{11}\,. \qquad \textbf{(6.1.12.7)}$$

We supplement this equation with the normalization condition

$$\varrho_{11} + \varrho_{22} = 1. \qquad \textbf{(6.1.12.8)}$$

6.1.13. Equation for Non-diagonal Density Matrix

Firstly, substituting formulas (6.1.10.8), (6.1.10.9), (6.1.12.4) and (6.1.12.5) into equation (11.5).

$$\dot{\varrho}_{12} = \mathrm{i}\,\tilde{\omega}\,\varrho_{12} - \Gamma\varrho_{12} + p\,\varrho_{21} - p/2\,(\mathrm{e}^{\beta\hbar\tilde{\omega}/2} + \mathrm{e}^{-\beta\hbar\tilde{\omega}/2})\,\varrho_{12} +$$

$$+ (A + Q - B\,W)\, \varrho_{12} + (B\,W - 2\sqrt{AQ})\, \varrho_{21} \,.$$

(6.1.13.1)

To solve this equation, the condition

$$\varrho_{21} = \varrho_{12}^{*} \qquad \textbf{(6.1.13.2)}$$

should be used. It should be remembered that in all three equations (6.1.12.7). The next work will be devoted to solving this system of equations.

6.1.14. Kinetics of Radiation

The laser was opened to electromagnetic radiation, which is characterized by the fact that it
travels almost in the same direction and has almost the same frequency. One of the tasks of the theory of laser is to find the spectral density of the laser radiation.

Because spontaneous migration of atoms is accompanied by the emission of photons moving in all directions, the number N_{photon} of photons travelling along the axis of the laser will satisfy the equation ,

$$\mathrm{d}N_{\text{photon}}/\mathrm{dt} = B\,W(N_2 - N_1) \,. \qquad \textbf{(6.1.14.1)}$$

The product

$$\Delta(t, \omega) = \mu\, N_{\text{photon}} \qquad \textbf{(6.1.14.2)}$$

is equal to the energy spectral density of these photons. Here μ is the coefficient which provides the desired dimension.

The spectral density of radiation energy *W* is equal to the sum of the Planck density and the photon density,

$$W(t, \omega) = W^{(\text{Planck})}(t, \omega) + \Delta(t, \omega) \,, \qquad \textbf{(6.1.14.3)}$$

When installed in the laser inverse population of levels ($N_2 > N_1$), we can assume the validity of inequalities,

$$\Delta(t,\omega) \gg W^{(\mathrm{Planck})}(t,\omega)\,. \tag{6.1.14.4}$$

In this case, we can write the equation as,

$$\mathrm{d}\Delta/\mathrm{d}t = \mu\, B\, \Delta\,(\,N_2 - N_1)\,. \tag{6.1.14.5}$$

The density Δ consists of two types; density Δ^+ and Δ^-. The density Δ^+ characterizes the radiation that flies to the right (see Fig. **6.1.1**), and the density Δ^- is the radiation, which travels in the opposite direction:

$$\Delta = \Delta^+ + \Delta^-. \tag{6.1.14.6}$$

These densities satisfy the equations:

$$\mathrm{d}\Delta^+/\mathrm{d}t = \mu\, B\, \Delta^+\,(\,N_2 - N_1)\,, \tag{6.1.14.7}$$

$$\mathrm{d}\Delta^-/\mathrm{d}t = \mu\, B\, \Delta^-\,(\,N_2 - N_1)\,. \tag{6.1.14.8}$$

Let the left mirror in Fig. (**1**) perfectly reflect the light falling on it and the right mirror has the reflection coefficient *R*. These mirrors are attached to the boundary conditions:

$$\Delta^-(x=0) = \Delta^+(x=0)\,, \tag{6.1.14.9}$$

$$\Delta^-(x=l) = R\,\Delta^+(x=l)\,. \tag{6.1.14.10}$$

The spectral energy density of the exciting laser radiation will be equal

$$\Delta^+_{\mathrm{output}}\,(\,x = l\,) = (\,1 - R\,)\,\Delta^+(\,x = l\,)\,. \tag{6.1.14.11}$$

In the following work, we will find this value

$$\Delta^+_{\mathrm{output}}(\,x = l\,) = (\,1 - R\,)\,(\mathrm{e}^{2\,\alpha\, l} - 1\,)/(\,1\; - \;R\,\mathrm{e}^{2\,\alpha\, l}\,)\; N\,\Omega^2\, M(l)/[\,\mu\,\omega_o^2\,(\,\delta\omega^2\; + \;\Gamma^2)\,], \tag{6.1.14.12}$$

where $M(l)$ is the coefficient,

$$\alpha = \mu\, NB\,(\varrho_{22} - \varrho_{11})/(2\,c)\,, \qquad \delta\omega = \omega - \omega_0\,.$$

6.1.15. Conclusion

In this work, we show that the method of density matrix is applicable to the theory of lasers. As we see from the exact formula (6.1.6.18), even the zero approximation of the density matrix is mixed, *i.e.* not equal to the product of the wave functions. To solve the other exact equations (6.1.12.7) and (6.1.13.4) and (6.1.14.1), it is necessary to use a computer. The theory of lasers described in this work, is based on the method of density-matrix and equation (6.1.4.1) obtained in works of the author.

REFERENCES

[1] B.V. Bondarev, "Vestnik Donetsk National University", *Ser. A: Natural Sciences*, vol. 4, pp. 54-68, 2017.
[2] N. G. Basov, A. M. Prokhorov, *Molecular oscillator and amplifier*, UFN, 1955, vol. 57, pp. 485-501; UFN, v. 93, p. 572-584, 1967.
[3] J.P. Gordon, H.J. Zeiger, C.H. Townes, "The Maser New Type of Microwave Amplifier, Frequency Standard, and Spectrometer", *Phys. Rev.*, vol. 99. pp. 1264-1274, 1955.

CHAPTER 7

Dissipative Operator

7.1. LINDBLAD EQUATION FOR HARMONIC OSCILLATOR UNCERTAINTY RELATION DEPENDING ON TEMPERATURE

Specific nonequilibrium states of the quantum harmonic oscillator described by the Lindblad equation have been hereby suggested. This equation makes it possible to determine time-varying effects produced by the statistical operator or statistical matrix. Thus, respective representation-varied equilibrium statistical matrixes and specific mean value equations have been found, and their equilibrium solutions have been obtained.

7.1.1. Lindblad Equation

Statistical operator $\hat{\varrho}$ or density matrix is basically applied as the quantum mechanics; any information of the nonequilibrium process proceeding within the tested system may be gained from the study [1]. When the process concerned proceeds within the system which fails to interact with its environment, statistical operator $\hat{\varrho}$ will satisfy the Liouville-von Neumann equation as follows:

$$i\hbar\dot{\hat{\varrho}} = [\hat{H}\hat{\varrho}]$$

(7.1.1.1)

With provision for the fact that the system interacts with any environment, a new equation shall be produced [1]. Lindblad is the first one who offered the equation describing the interaction of the system with a thermostat. This work is devoted to the Markovian equation [4], which hereby describes nonequilibrium quantum harmonic oscillator performance.

We will write the kinetic equation for a quantum harmonic oscillator as follows:

$$i\hbar\dot{\hat{\varrho}} = [\hat{H}\hat{\varrho}] + i\hbar A([\hat{a}\hat{\varrho}, \hat{a}^+] + [\hat{a}, \hat{\varrho}\hat{a}^+]) + i\hbar B([\hat{a}^+\hat{\varrho}, \hat{a}] + [\hat{a}^+, \hat{\varrho}\hat{a}]),$$

(7.1.1.2)

Boris V. Bondarev

where

$$\hat{H} = \hbar\,\omega\,(\,\hat{a}^{+}\hat{a} + 1/2\,)\,,$$

(7.1.1.3)

A and B are constants. Operator $\hat{a}$ is formulated as follows:

$$\hat{a} = (\,\mathrm{i}\ \hat{p}/\sqrt{m} + \sqrt{\kappa}\,\hat{x}\,)/\sqrt{2\,\hbar\,\omega}\,,$$

(7.1.1.4)

where

$$\omega = \sqrt{\kappa/m}\,.$$

Equation (7.1.1.2) is very precise to describe the time-varying state of the thermostat-interacted quantum harmonic oscillator and its equilibrium state.

7.1.2. Energy Representation

Now, we will define the wave functions, describing specific energy state $\varphi_n(x)$ which will satisfy the equation as follows:

$$\hat{H}\,\varphi_n(x) = E_n\ \varphi_n(x),$$

(7.1.2.1)

where

$$E_n = \hbar\,\omega\,(\,n + 1/2\,)\,,$$

(7.1.2.2)

$$n = 0,\ 1,\ 2,\ \ldots$$

As referred to energy representation, the matrix elements of statistical operator $\hat{\varrho}$ will be formulated by the equation as follows:

$$\varrho_{nn'} = \int \varphi_n^*(x)\ \hat{\varrho}\ \varphi_{n'}(x)\ \mathrm{d}x\,.$$

(7.1.2.3)

Wave functions satisfy the following equations:

$$\hat{a}\,\varphi_n = \sqrt{n}\,\varphi_{n-1}, \qquad \hat{a}^{+}\varphi_n = \sqrt{n+1}\,\varphi_{n+1}\,.$$

(7.1.2.4)

With provision for the above formulas, the following matrix-formed Equation (1.2) is derived as:

$$\dot{\varrho}_{nn'} = -\,\mathrm{i}\,\omega\,(n-n')\,\varrho_{nn'} + A\,[\,2\sqrt{(n+1)(n'+1)}\,\varrho_{n+1,n'+1} - (n+n')\,\varrho_{nn'}] + B\,[\,2\sqrt{n\,n'}\,\varrho_{n-1,n'-1} - (n+n'+2)\,\varrho_{nn'}]\,.$$

(7.1.2.5)

Now, we will write the equation for diagonal elements of the density matrix $\varrho_{nn} = w_n$, where w_n is the probability referred to oscillator state φ_n. The equation produced has the form as follows:

$$\dot{w}_n = 2\,A\,[\,(n+1)\,w_{n+1} - nw_n\,] + 2\,B\,[n\,w_{n-1} - (n+1)\,w_n]\,.$$

(7.1.2.6)

This kinetic equation describes particular harmonic oscillator state transitions. In this case, there may be gained coefficients *A* and *B* as follows:

$$A = 1/2\,P\exp\,(\,\beta\,\hbar\,\omega/2\,)\,, \qquad B = 1/2\,P\exp\,(\,-\,\beta\,\hbar\,\omega/2\,)\,,$$

(7.1.2.7)

where P is probability of transition per unit time; $\beta = 1/(k_{\mathrm{B}}T)$ is reciprocal temperature.

Equation (7.1.2.6) has specific oscillator state equilibrium distribution, which satisfies the following equation:

$$A\,[\,(n+1)\,w_{n+1} - n\,w_n] + B\,[\,n\,w_{n-1} - (n+1)\,w_n] = 0\,.$$

(7.1.2.8)

This equation is solved by the method as follows:

$$w_n = (1 - q)\, q^n$$

(7.1.2.9)

under the following condition

$$q = B/A = \exp(-\beta \hbar \omega)$$

(7.1.2.10)

7.1.3. Mean Value of the Coordinate

Mean value $\bar{b}$ assigned by operator $\hat{b}$ is defined as

$$\bar{b} = \mathrm{Tr}(\hat{b}\,\hat{\varrho})$$

(7.1.3.1)

For gaining mean value $\bar{a}$, the respective equation may be derived from formula (7.1.1.2). Using the equality of:

$$\hat{a}\,\hat{a}^{+} - \hat{a}^{+}\hat{a} = 1,$$

(7.1.3.2)

we will get the equation as follows:

$$\dot{\bar{a}} = (-\mathrm{i}\,\omega - A + B)\,\bar{a}\,.$$

(7.1.3.3)

Now, we can find the derivatives from mean values $\bar{x}$ and $\bar{p}$. By applying formula (7.1.1.4), we will get:

$$\mathrm{i}\,\dot{\bar{p}}/\sqrt{m} + \sqrt{\kappa}\,\dot{\bar{x}} = (-\mathrm{i}\,\omega - A + B)(\mathrm{i}\,\bar{p}/\sqrt{m} + \sqrt{\kappa}\,\bar{x})\,.$$

Then, we will try to equate both the real and imaginary parts of this equation:

$$\begin{cases} \dot{\bar{x}} = -(A-B)\,\bar{x} + \bar{p}/m, \\ \dot{\bar{p}} = -\kappa\,\bar{x} - (A-B)\,\bar{p}. \end{cases}$$

(7.1.3.4)

If we eliminate $\bar{p}$ from this set of equations, we can obtain the mean coordinate equation

$$\ddot{\bar{x}} + 2\,(A-B)\,\dot{\bar{x}} + [\,\omega^2 + (A-B)^2]\,\bar{x} = 0.$$

(7.1.3.5)

The above Equation (7.1.3.5) provides the following solution:

$$\bar{x}(t) = (\,C_1 \cos\omega\, t + C_2 \sin\omega\, t\,)\exp\,[\,-(\,A-B\,)\,t\,],$$

(7.1.3.6)

where C_1 and C_2 are arbitrary constants.

7.1.4. Mean Oscillator Energy

Now, we will find the time derivative from $\overline{a^+a}$. By applying the above equality (7.1.3.2), we will produce the following derivative from Equation (7.1.1.2):

$$\dot{\overline{a^+a}} + 2\,(\,A-B\,)\,\overline{a^+a} = 2\,B\,.$$

(7.1.4.1)

We can define harmonic oscillator time-varying energy effects inserting the following formula in Equation (7.1.4.1):

$$\overline{a^+a} = \bar{H}/(\hbar\,\omega) - 1/2\,.$$

Thus, the following differential equation is derived:

$$\dot{\bar{H}} + 2\,(\,A-B\,)\,\bar{H} = \hbar\,\omega\,(\,A+B\,)\,.$$

(7.1.4.2)

The solution of the equation is:

$$\overline{H}(t) = C \exp\left[-2\,(A - B)\,t \right] + \left[\hbar\,\omega\,(A + B) \right] / \left[2\,(A - B) \right],$$

(7.1.4.3)

where C is an arbitrary constant. The Equation (7.1.4.2) has specific stationary solution:

$$\overline{H} = \left[\hbar\,\omega\,(A + B) \right] / \left[2\,(A - B) \right].$$

(7.1.4.4)

Since constants A and B are related (7.1.2.7), the stationary solution obeys the formula as follows:

$$\overline{H} = \left[\hbar\,\omega\,(e^{\beta\,\hbar\,\omega} + 1) \right] / \left[2\,(e^{\beta\,\hbar\,\omega} - 1 \right] = \hbar\,\omega/2\ \mathrm{cth}\,(\beta\,\hbar\,\omega/2).$$

(7.1.4.5)

If $T = 0$, then $\overline{H} = \hbar\,\omega/2$. If T increases to infinity, then $\overline{H} = k_{\mathrm{B}}T$.

7.1.5. Kinetic Equation Expressed in Terms of Coordinate and Momentum Operators

Let us express the Equation (7.1.1.2) in terms of operators $\hat{x}$ and $\hat{p}$.For this purpose, we will first write the Equation (7.1.1.2) as follows:

$$\mathrm{i}\,\hbar\,\dot{\hat{\varrho}} = \widehat{H}\,\hat{\varrho} - \hat{\varrho}\,\widehat{H} + \mathrm{i}\,\hbar\,A\,(2\,\hat{a}\,\hat{\varrho}\,\hat{a}^{+} - \hat{a}^{+}\hat{a}\,\hat{\varrho} - \hat{\varrho}\,\hat{a}^{+}\hat{a}) + \mathrm{i}\,\hbar\,B\,(2\,\hat{a}^{+}\hat{\varrho}\,\hat{a} - \hat{a}\,\hat{a}^{+}\hat{\varrho} - \hat{\varrho}\,\hat{a}\,\hat{a}^{+}).$$

(7.1.5.1)

Since the energy operator is equal to:

$$\widehat{H} = \hat{p}^2/(2m) + \kappa\,\hat{x}^2/2,$$

(7.1.5.2)

we will insert it in Equation (7.1.5.1) along with formula (7.1.1.4) to obtain the following one:

$$\mathrm{i}\,\hbar\,\dot{\hat{\varrho}} = [\,\hat{p}^2/(2m) + \kappa\,\hat{x}^2/2\,]\,\hat{\varrho} - \hat{\varrho}\,[\,\hat{p}^2/(2m) + \kappa\,\hat{x}^2/2\,] -$$

$$-\mathrm{i}\,(A+B)/(2\omega)\,[\,(\hat{p}^2\hat{\varrho} - 2\,\hat{p}\,\hat{\varrho}\,\hat{p} + \hat{\varrho}\,\hat{p}^2)/m + \kappa\,(\hat{x}^2\hat{\varrho} - 2\,\hat{x}\,\hat{\varrho}\,\hat{x} + \hat{\varrho}\,\hat{x}^2)] -$$

$$-\,(A - B)\,(\,\hat{x}\,\hat{\varrho}\,\hat{p} - \hat{p}\,\hat{\varrho}\,\hat{x} + \mathrm{i}\,\hbar\,\hat{\varrho}\,).$$

(7.1.5.3)

7.1.6. Coordinate Representation

In coordinate representation, the density matrix looks like this: $\varrho = \varrho(t, x, x')$. The coordinate and momentum operators are:

$$\hat{x} = x, \qquad \hat{p} = -\,\mathrm{i}\,\hbar\,\partial_x\,.$$

Using the above values, we can write Equation (5.3) by the formula as follows:

$$\partial_t\varrho = \mathrm{i}\hbar/(2\,m)\,(\,\partial_x^2 - \partial_{x'}^2)\,\varrho - \mathrm{i}\,\kappa/(2\,\hbar)\,(\,x^2 - x'^2)\,\varrho +$$

$$+\,(A+B)/(2\,\hbar\,\omega)\,[\,\hbar^2(\,\partial_x - \partial_{x'})^2/m - \kappa\,(x - x')^2]\,\varrho + (\,A - B\,)\,(\,1 + x\,\partial_{x'} + x'\partial_x)\varrho.$$

(7.1.6.1)

The physical interpretation of the density matrix implies that the following expression is the probability density.

$$w(t, x) = \varrho(\,t, x, x)$$

(7.1.6.2)

Let us introduce new variables:

$$x_1 = (\,x + x')/2 \quad x_2 = x - x'.$$

(7.1.6.3)

In this case,

$$\partial_x = \partial_1/2 + \partial_2 , \qquad \partial_{x'} = \partial_1/2 - \partial_2 .$$

Referring to density matrix $\varrho(t, x_1, x_2)$ and using the above new variables, we will get the equation as follows:

$$\partial_t \varrho = \mathrm{i}\, \hbar/m\ \partial_1 \partial_2 \varrho - \mathrm{i}\, \kappa/\hbar\ x_1 x_2\ \varrho + (A + B)/(\hbar\, \omega)\ [\hbar^2/(2\, m)\ \partial_1^2 \varrho - \kappa/2\ x_2^2\ \varrho + \varepsilon\, (1 + x_1 \partial_1 - x_2 \partial_2)\, \varrho\,] ,$$

(7.1.6.4)

where

$$\varepsilon = \hbar\, \omega\, (A - B)/(A + B) = \hbar\, \omega\ \mathrm{th}\, (\beta\, \hbar\, \omega\, /2).$$

(7.1.6.5)

In this case,

$$\varrho(t, x, 0) = w(t, x).$$

(7.1.6.6)

We will find the solution of Equation (7.1.6.4) as follows:

$$\varrho(t, x_1, x_2) = 1/(2\, \pi) \int f(t, k, x_2) \exp\, (\mathrm{i}\, k\, x_1)\, \mathrm{d}k .$$

(7.1.6.7)

Reciprocal transformation

$$f(t, k, x_2) = \int \varrho(t, x_1, x_2) \exp\, (-\, \mathrm{i}\, k\, x_1)\, \mathrm{d}x_1 .$$

(7.1.6.8)

Taking into account (7.1.6.6), we will obtain:

$$f(t, 0, 0) = \int \varrho(t, x, 0)\, \mathrm{d}x = \int w(t, x)\, \mathrm{d}x = 1 .$$

(7.1.6.9)

Thus, in view of function (7.1.6.8), the following equation is formed:

$$\partial_t f = -\hbar\, k/m\ \partial_x f + \kappa/\hbar\, x\, \partial_k f -$$

$$-\ (A+B)/(\hbar\,\omega)\,[\,\hbar^2 k^2/(2\,m) + \kappa\, x^2/2 + \varepsilon\,(k\,\partial_k + x\,\partial_x)\,]\, f.$$

(7.1.6.10)

This equation has an equilibrium solution which satisfies both equations as follows:

$$-\hbar\, k/m\ \partial_x f + \kappa/\hbar\, x\, \partial_k f = 0\,,$$

(7.1.6.11)

$$\hbar^2 k^2 f/(2\,m) + \kappa\, x^2 f/2 + \varepsilon\,(k\,\partial_k + x\,\partial_x) f = 0\,.$$

(7.1.6.12)

We will write the performance equation of the above formula (7.1.6.11):

$$-\,m\,\mathrm{d}x\,/(\hbar^2 k) = x\,\mathrm{d}k/\kappa.$$

This equation has a solution as follows:

$$\hbar^2 k^2/(2\,m) + \kappa\, x^2/2 = \text{const}\,.$$

This formula implies that the general solution of Equation (7.1.6.11) takes the form as follows:

$$f = f(E)\,,$$

where

$$E = \hbar^2 k^2/(2\,m) + \kappa\, x^2/2\,.$$

We put this function into Equation (7.1.6.12) to get the following formula:

$$\mathrm{d}f/\mathrm{d}E + f/(2\,\varepsilon) = 0\,.$$

Taking into account condition (7.1.6.9), this equation has the following solution:

$$f(E) = \exp\,[\,-\,E/(2\,\varepsilon)\,]\,.$$

Thus, the equilibrium solution of Equation (7.1.6.10) takes the form as follows:

$$f(k, x) = \exp \{ - [\hbar^2 k^2/(2 m) + \kappa x^2/2]/(2 \varepsilon) \} . \tag{7.1.6.13}$$

We will find the equilibrium density matrix by Formula (7.1.6.7):

$$\varrho(x_1, x_2) = 1/(2 \pi) \int \exp \{ - [\hbar^2 k^2/(2m) + \kappa x_2^2/2]/(2 \varepsilon)\} e^{i k x_1} dk .$$

Integration brings us to the formula:

$$\varrho(x_1, x_2) = \sqrt{\alpha/\pi} \exp [- \alpha x_1^2 - \sigma^2 x_2^2/(4 \alpha)], \tag{7.1.6.14}$$

where

$$\alpha = \sigma \, \mathrm{th} (\beta \hbar \omega/2), \qquad \sigma = m \omega/\hbar .$$

Using Formula (7.1.6.3), we will obtain the following equation:

$$\varrho(x, x') = \sqrt{\alpha/\pi} \exp [- \alpha (x + x')^2/4 - \sigma^2 (x - x')^2/(4 \alpha)]. \tag{7.1.6.15}$$

Using Formula (7.1.6.2), we will obatin equilibrium probability density [19]

$$w(x) = \sqrt{\alpha/\pi} \exp (- \alpha x^2) . \tag{7.1.6.16}$$

7.1.7. Momentum Representation

In momentum representation, the coordinate and momentum operators are:

$$\hat{x} = i \hbar \partial_p, \qquad \hat{p} = p .$$

In this representation, the density matrix looks like this: $\varrho = \varrho(t, p, p')$. This enables to write Equation (7.1.5.3) as follows:

$$\partial_t \varrho = - \mathrm{i}/(2\,\hbar\, m)\,(p^2 - p'^2)\,\varrho + \mathrm{i}\,\hbar\,\kappa/2\,(\partial_p^2 - \partial_{p'}^2)\,\varrho -$$

$$- (A + B)/(2\,\hbar\,\omega)\,[\,(p - p')^2/m - \kappa\,\hbar^2\,(\partial_p + \partial_{p'})^2\,]\,\varrho -$$

$$- (A - B)\,(1 + p\,\partial_{p'} + p'\partial_p)\,\varrho\,.$$

(7.1.7.1)

The physical interpretation of density matrix $\varrho\,(t, p, p')$ implies that the expression

$$w(t, p) = \varrho\,(t, p, p)$$

(7.1.7.2)

is the probability density to detect the state when an oscillator has impulse p.

7.1.8. Wigner Function

In order to better understand the physical meaning of various kinetic state summands, we will derive the equation for Wigner function $w = w\,(t, x, p)$, which is a quantum analog of the classical distribution function and can be defined with the use of density matrix $\varrho\,(t, x, x')$ by the relation:

$$w\,(t, x, p) = 1/(2\,\pi) \int \varrho\,(t, x + \hbar\, q/2, x - \hbar\, q/2)\,\exp\,(-\,\mathrm{i}\, p\, q)\,\mathrm{d}q\,.$$

(7.1.8.1)

If the density matrix depends on x_1 andx_2, than

$$w\,(t, x, p) = 1/(2\,\pi\,\hbar) \int \varrho\,(t, x_1 = x, x_2)\,\exp\,(-\,\mathrm{i}\, p\, x_2/\hbar)\,\mathrm{d}x_2\,.$$

(7.1.8.2)

Reciprocal transformation

$$\varrho\,(t, x_1, x_2) = \int w(t, x_1, p)\,\exp\,(\mathrm{i}\, p\, x_2/\hbar)\,\mathrm{d}p\,.$$

(7.1.8.3)

Since there is Formula (7.1.6.9)

$$\int \varrho(t, x_1, 0)\, dx_1 = 1 ,$$

Wigner function satisfies the normalization requirement

$$\int w(t, x, p)\, dx\, dp = 1 .$$

(7.1.8.4)

From Equation (7.1.6.4) for density matrix $\varrho(t, x_1, x_2)$, we will get the equation for Wigner function

$$\partial_t w = -\, p/m\, \partial_x w + \kappa\, x\, \partial_p w + (A + B)/(2\, \hbar\, \omega)\, [\, \hbar^2/(2\, m)\, \partial_x^2 w + \hbar^2 \kappa/2\, \partial_p^2 w + \varepsilon\, (\, 2 + x\, \partial_x + p\, \partial_p)\, w\,] .$$

(7.1.8.5)

The equation obtained is much different from its quantum analog of the Fokker – Planck equation. Summands containing derivatives $\partial_x^2 w$ and $\partial_p^2 w$ can be interpreted as those describing phase space diffusion. And still, it is necessary to add that it is rather hard to find the physical meaning of the formula in parentheses that follows coefficient ε.

The equilibrium solution of Equation (7.1.8.5) should, at the same time, be a solution for the following equations:

$$-\, p/m\, \partial_x w + \kappa\, x\, \partial_p w = 0 ,$$

(7.1.8.6)

$$\hbar^2/(2\, m)\, \partial_x^2 w + \hbar^2 \kappa/2\, \partial_p^2 w + \varepsilon\, (\, 2 + x\, \partial_x + p\, \partial_p)\, w = 0 .$$

(7.1.8.7)

General solution of the Equation (7.1.8.6) is the function:

$$w = w\, (E) ,$$

(7.1.8.8)

where

$$E = p^2/(2\,m) + \kappa\, x^2/2\,.$$

We will insert this function in Equation (7.1.8.7) and get the following equation:

$$E\, \mathrm{d}^2 w/\mathrm{d}E^2 + (\,1 + \mu\, E\,)\, \mathrm{d}w/\mathrm{d}E + \mu\, w = 0\,,$$

(7.1.8.9)

where

$$\mu = 2/(\hbar\,\omega)\ \mathrm{th}\,(\,\beta\,\hbar\,\omega/2\,)\,.$$

(7.1.8.10)

The solution of this equation is the function:

$$w(E) = C\,\mathrm{e}^{-\mu E}.$$

(7.1.8.11)

Wigner function can be obtained by Formula (7.1.8.2), inserting in its equilibrium function (7.1.6.14). We have:

$$w(x, p) = 1/(2\,\pi\,\hbar)\,\sqrt{\alpha/\pi}\int \exp\,[\,-\,\alpha\, x^2 - \sigma^2 x_2^2/(4\,\alpha)]\,\exp\,(\,-\,i\,p\,x_2/\hbar)\,\mathrm{d}x_2\,.$$

(7.1.8.12)

Integration brings us to the equilibrium function

$$w(x, p) = \mu\,\omega/(2\pi)\,\exp\{\,-\,\mu\,[\,p^2/(2\,m) + \kappa\, x^2/2]\,\}.$$

(7.1.8.13)

This function can be represented as:

$$w(x, p) = \sqrt{\alpha/\pi}\,\exp\,(\,-\,\alpha\, x^2\,)\,\sqrt{\gamma/\pi}\,\exp\,(\,-\,\gamma\, p^2)\,,$$

(7.1.8.14)

where

$$\gamma = \mu/(2\,m)\,.$$

(7.1.8.15)

Let us find the mean value

$$\overline{x^2} \cdot \overline{p^2} = \int x^2\, p^2 w(x, p)\, \mathrm{d}x\, \mathrm{d}p\,.$$

(7.1.8.16)

Calculation gives the following formula:

$$\overline{x^2} \cdot \overline{p^2} = \hbar^2/4\,[\,(e^{\beta\,\hbar\,\omega} + 1)/(e^{\beta\,\hbar\,\omega} - 1)\,]^2.$$

(7.1.8.17)

This formula leads to the following result. If $T = 0$, the uncertainty is equal to

$$\overline{x^2} \cdot \overline{p^2} = \hbar^2/4\,.$$

If$T \to \infty$, than

$$\overline{x^2} \cdot \overline{p^2} = (\,k_{\mathrm{B}}\,T/\omega\,)^2.$$

7.1.9. Lindblad Equation is First-order Approximation

It was proved that the Lindblad equation could be derived from the quantum equation for a small system that interacts with the equilibrium system from the equation of Liouville – von Neumann [10]. The Lindblad equation can be written as

$$\mathrm{i}\,\hbar\,\dot{\hat{\varrho}} = [\,\hat{H}\,\hat{\varrho}\,] + \lambda\,\mathrm{i}\,\hbar\,\hat{D}(\hat{\varrho}) + \ldots,$$

(7.1.9.1)

here λ is the order parameter. Statistical operator ϱ write as well

$$\hat{\varrho} = \hat{\varrho}_0 + \lambda\, \hat{\varrho}_1 + \dots$$

(7.1.9.2)

Let us substitute the Operator (7.1.9.2) in Equation (7.1.9.1). We have

$$i\hbar\,(\dot{\hat{\varrho}}_0 + \lambda\,\dot{\hat{\varrho}}_1 + \dots) = [\,\hat{H}, \hat{\varrho}_0 + \lambda\,\hat{\varrho}_1 + \dots\,] + \lambda\, i\hbar\,\hat{D}(\hat{\varrho}_0 + \lambda\,\hat{\varrho}_1 + \dots)\,.$$

(7.1.9.3)

If the order parameter $\lambda = 0$, we get the unperturbed statistical operator $\hat{\varrho}_o$:

$$i\hbar\,\dot{\hat{\varrho}}_0 = [\,\hat{H}\,\hat{\varrho}_0\,]\,.$$

(7.1.9.4)

The first value of λ, gives

$$i\hbar\,\dot{\hat{\varrho}}_1 = [\,\hat{H}\,\hat{\varrho}_1] + i\hbar\,\hat{D}(\hat{\varrho}_0)\,,$$

(7.1.9.5)

Equilibrium values obey the equations:

$$[\,\hat{H}\,\hat{\varrho}_0\,] = 0\,,$$

(7.1.9.6)

$$[\,\hat{H}\,\hat{\varrho}_1] + i\hbar\,\hat{D}(\hat{\varrho}_0) = 0\,.$$

(7.1.9.7)

Equation (7.1.9.7) is equivalent to the equation

$$\hat{D}(\hat{\varrho}_o) = 0\,.$$

(7.1.9.8)

Therefore, the equilibrium values satisfy the following equations

$$\begin{cases} [\hat{H}\,\hat{\varrho}_o] = 0\,, \\ \hat{D}(\hat{\varrho}_o) = 0\,. \end{cases}$$

(7.1.9.9)

Examples of such equations are the equations (7.1.6.11), (7.1.6.12) and (7.1.8.6), (7.1.8.7).

7.1.10. Conclusion

We considered the equation proposed by Lindblad for the statistical operator describing nonequilibrium state of the quantum harmonic oscillator. From this equation, first, we obtained the density matrix equation in energy representation and the equation for the diagonal elements of this matrix. We formulated the expressions defining the physical meaning of Lindblad equation coefficients. Then, we derived the equation for the mean value of a coordinate and found its general solution. We demonstrated that the mean coordinate value exponentially decreases in time.We obtained the equation for the mean oscillator energy and its general solution. We found the equilibrium mean energy value. This value is a monotonic decreasing function of temperature, which can formulate the Lindblad equation using coordinate and momentum operators. We obtained the density matrix equation in the coordinate representation. From this equation, we derived the formula for the equilibrium density matrix. We wrote the density matrix equation in the momentum representation. We obtained the Wigner function equation and found the respective equilibrium state function. We found the uncertainty relation for various temperatures by applying a Wigner equilibrium function.

7.2. STATISTICAL OPERATOR IN THEORY OF QUANTUM OSCILLATOR. DISSIPATIVE OPERATOR DAMPING

The Lindblad equation for the statistical operator is applied to the quantum oscillator in which the attenuation is present [2]. Description of attenuation is carried out by means of a dissipative operator. It is shown that the equation of damped oscillations for the mean value of the coordinate can be found from this Equation [3].

7.2.1. Introduction

In quantum mechanics, the most common description is the statistical operator or density matrix. The equation for the statistical operator was first obtained by Lindblad [2]. This equation has the form

$$\mathrm{i}\,\hbar\,\dot{\hat{\varrho}} = [\,\hat{H}\;\hat{\varrho}\,] + \mathrm{i}\,\hbar\,\hat{D}\,, \tag{7.2.1.1}$$

where $\hat{H}$ – Hamiltonian,

$$\hat{D} = \sum_{jk} C_{jk}\{\,[\,\hat{a}_j\,\hat{\varrho}\,,\,\hat{a}_k^{+}] + [\hat{a}_j\,,\,\hat{\varrho}\,\hat{a}_k^{+}]\,\}\,, \tag{7.2.1.2}$$

C_{jk} are some numbers, $\hat{a}_j$ is an arbitrary operator. The $\hat{D}$ operator is called the dissipative operator. The statistical operator must be normalized at any time

$$\mathrm{Tr}\,\hat{\varrho} = 1\,, \tag{7.2.1.3}$$

self-adjoint

$$\hat{\varrho}^{+} = \hat{\varrho} \tag{7.2.1.4}$$

and positive definite. A correct equation describing the evolution of a statistical operator must ensure that these properties are preserved over time. This equation was phenomenologically derived by Lindblad.

A one-dimensional oscillatory system, which acts on the braking force, can be described by the equation

$$\ddot{\bar{x}} + 2\,\alpha\,\dot{\bar{x}} + \omega_0^2\,\bar{x} = 0\,, \tag{7.2.1.5}$$

where $\bar{x}$ is the mean value of the coordinate, α and ω_0 are constants. Equation (7.2.1.5) has a solution

$$\bar{x}(t) = A\, e^{-\alpha t} \cos\,(t + \varphi),$$

(7.2.1.6)

which describes the so-called damped oscillations. Here, A and φ are constant integrations,

$$\omega = \sqrt{\omega_0^2 - \alpha^2}\,,$$

(7.2.1.7)

where $\omega_0 > \alpha$. Our task is to obtain the Equation (1.5) from the laws of quantum mechanics.

7.2.2. Equation for Statistical Operator that Describes Damped Oscillations

The author considers the Lindblad equation in the simplest form when it has only one operator $\hat{a}$:

$$\mathrm{i}\,\hbar\,\dot{\hat{\varrho}} = [\,\hat{H}\,\hat{\varrho}\,] - \mathrm{i}\,\gamma/\hbar\,\{2\,\hat{a}\,\hat{\varrho}\,\hat{a}^{+} - (\,\hat{a}^{+}\hat{a}\,\hat{\varrho} + \hat{\varrho}\,\hat{a}^{+}\hat{a}\,)\,\}\,,$$

(7.2.2.1)

where the operator $\hat{a}$ is equal to

$$\hat{a} = \hat{x} + \mathrm{i}\,\hbar\,\beta\,\hat{p}/(\,4\,m\,),$$

(7.2.2.2)

β and γ are some numbers, m is the mass of the particle, $\hat{x}$ and $\hat{p}$ are the coordinate and momentum operators of the particle.Since the operators $\hat{x}$ and $\hat{p}$ are self-conjugated, the operator $\hat{a}^{+}$ is

$$\hat{a}^{+} = \hat{x} - \mathrm{i}\,\hbar\,\beta\,\hat{p}/(\,4\,m\,)\,,$$

(7.2.2.3)

If the particle oscillates, its Hamiltonian operator can be considered equal to

$$\hat{H} = \hat{p}^2/(2\,m) + k\,\hat{x}^2/2\,,$$

(7.2.2.4)

where k is the power constant.

Substitution of operators (7.2.2.2) – (7.2.2.4) in equation (7.2.2.1) gives

$$\dot{\hat{\varrho}} = -\,\mathrm{i}/\hbar\,[\,\hat{p}^2/(2\,m) + k\,\hat{x}^2/2\,,\hat{\varrho}\,] - 2\,\gamma/\hbar^2\{\,[\,\hat{x}\,[\,\hat{x}\,\hat{\varrho}\,]\,] + \mathrm{i}\,\hbar\,\beta/(2\,m)\,(\,\hat{x}\,(\,\hat{p}\,\hat{\varrho} + \hat{\varrho}\,\hat{p}\,)\,) + (\,\hbar\,\beta/(4\,m)\,)^2\,[\,\hat{p}\,[\,\hat{p}\hat{\varrho}\,]\,]\,\}.$$

(7.2.2.5)

Now, we prove that this quantum equation describes the damped oscillations.

7.2.3. Average Values of Position and Momentum

To understand the physical meaning of the summands in the right part of Equation (2.5), we find time derivatives of the mean values of $\bar{x}$ and $\bar{p}$ of operators $\hat{x}$ and $\hat{p}$. By definition, the mean values are

$$\bar{x} = \mathrm{Tr}\,(\,\hat{x}\,\hat{\varrho}\,), \qquad \bar{p} = \mathrm{Tr}\,(\,\hat{p}\,\hat{\varrho}\,).$$

(7.2.3.1)

The time derivatives of these averages will be

$$\dot{\bar{x}} = \mathrm{Tr}\,(\,\hat{x}\,\dot{\hat{\varrho}}\,), \qquad \dot{\bar{p}} = \mathrm{Tr}\,(\,\hat{p}\,\dot{\hat{\varrho}}\,).$$

(7.2.3.2)

To find the derivative of $\dot{\bar{x}}$, substitute the derivative $\dot{\hat{\varrho}}$ from Equation (2.5) and obtain

$$\dot{\bar{x}} = -\,\mathrm{i}/\hbar\,\mathrm{Tr}\,(\,\hat{x}\,[\,\hat{p}^2/(2\,m) + k\,\hat{x}^2/2\,,\hat{\varrho}\,]\,) -$$
$$-\,2\,\gamma/\hbar^2\,\{\,\mathrm{Tr}\,(\,\hat{x}\,[\,\hat{x}\,[\,\hat{x}\,\hat{\varrho}\,]\,]\,) + \mathrm{i}\,\hbar\,\beta/(2\,m)\,\mathrm{Tr}\,(\,\hat{x}\,(\,\hat{x}\,(\,\hat{p}\,\hat{\varrho} + \hat{\varrho}\,\hat{p}\,)\,)\,) +$$
$$+\,(\,\hbar\,\beta/(4\,m)\,)^2\,\mathrm{Tr}\,(\,\hat{x}\,[\,\hat{p}\,[\,\hat{p}\hat{\varrho}\,]\,]\,)\,\}\,.$$

(7.2.3.3)

Calculate

$$\mathrm{Tr}\,(\hat{x}\,[\,\hat{p}^2\,\hat{\varrho}\,]\,) = \mathrm{Tr}\,(\,\hat{x}\,(\,\hat{p}^2\hat{\varrho} - \hat{\varrho}\,\hat{p}^2\,)\,) = \mathrm{Tr}\,(\,\hat{x}\,\hat{p}^2\hat{\varrho} - \hat{x}\,\hat{\varrho}\,\hat{p}^2)\,.$$

Now use the formula

$$\mathrm{Tr}\,(\,\hat{a}\,\hat{b}\,) = \mathrm{Tr}\,(\,\hat{b}\,\hat{a}\,)\,.$$

(7.2.3.4)

Obtain

$$\mathrm{Tr}\,(\,\hat{x}\,[\,\hat{p}^2\,\hat{\varrho}\,]\,) = \mathrm{Tr}\,(\,\hat{x}\,\hat{p}^2\hat{\varrho} - \hat{p}^2\hat{x}\,\hat{\varrho}\,) = \mathrm{Tr}\,(\,(\,\hat{x}\,\hat{p}\,)\,\hat{p}\,\hat{\varrho} - \hat{p}\,(\,\hat{p}\,\hat{x}\,)\hat{\varrho}\,)\,.$$

Let's use the formula

$$\hat{x}\,\hat{p} - \hat{p}\,\hat{x} = \mathrm{i}\,\hbar\,.$$

(7.2.3.5)

We will have

$$\mathrm{Tr}\,(\,\hat{x}\,[\,\hat{p}^2\,\hat{\varrho}\,]\,) = \mathrm{Tr}\,(\,(\,\mathrm{i}\,\hbar + \hat{p}\,\hat{x}\,)\,\hat{p}\,\hat{\varrho} - \hat{p}\,(\,\hat{x}\,\hat{p} - \mathrm{i}\,\hbar\,)\,\hat{\varrho}\,) = 2\,\mathrm{i}\,\hbar\,\mathrm{Tr}\,(\,\hat{p}\,\hat{\varrho}\,) = 2\,\mathrm{i}\,\hbar\,\bar{p}\,.$$

(7.2.3.6)

Convert the following member

$$\mathrm{Tr}\,(\,\hat{x}\,[\,\hat{x}^2\hat{\varrho}\,]\,) = \mathrm{Tr}\,(\,\hat{x}^3\hat{\varrho} - \hat{x}\,\hat{\varrho}\,\hat{x}^2\,)$$

According to the Formula (7.2.3.4), we will have

$$\mathrm{Tr}\,(\,\hat{x}\,[\,\hat{x}^2\hat{\varrho}\,]\,) = 0\,.$$

The following term

$$\mathrm{Tr}\,(\,\hat{x}\,[\,\hat{x}\,[\hat{x}\,\hat{\varrho}\,]\,]\,) = \mathrm{Tr}\,(\,\hat{x}\,[\,\hat{x}\,(\,\hat{x}\,\hat{\varrho} - \hat{\varrho}\,\hat{x}\,)\,]\,) = \mathrm{Tr}\,\{\,\hat{x}\,(\hat{x}^2\hat{\varrho} - 2\,\hat{x}\,\hat{\varrho}\,\hat{x} + \hat{\varrho}\,\hat{x}^2\,)\,\}\,.$$

Applying the formula (3.4), we obtain

$$\mathrm{Tr}\,(\,\hat{x}\,[\,\hat{x}\,[\hat{x}\,\hat{\varrho}\,]\,]) = 0\,.$$

The penultimate member

$$\mathrm{Tr}\,(\hat{x}\,[\,\hat{x}\,(\hat{p}\,\hat{\varrho}+\hat{\varrho}\,\hat{p}\,)\,]\,)=\mathrm{Tr}\,(\,\hat{x}^2\,(\hat{p}\,\hat{\varrho}+\hat{\varrho}\,\hat{p}\,)-\hat{x}\,(\,\hat{p}\,\hat{\varrho}+\hat{\varrho}\,\hat{p}\,)\,\hat{x}\,)=$$

$$=\mathrm{Tr}\,(\,\hat{x}^2\hat{p}\,\hat{\varrho}+\hat{x}^2\hat{\varrho}\,\hat{p}-\hat{x}\,\hat{p}\,\hat{\varrho}\,\hat{x}-\hat{x}\,\hat{\varrho}\,\hat{p}\,\hat{x}\,)=\mathrm{Tr}\,(\,\hat{x}^2\hat{p}\,\hat{\varrho}+\hat{p}\,\hat{x}^2\hat{\varrho}-\hat{x}^2\hat{p}\,\hat{\varrho}\,--\hat{p}\,\hat{x}^2\hat{\varrho}\,)$$
$$=0\,.$$

And finally

$$\mathrm{Tr}\,(\,\hat{x}\,[\,\hat{p}\,[\,\hat{p}\,\hat{\varrho}\,]\,]\,)=\mathrm{Tr}\,(\,\hat{x}\,(\,\hat{p}^2\hat{\varrho}-2\,\hat{p}\,\hat{\varrho}\,\hat{p}+\hat{\varrho}\,\hat{p}^2)\,)=$$

$$=\mathrm{Tr}\,(\,\hat{x}\,\hat{p}^2\hat{\varrho}-2\,\hat{x}\,\hat{p}\,\hat{\varrho}\,\hat{p}+\hat{x}\,\hat{\varrho}\,\hat{p}^2)=\mathrm{Tr}\,(\,\hat{x}\,\hat{p}^2\hat{\varrho}-2\,\hat{p}\,\hat{x}\,\hat{p}\,\hat{\varrho}+\hat{p}^2\hat{x}\,\hat{\varrho}\,)=$$

$$=\mathrm{Tr}\,(\,(\,\hat{x}\,\hat{p}\,)\,\hat{p}\,\hat{\varrho}-2\,\hat{p}\,\hat{x}\,\hat{p}\,\hat{\varrho}+\hat{p}\,(\,\hat{p}\,\hat{x}\,)\,\hat{\varrho}\,)\,.$$

Apply formula (7.2.3.5). We will have

$$\mathrm{Tr}\,(\,\hat{x}\,[\,\hat{p}\,[\,\hat{p}\,\hat{\varrho}\,]\,]\,)=\mathrm{Tr}\,(\,(\,\mathrm{i}\,\hbar+\hat{p}\,\hat{x}\,)\,\hat{p}\,\hat{\varrho}-2\,\hat{p}\,\hat{x}\,\hat{p}\,\hat{\varrho}+\hat{p}\,(\,\hat{x}\,\hat{p}-\mathrm{i}\,\hbar\,)\,\hat{\varrho}\,)=0\,.$$

Substituting the Formula (7.2.3.6) into the derivative (7.2.3.3), we obtain an almost trivial equation

$$\dot{\bar{x}}=\bar{p}/m\,.$$

(7.2.3.7)

Now let's make a derivative of the average pulse value:

$$\dot{\bar{p}}=-\,\mathrm{i}/\hbar\;\mathrm{Tr}\,(\,\hat{p}\,[\,\hat{p}^2/(2\,m)+k\,\hat{x}^2/2\,,\hat{\varrho}\,]\,)-$$

$$-\,2\,\gamma/\hbar^2\,\{\,\mathrm{Tr}\,(\,\hat{p}\,[\,\hat{x}\,[\,\hat{x}\,\hat{\varrho}\,]\,]\,)+\mathrm{i}\,\hbar\,\beta/(2\,m)\;\mathrm{Tr}\,(\,\hat{p}\,(\,\hat{x}\,(\,\hat{p}\,\hat{\varrho}+\hat{\varrho}\,\hat{p}\,)\,)\,)+$$

$$+\,(\,\hbar\,\beta/(4\,m)\,)^2\;\mathrm{Tr}\,(\,\hat{p}\,[\,\hat{p}\,[\,\hat{p}\hat{\varrho}\,]\,]\,)\,\}\,.$$

(7.2.3.8)

According to Formula (7.2.3.4), the first sum will be equal to zero:

$$\mathrm{Tr}\,(\,\hat{p}\,[\,\hat{p}^2\hat{\varrho}\,]\,)=0\,.$$

Following amount

$$\mathrm{Tr}\,(\hat{p}\,[\,\hat{x}^2\hat{\varrho}\,]\,) = \mathrm{Tr}\,(\hat{p}\,(\,\hat{x}^2\hat{\varrho} - \hat{\varrho}\,\hat{x}^2\,)\,) = \mathrm{Tr}\,(\hat{p}\,\hat{x}^2\hat{\varrho} - \hat{p}\,\hat{\varrho}\,\hat{x}^2\,)\,.$$

Now we use the Formula (7.2.3.4):

$$\mathrm{Tr}\,(\hat{p}\,[\,\hat{x}^2\hat{\varrho}\,]\,) = \mathrm{Tr}\,(\hat{p}\,\hat{x}^2\hat{\varrho} - \hat{x}^2\hat{p}\,\hat{\varrho}\,) = \mathrm{Tr}\,(\,(\hat{p}\,\hat{x}\,)\,\hat{x}\,\hat{\varrho} - \hat{x}\,(\,\hat{x}\,\hat{p}\,)\,\hat{\varrho}\,)\,.$$

According to the Formula (7.2.3.5):

$$\mathrm{Tr}\,(\hat{p}\,[\,\hat{x}^2\hat{\varrho}\,]\,) = \mathrm{Tr}\,(\,(\,\hat{x}\,\hat{p} - \mathrm{i}\,\hbar\,)\,\hat{x}\,\hat{\varrho} - \hat{x}\,(\,\hat{p}\,\hat{x} + \mathrm{i}\,\hbar\,)\,\hat{\varrho}\,) =$$

$$= -\,2\,\mathrm{i}\,\hbar\,\mathrm{Tr}\,(\,\hat{x}\,\hat{\varrho}\,) = -\,2\,\mathrm{i}\,\hbar\,\bar{x}\,. \qquad \textbf{(7.2.3.9)}$$

Calculate the third term

$$\mathrm{Tr}\,(\hat{p}\,[\,\hat{x}\,[\,\hat{x}\,\hat{\varrho}\,]\,]\,) = \mathrm{Tr}\,(\hat{p}\,(\,\hat{x}^2\hat{\varrho} - 2\,\hat{x}\,\hat{\varrho}\,\hat{x} + \hat{\varrho}\,\hat{x}^2\,)\,) =$$

$$= \mathrm{Tr}\,(\hat{p}\,\hat{x}^2\hat{\varrho} - 2\,\hat{p}\,\hat{x}\,\hat{\varrho}\,\hat{x} + \hat{p}\,\hat{\varrho}\,\hat{x}^2\,) = \mathrm{Tr}(\,\hat{p}\,\hat{x}^2\hat{\varrho} - 2\,\hat{x}\,\hat{p}\,\hat{x}\,\hat{\varrho} + \hat{x}^2\hat{p}\,\hat{\varrho}\,) =$$

$$= \mathrm{Tr}\,(\,(\,\hat{p}\,\hat{x}\,)\,\hat{x}\,\hat{\varrho} - 2\,\hat{x}\,\hat{p}\,\hat{x}\,\hat{\varrho} + \hat{x}\,(\,\hat{x}\,\hat{p}\,)\,\hat{\varrho}\,)\,.$$

Apply Formula (7.2.3.5), we will obtain

$$\mathrm{Tr}\,(\hat{p}\,[\,\hat{x}\,[\,\hat{x}\,\hat{\varrho}\,]\,]\,) = \mathrm{Tr}\,(\,(\,\hat{x}\,\hat{p} - \mathrm{i}\,\hbar\,)\,\hat{x}\,\hat{\varrho} - 2\,\hat{x}\,\hat{p}\,\hat{x}\,\hat{\varrho} + \hat{x}\,(\,\hat{p}\,\hat{x} + \mathrm{i}\,\hbar\,)\,\hat{\varrho}\,) = 0\,.$$

The penultimate term

$$\mathrm{Tr}\,(\hat{p}\,[\,\hat{x}\,(\,\hat{p}\,\hat{\varrho} + \hat{\varrho}\,\hat{p}\,)\,]\,) = \mathrm{Tr}\,(\,\hat{p}\,\hat{x}\,(\,\hat{p}\,\hat{\varrho} + \hat{\varrho}\,\hat{p}\,) - \hat{p}(\,\hat{p}\,\hat{\varrho} + \hat{\varrho}\,\hat{p}\,)\,\hat{x}\,) =$$

$$= \mathrm{Tr}\,(\,\hat{p}\,\hat{x}\,\hat{p}\,\hat{\varrho} + \hat{p}\,\hat{x}\,\hat{\varrho}\,\hat{p} - \hat{p}^2\hat{\varrho}\,\hat{x} - \hat{p}\,\hat{\varrho}\,\hat{p}\,\hat{x}\,) =$$

$$= \mathrm{Tr}\,(\,\hat{p}\,\hat{x}\,\hat{p}\,\hat{\varrho} + \hat{p}^2\hat{x}\,\hat{\varrho} - \hat{x}\,\hat{p}^2\hat{\varrho} - \hat{p}\,\hat{x}\,\hat{p}\,\hat{\varrho}\,) = \mathrm{Tr}\,(\,\hat{p}\,(\,\hat{p}\,\hat{x}\,)\,\hat{\varrho} - (\,\hat{x}\,\hat{p}\,)\,\hat{p}\,\hat{\varrho}\,)\,.$$

Applying the Formula (7.2.3.5), we will obtain

$$\mathrm{Tr}\,(\,\hat{p}\,[\,\hat{x}\,(\,\hat{p}\,\hat{\varrho} + \hat{\varrho}\,\hat{p}\,)\,]\,) =$$

$$= \mathrm{Tr}\,(\,\hat{p}\,(\,\hat{x}\,\hat{p} - \mathrm{i}\,\hbar\,)\,\hat{\varrho} - (\,\hat{p}\,\hat{x} + \mathrm{i}\,\hbar\,)\,\hat{p}\,\hat{\varrho}\,) = -\,2\,\mathrm{i}\,\hbar\,\mathrm{Tr}\,(\,\hat{p}\,\hat{\varrho}\,) = -\,2\,\mathrm{i}\,\hbar\,\bar{p}\,. \qquad \textbf{(7.2.3.10)}$$

Last amount

$$\mathrm{Tr}\,(\,\hat{p}\,[\,\hat{p}\,[\,\hat{p}\,\hat{\varrho}\,]\,]\,) = 0\,.$$

Now we will deliver formulas (7.2.3.9) and (7.2.3.10) to the derivative (3.8) and receive

$$\dot{\bar{p}} = -\,k\,\bar{x} - 2\,\beta\,\gamma/m\,\bar{p}\,.$$

(7.2.3.11)

Equations (7.2.3.7) and (7.2.3.11) form a system

$$\begin{cases} \dot{\bar{x}} = \bar{p}/m\,, \\ \dot{\bar{p}} = -\,k\,\bar{x} - 2\,\beta\,\gamma\,\bar{p}/m\,. \end{cases}$$

(7.2.3.12)

Eliminate the impulse from this system and we will obtain

$$\ddot{\bar{x}} + 2\,\beta\,\gamma/m\,\dot{\bar{x}} + k/m\,\bar{x} = 0\,,$$

(7.2.3.13)

where

$$\alpha = \beta\,\gamma/m\,, \qquad \omega_0 = \sqrt{k/m}\,.$$

(7.2.3.14)

We proved that the Quantum Equation (7.2.2.5) for the statistical operator ϱ yields the equation of damped oscillations $\bar{x} = \bar{x}(t)$.

7.2.4. The Average Values of Squares of Position and Momentum

For the mean values $\overline{\hat{x}^2}$, $\overline{\hat{p}^2}$ and $\overline{\hat{x}\,\hat{p} + \hat{p}\,\hat{x}}$ of the operators $\hat{x}^2$, $\hat{p}^2$ and $\hat{x}\,\hat{p} + \hat{p}\,\hat{x}$, we find the formulas

$$\overline{\hat{x}^2} = \mathrm{Tr}\,(\,\hat{x}^2\hat{\varrho}\,),\ \overline{\hat{p}^2} = \mathrm{Tr}\,(\,\hat{p}^2\hat{\varrho}\,), \qquad \overline{\hat{x}\,\hat{p} + \hat{p}\,\hat{x}} = \mathrm{Tr}\,(\,(\hat{x}\,\hat{p} + \hat{p}\,\hat{x}\,)\,\hat{\varrho}\,)\,.$$

(7.2.4.1)

The time derivatives of these averages will be

$$\dot{\overline{\hat{x}^2}} = \mathrm{Tr}\,(\,\hat{x}^2\dot{\hat{\varrho}}\,)\,,\, \dot{\overline{\hat{p}^2}} = \mathrm{Tr}\,(\,\hat{p}^2\dot{\hat{\varrho}}\,)\,, \qquad \dot{\overline{\hat{x}\,\hat{p} + \hat{p}\,\hat{x}}} = \mathrm{Tr}\,(\,(\hat{x}\,\hat{p} + \hat{p}\,\hat{x}\,)\,\dot{\hat{\varrho}}\,)\,.$$

(7.2.4.2)

By substituting $\dot{\hat{\varrho}}$ from Equation (7.2.2.5) here, we obtained the equations for these averages by introducing the notation

$$X = \overline{\hat{x}^2},\; Y = \overline{\hat{p}^2}, \qquad Z = \overline{\hat{x}\,\hat{p} + \hat{p}\,\hat{x}}\,.$$

(7.2.4.3)

We will have

$$\begin{cases} \dot{X} = 1/m\; Z + A\,, \\ \dot{Y} = -\,k\,Z + 4\,\gamma - 2\,B\,Y\,, \\ \dot{Z} = 2/m\; Y - 2\,k\,X - B\,Z\,, \end{cases}$$

(7.2.4.4)

where

$$A = \gamma/4\,(\,\hbar\,\beta/m\,)^2, \qquad B = 2\,\beta\,\gamma/m\,.$$

The solution of this inhomogeneous system is equal to the sum of the total solution of a homogeneous system

$$\begin{cases} \dot{X} = 1/m\; Z\,, \\ \dot{Y} = -\,k\,Z - 2\,B\,Y\,, \\ \dot{Z} = 2/m\; Y - 2\,k\,X - B\,Z\,, \end{cases}$$

(7.2.4.5)

and stationary solutions

$$\begin{cases} 0 = 1/m\; Z + A\,, \\ 0 = -\,k\,Z + 4\,\gamma - 2\,B\,Y\,, \\ 0 = 2/m\; Y - 2\,k\,X - B\,Z\,, \end{cases}$$

(7.2.4.6)

The roots of the characteristic equation are

$$\lambda_1 = -2, \lambda_{23} = -2\,(\,\alpha \pm \sqrt{\omega_0^2 - \alpha^2}\,)\,.$$

Therefore, the general solution of a homogeneous equation tends to be zero over time.

It is not difficult to make $A = 0$. To do this, add the dissipative operator to the right side of Equation (7.2.2.5)

$$-\,(\,\hbar\,\beta/(4\,m)\,)^2\,[\,\hat{p}\,[\,\hat{p}\hat{\varrho}\,]\,.$$

Herewith, the term proportional β^2 disappears in Equation (7.2.2.5). In this case, the System (7.2.4.6) takes the form

$$\begin{cases} 0 = Z, \\ 0 = -\,k\,Z + 4\,\gamma - 2\,B\,Y\,, \\ 0 = 2/m\;Y - 2\,k\,X - B\,Z\,, \end{cases}$$

Thus, we obtain

$$\begin{cases} X = Y/(k\,m)\,, \\ Y = 2\,\gamma/B\,, \\ Z = 0\,, \end{cases}$$

(7.2.4.7)

Now find

$$\overline{\hat{x}^2}\cdot\overline{\hat{p}^2} = X\,Y = Y^2/(k\,m) = (k_B T/\omega_0)^2\,.$$

In this case, the ratio $\overline{\hat{x}^2}\cdot\overline{\hat{p}^2}$increases with temperature and tends to infinity:

$$\overline{\hat{x}^2}\cdot\overline{\hat{p}^2} \to \infty\,.$$

(7.2.4.8)

7.2.5. Conclusion

We have shown that the quantum equation for the statistical operator, where a special dissipative operator is chosen, yields the exact Newton equation for the damped oscillator.

7.3. PARTICLE IN A STOCHASTIC ENVIRONMENT DISSIPATIVE MATRIX

7.3.1. Particle in a Homogeneous Isotropic Continuum

Let $\hat{\varrho}$ be the statistical operator describing the motion of a particle of mass m in a stochastic continuous medium. If the medium is homogeneous and isotropic, Lindblad equation for the single-particle is

$$\mathrm{i}\,\hbar\,\partial\hat{\varrho}/\partial t = [\,\hat{H}\,\hat{\varrho}\,] +$$

$$+\,\mathrm{i}\,D/\hbar\,\{\,2\,\hat{a}\,\hat{\varrho}\,\hat{a}^{+} - [\,\hat{a}^{+}\hat{a}\,,\hat{\varrho}\,]_{+}\,\} + \mathrm{i}\,\gamma/\hbar\,\{\,2\,\hat{b}\,\hat{\varrho}\,\hat{b}^{+} - [\,\hat{b}^{+}\hat{b}\,,\hat{\varrho}\,]_{+}\,\}\,, \tag{7.3.1}$$

where the terms with coefficients D and γ are dissipative operators,

$$\hat{H} = \hat{\mathbf{p}}^{2}/(2\,m) + \hat{U} \tag{7.3.2}$$

is the particle Hamiltonian; $\hat{\boldsymbol{r}}$, $\hat{\mathbf{p}}$ and $\hat{U}$ are the operators of the coordinates, momentum and potential energy, respectively,

$$\boldsymbol{F} = -\,\nabla U$$

is the external force acting on the particle;

$$\beta = 1/(k_{\mathrm{B}}T)$$

is the reverse temperature; D is the diffusion coefficient; $\alpha = \beta\,\gamma$ is the friction coefficient;

$$\hat{\boldsymbol{a}} = \hat{\mathbf{p}} + \mathrm{i}\,\hbar\,\beta\,\hat{\mathbf{F}}/4\,, \qquad \hat{\boldsymbol{b}} = \hat{\boldsymbol{r}} + \mathrm{i}\,\hbar\,\beta\,\hat{\mathbf{p}}/(4\,m)\,.$$

(7.3.3)

Substituting the operators (7.3.2) and (7.3.3) into Eq. (7.3.1) and performing a series of simple transformations, we obtain the following equation:

$$\mathrm{i}\,\hbar\,\partial\hat{\varrho}/\partial t = [\,\hat{H}\,\hat{\varrho}\,] +$$

$$+\mathrm{i}\,D/\hbar\,\{\,[\,\hat{\boldsymbol{p}}\,[\,\hat{\boldsymbol{p}}\,\hat{\varrho}\,]\,] + \mathrm{i}\,\hbar\,\beta/2\,[\,\hat{\boldsymbol{p}}\,[\,\hat{\boldsymbol{F}}\,\hat{\varrho}\,]_{+}\,] + (\hbar\,\beta/4)^{2}\,[\,\hat{\boldsymbol{F}}\,[\,\hat{\boldsymbol{F}}\,\hat{\varrho}\,]\,]\,\} - \mathrm{i}\,\gamma/\hbar\,\{\,[\,\hat{\boldsymbol{r}}\,[\,\hat{\boldsymbol{r}}\,\hat{\varrho}\,]\,] + \mathrm{i}\,\hbar\,\beta/(2\,\mathrm{m})\,[\,\hat{\boldsymbol{r}}\,[\,\hat{\boldsymbol{p}}\,\hat{\varrho}\,]_{+}\,] + \{\,\hbar\,\beta/(4\,m)\,\}^{2}\,[\,\hat{\boldsymbol{p}}\,[\,\hat{\boldsymbol{p}}\,\hat{\varrho}\,]\,]\,\}. \quad \mathbf{(7.3.4)}$$

This equation is a quantum analogue of the Fokker – Planck equation for one particle. If we assume that $\beta = 0$ in Eq. (7.3.4), we shall obtain the equation counted in the quantum theory of continual measurements.

7.3.2. Equation for the Density Matrix

In the coordinate representation, Eq. (7.3.4) takes on the form

$$\mathrm{i}\,\hbar\,\partial\varrho/\partial t = -\,\hbar^{2}/(2\,m)\,(\,\nabla^{2} - \nabla'^{2})\,\varrho + (\,U - \mathrm{U}'\,)\,\varrho + \mathrm{i}\,\hbar\,D\,\{\,(\,\nabla + \nabla'\,)^{2}\,\varrho - \beta/2\,(\,\nabla + \nabla'\,)\,(\,\boldsymbol{F} + \boldsymbol{F}'\,)\,\varrho - (\beta/4)^{2}\,(\,\boldsymbol{F} - \boldsymbol{F}'\,)^{2}\,\varrho\,\} - \mathrm{i}\,\gamma/\hbar\,\{\,(\,\boldsymbol{r} - \boldsymbol{r}'\,)^{2}\,\varrho + \hbar^{2}\,\beta/(2\,\mathrm{m})\,(\,\boldsymbol{r} - \boldsymbol{r}'\,)\,(\,\nabla - \nabla'\,)\,\varrho + \{\,\hbar^{2}\,\beta/(4\,m)\,\}^{2}\,(\,\nabla + \nabla'\,)^{2}\,\varrho\,\},$$

(7.3.5)

where

$$\varrho = \varrho\,(t, \boldsymbol{r}, \boldsymbol{r}'\,)\,.$$

Eqs. (7.3.4) and (5) have the property that ϱ is self-adjoint and completely positive for all t.

It is not difficult to make sure that the function

$$\varrho(\boldsymbol{r}, \boldsymbol{r}'\,) = n\,\exp\,\{\,-\,m/(2\,\hbar^{2}\beta)\,(\,-\boldsymbol{r}'\,)^{2} + \mathrm{i}\,m/(\hbar\,\alpha)\,\boldsymbol{F}\cdot(\,\boldsymbol{r} - \boldsymbol{r}'\,)\}$$

(7.3.6)

represents the stationary solution of Eq. (7.3.5), which describes a system of noninteracting particles in a constant homogeneous force field:

$$U = -\,\boldsymbol{F}\cdot\boldsymbol{r}\,, \qquad \boldsymbol{F} = \text{const.};$$

n = const. is the particle concntration.

7.3.3. Equation for the Wigner Function

The construction of the dissipative part has no direct physical interpretation. To understand its meaning better, let us derive an equation for the Wigner function

$$w(t, \boldsymbol{r}, \boldsymbol{p}) = 1/(2\,\pi\,\hbar)^3 \int \varrho(t, \boldsymbol{r} + \boldsymbol{r}'/2, \boldsymbol{r} - \boldsymbol{r}'/2) \exp(-\mathrm{i}\,\boldsymbol{p}\,\boldsymbol{r}'/\hbar)\, \mathrm{d}\boldsymbol{r}',$$

(7.3.7)

which is a quantum analogue of the classical distribution function. Using Eq.

(7.3.5), we find that at $\boldsymbol{F}$ = const, the desired equation is as follows:

$$\partial w/\partial t = -\,\boldsymbol{p}/m\,\nabla w - \boldsymbol{F}\,\nabla_{\boldsymbol{p}} w + D\,\nabla\,(\nabla - \beta\,\boldsymbol{F})\,w + \gamma\,\{\,\nabla_{\boldsymbol{p}}\,(\nabla_{\boldsymbol{p}} + \beta/m\,\boldsymbol{p}) + [\,\hbar\,\beta/(4\,m)\,\nabla\,]^2\,\}\,w\,,$$

(7.3.8)

This equation is similar to the Fokker – Planck equation for a Brownian particle. However, Eq. (3.8) contains additional terms of a quantum mechanical nature. Substituting expression (7.3.6) into formula (7.3.7) yields the equilibrium solution of Eq. (7.3.8),

$$w(\boldsymbol{p}) = n\,[\,\beta/(2\,\pi\,m)\,]^{3/2} \exp\{-\beta/(2\,m)\}\,[\,\boldsymbol{p} - m\,/\,\alpha\,\boldsymbol{F}\,]^2\,\}\,.$$

(7.3.9)

This is none other than the Maxwell distribution function.

7.3.4. Conclusion

There can be no doubt that Eq. (7.3.4) correctly describes the motion of the single-particle in a stochastic medium. As we can see, correctly chosen dissipative operators lead to equations known from classical mechanics with absolute accuracy.

7.4. DENSITY MATRIX METHOD IN QUANTUM THEORY OF LIGHT EMITTING DIODE (LED)

Considering the work of LED, the Lindblad equation for the statistical operation is given. This equation presents the dissipative operator, which describes the diffusion of particles and attenuation. An equation for the density matrix is also provided, which shows that the Wigner equation obtained from the equation for the density matrix coincides with the Fokker – Planck equation. Equations for holes and electrons are also given. These equations contain a term that is responsible for the relaxation of holes and electrons in the *p-n* junction.

7.4.1. Introduction

The first known report of light emission by a solid-state diode was made in 1907 by the British experimenter Henry Joseph Round [5]. He first discovered and described the electroluminescence when studying the passage of current in a pair of metal-silicon carbide and noted the yellow, green and orange glow at the cathode. These experiments were later repeated in 1923 by Oleg Vladimirovich Losev, who experimented in the Nizhny Novgorod radio laboratory with a rectifying contact of a pair of carborundum - steel wire and found a weak glow-the electroluminescence of the semiconductor transition at the point of contact of two dissimilar materials [9].

In 1961, Robert Bayard and Gary Pittman discovered infrared led technology. The world's first practical led operating in the red range was developed by Nick Holonyak in 1962. In 1972, his former student, George Craford, invented the world's first yellow led and improved the brightness of red and red-orange LEDs by a factor of 10. In 1976, T. Pearsall created the world's first high-efficiency high-brightness led for telecommunication applications, specially adapted to data transmission over fiber-optic communication lines.

In the early 1990s, Isama Akasaki, Hiroshi Amano, and Suji Nakamura invented the cheap blue led. In 2014, the three were awarded the Nobel prize in physics for their discovery of the blue led. The first commercial blue led was made by Nakamura in early 1994.

7.4.2. Light-diode

Luminescence in a semiconductor crystal occurs when electrons and holes recombine in the p-n transition region. The *p-n* junction region is formed by the contact of two semiconductors with different types of conductivity. To do this, the contact layers of the semiconductor crystal are doped with different impurities: on one side acceptor, on the other – donor.

донорскими.

In order for the *p-n* junction to emit light, the width of the band gap in the active region of the led must be close to the energy of the light quanta of the visible range. The semiconductor crystal should contain few defects, due to which recombination occurs without radiation. To meet both conditions, often one *p-n* transition in the crystal is not enough, and manufacturers are forced to utilizemanufacture multilayer semiconductor structures, the so-called heterostructures.

The light diode consists of a semiconductor crystal on a current-nonconducting substrate, a housing with contact terminals and an optical system. To increase vitality, the space between the crystal and the plastic lens is filled with transparent silicone. The aluminum base serves to remove excess heat.

The color of the led depends on the width of the band gap in which the electrons and holes recombine, that is, on the semiconductor material and alloying impurities. The light emitted by the led lies in a narrow range of the spectrum. The higher the frequency of the led, the higher the energy of the quanta, which means that the greater the width of the band gap should be. The range of led radiation depends largely on the chemical composition of the semiconductors used. Diodes made of non-direct-band semiconductors (such as silicon, germanium, or silicon carbide) emit little or no light. In terms of light output, LEDs have overtaken conventional incandescent bulbs. In terms of durability, reliability, safety, they have also surpassed them.

7.4.3. Connect LED

The wiring diagram of the LED is shown in Fig. (**7.4.1**).Zack Ohm for this chain is given as follows:

$$(R + R_d)I = V, \tag{7.4.3.1}$$

here R_d is the resistance of the diode. Unfortunately, this resistance is unknown to us.

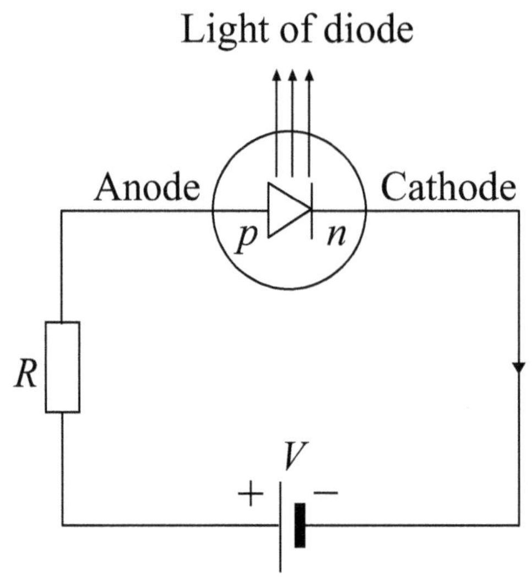

Fig. (7.4.1). LED connection diagram.

An important feature of the diode is the volt-ampere characteristic, *i.e.* the dependence of the current strength on the voltage. The graph of this function is shown in Fig. (**7.4.2**). The volt-ampere characteristic of the led is a nonlinear function.

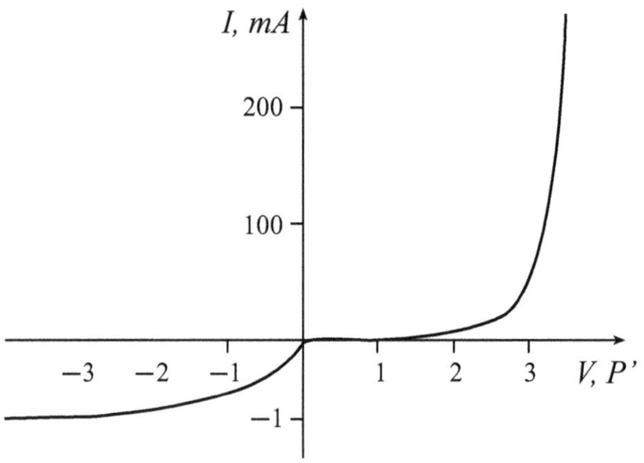

Fig. (7.4.2). The volt-ampere characteristic of the diode.

7.4.4. Equilibrium Distribution of Electron Energy

The equilibrium distribution of electrons in States when the temperature is not zero: $T \neq 0$ is described by the Fermi-Dirac function

$$\bar{N}(E) = 1/\{\, 1 + \exp\,[\, \beta\, (\, E - E_F\,)\,]\, \}\,, \tag{7.4.4.1}$$

where $\beta = 1/(k_B T)$ is the inverse temperature, k_B is the Boltzmann constant, E_F is the Fermi energy. In this formula, $\bar{N}(E)$ is the average number of electrons in a state with energy E, or the probability of filling that state with an electron.

The Fermi-Dirac function graph is shown in Fig. (**7.4.3**). The dotted line shows the graph of dependence $\bar{N}(E)$for $T = 0$. For $E = E_F$, the Fermi-Dirac function takes the value $1/2$: $\bar{N}(E_F)$ = 0,5.It can be seen that with $E < E_F$, all States are filled with probability 1, and all States $E > E_F$ are absolutely free. As the temperature increases, the "break" in the Fermi – Dirac function graph becomes more gentle.

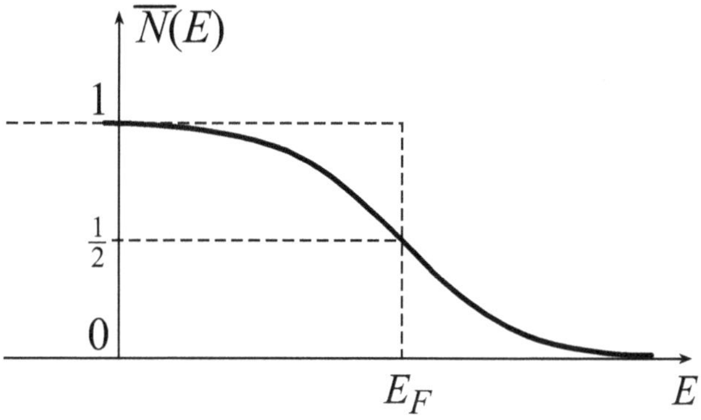

Fig. (7.4.3). Fermi - Dirac Function.

When the temperature $T > 0$, the values of the Fermi - Dirac function at $E > E_F$ are no longer zero, *i.e.* the probability of filling States with electrons whose energy is greater than the Fermi energy is not zero. In this case, the Fermi – Dirac function values corresponding to the energies $E < E_F$ are not equal to 1, *i.e.* not all States whose energy is lower than the Fermi energy are filled with electrons.

7.4.5. Impurity Semiconductors of *p* -Type

Foreign atoms that replace some of the atoms of a given semiconductor (say, germanium Ge) in the lattice nodes are called impurity atoms. Let us consider the case when atoms of trivalent indium In are present as impurities in a germanium Ge crystal (Fig. **7.4.4**).

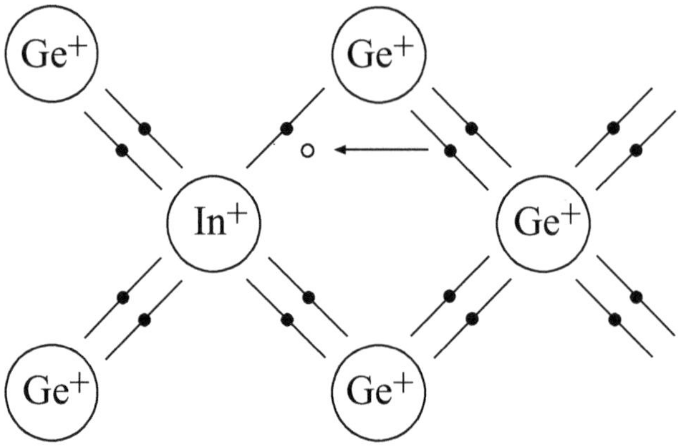

Fig. (7.4.4). Indium atom In in Germanium Ge crystal.

The three outer electrons of the indium atom form covalent bonds with three of the four neighboring germanium Ge atoms. In this case, one state near the atom of indium remains free. The energy of the electron in this state is only a relatively small amount δE_A above the ceiling of the valence band. The energies of these States are called acceptor levels. At $T = 0$, all these States are free and the Fermi level is located between the ceiling of the valence band and the acceptor levels (Fig. **7.4.5**).

As the temperature rises, electrons with energies from the valence band (*i.e.*, external electrons that bond between Ge atoms) begin to fill free States with higher energies. As seen in Fig. (**5**), the energy of electrons detached from germanium atoms will take the values marked on the energy axis by acceptor levels, *i.e.* these electrons will be captured by indium atoms. The impurity atoms, in this case, are called **acceptors**, and such an impurity semiconductor is called a p -type semiconductor.

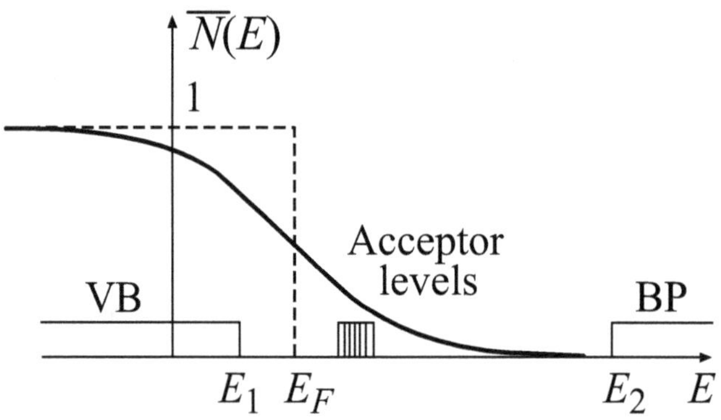

Fig. (7.4.5). Fermi - Dirac Function and energy levels of electrons in a p -type semiconductor.

When an electron is captured, the indium atom turns into a stationary negatively charged ion, and in place of the electron, the germanium atom forms a movable **hole**. The concentration of holes is denoted as $p = p^{(p)}$. Free electrons with energies in the conduction band are present in very small amounts. Their concentration is denoted by $n^{(p)}$. Thus, in a germanium crystal with impurities of indium atoms, the main current carriers are holes. Their concentration is greater than the concentration of electrons:

$$p = p^{(p)} \gg n^{(p)} \text{ in a } p \text{ -type semiconductor.} \tag{7.4.5.1}$$

7.4.6. *n* -type Impurity Semiconductors

Now as impurities in the germanium crystal, there are atoms of 5-valence phosphorus P (see Fig. **7.4.6**).

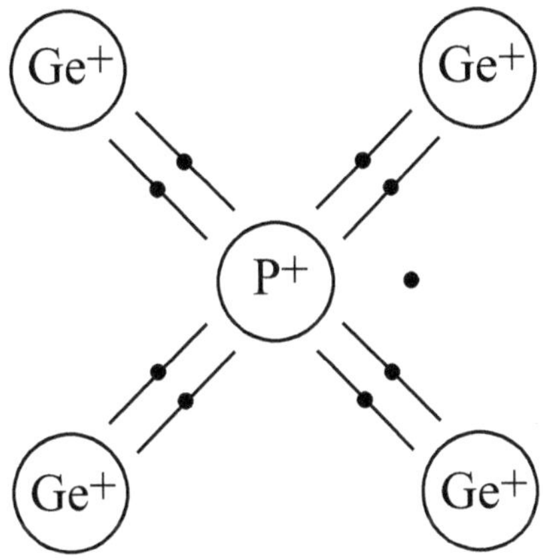

Fig. (7.4.6). Phosphorus atom P in germanium crystal.

Four of the five outer electrons of the phosphorus atom carry out its covalent bond with the four neighboring, surrounding germanium atoms, and the fifth outer electron in the phosphorus atom is as if "out of business". This electron is very weakly bound to the phosphorus atom, *i.e.* it is enough to give it a littleenergy and it will become free. This means that at low temperatures, the energy levels of these "extra" electrons in phosphorus atoms lie just below the bottom of the conduction band (see Fig. **7.4.7**).

Since the separation of the "extra" electron from the phosphorus atom requires significantly less energy than the separation of one electron from the germanium atom, phosphorus atoms will be the main "suppliers" of free electrons. Therefore, the phosphorus atoms in the germanium crystal are called free-electron donors, and the energy levels of the "extra" electrons are called donor levels. An atom of phosphorus, from which an electron has broken away, turns into a sedentary positively charged ion.

As long as there are still "extra" electrons around the phosphorus atoms, only a very small number of free electrons will occur when they are separated from the germanium atoms. This produces the same small number of holes. Thus, in a germanium crystal with impurities of phosphorus atoms, mobile charged particles (current carriers) are mainly free electrons. Therefore, such a crystal is called an *n*-type semiconductor. In donor semiconductors, the electron concentration $n = n^{(n)}$ is greater than the hole concentration $p^{(n)}$:

$$n^{(n)} \gg p^{(n)} \text{ in an } n\text{-type semiconductor.} \quad \textbf{(7.4.6.1)}$$

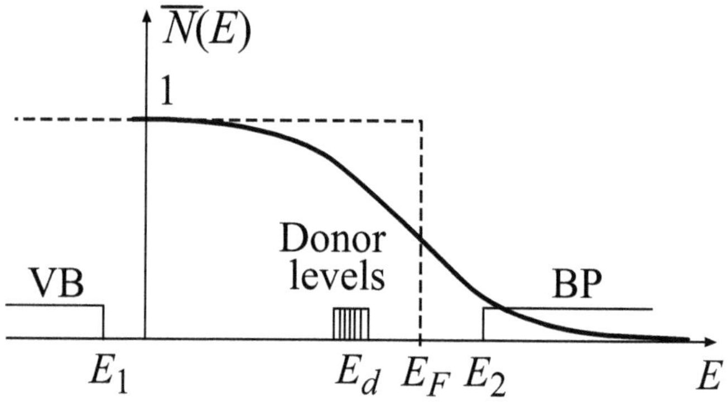

Fig. (7.4.7). Fermi - Dirac Function and energy levels of electrons in an *n*-type semiconductor.

7.4.7. *p-n* transition

Consider a crystal, one part of which, due to the presence of impurities in it, is a *p*-type semiconductor, and the other part is an n-type semiconductor. When the *p* - and *n* -regions are brought into contact, an equilibrium is established between them. This construction is called a *p-n* transition (Fig. **7.4.8**). In the future, we assume that the number of acceptors N_p in a p-conductor is equal to the number of holes N_n in an *n*-type semiconductor:

$$N_p = N_n = N_o \,. \quad \textbf{(7.4.7.1)}$$

At room temperature, most acceptor atoms capture electrons and holes appear. Their number will be equal to $N_p = N_0$. and most donors are released from electrons and they become free. So the number of free electrons in the n -region will be equal to $N_n = N_0$. On the basis of inequalities (5.1) and (6.1), we can neglect the concentrations of $p^{(n)}$ and $n^{(p)}$ of non-basic charge carriers and assume that the average concentrations of holes in the p-semiconductor are equal to the average concentrations of electrons in the n -semiconductor:

$$p = n \,. \tag{7.4.7.2}$$

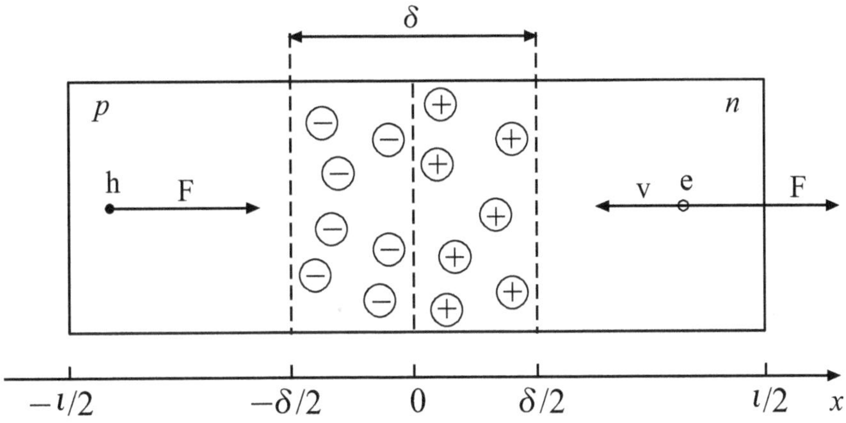

Fig. (7.4.8). *p-n* transition.

In an n -type semiconductor, mobile charged particles are free electrons whose charge is compensated by the charge of fixed positively charged ions formed from donor atoms. In a p-type semiconductor, the mobile charged particles are holes, and their charge is neutralized by the charge of negative ions formed from the acceptor atoms. When these semiconductors touch, electrons travel across the interface from an n -type semiconductor to a p -type semiconductor, where they recombine with holes. In this case, positive ions are exposed in the thin layer of the n -semiconductor at the interface, and negative ions are exposed in the layer of the p -semiconductor on the other side of the interface. These charges create an electric field that prevents the diffusion of electrons from the n -semiconductor to the p -semiconductor.

Over time, a dynamic equilibrium is established, in which the number of electrons diffusing through the interface of the n -semiconductor for some time is equal to the number of electrons flowing during this time in the opposite direction under the action of the electric field created by the ions. As a result of these processes, a double layer is formed at the interface of semiconductors, **depleted by current carriers**. Therefore, this layer has a high electrical resistance, which is proportional to its width δ.

Now draw a graph of the dependence of the electric charge density ρ on the p-n transition. These fixed charges were formed when electrons "stuck" to the impurity atoms in the left p -conductor, and electrons "left" from the right n -conductor. Suppose that these charges are evenly distributed over their volumes (see Fig. **7.4.9**).

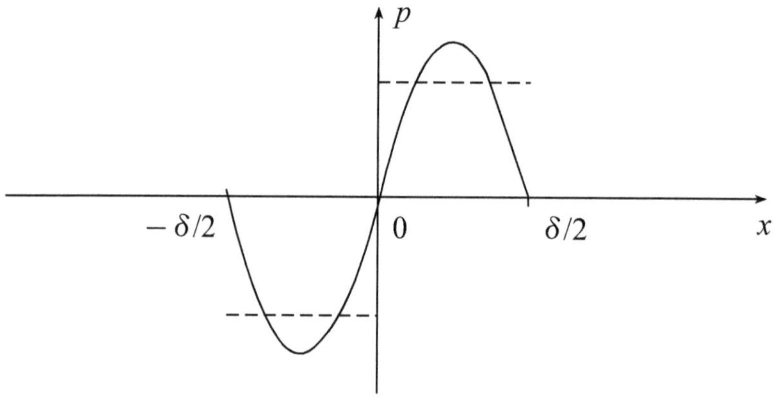

Fig. (7.4.9). Dependence of electric charge density ρ on p-n transition.

Now look at the circuit of the led in Fig. (**7.4.1**). Electrons start moving from the n -semiconductor to the p-semiconductor. When an electron flies into a p-n junction, it first encounters a positively distributed charge in its path. Under the action of this charge, the electron accelerates. Holes from the p -semiconductor move to meet the electrons. Upon entering the p-n junction, the hole is accelerated by a negative charge. Meeting, somewhere in the middle of the p-n junction, some of the accelerated electrodes and holes recombine. As a result of this recombination, a photon is born:

$$e^{-} + \text{"hole"} \rightarrow \hbar\,\omega\ .$$

This recombination does not always occur. Here the skill and tenacity of the experimenters were manifested.

Now let's draw how the electrons will be distributed according to their energies. To do this, connect the Figs. (**7.4.5** and **7.4.7**). First draw the Fermi level. Look at Fig. (**7.4.5**) and we see that the acceptor levels are located above the Fermi level. From Fig. (**7.4.7**), it can be seen that donor levels lie below this level. There is the *p-n* transition between the acceptor and donor levels (see Fig. **7.4.10**).

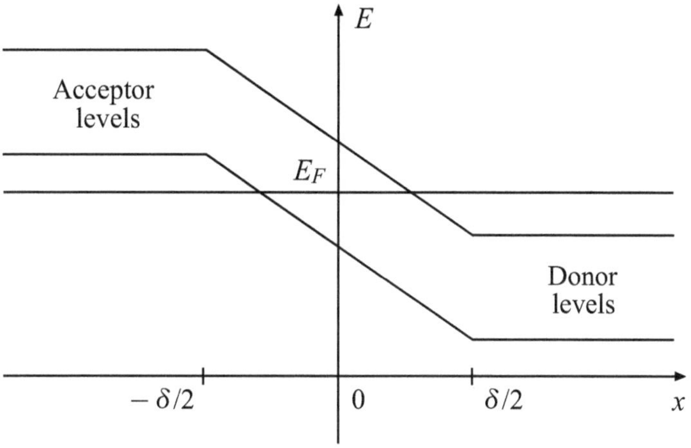

Fig. (7.4.10). The dependence of electron energy.

7.4.8. Lindblad Equation Statistical Operator

The statistical operator is the most informative tool in quantum physics

$$\hat{\varrho} = \hat{\varrho}\,(\,t, q)\,, \tag{7.4.8.1}$$

where t is time and the quantity q is the quantum variable that determines the state of system S. This operator determines the probability of a quantum system being in state q at time t. Therefore, it must obey three conditions at any given time. The statistical operator at time t must be normalized

$$\mathrm{Tr}\,\hat{\varrho} = 1\,, \tag{7.4.8.2}$$

self-adjoint

$$\hat{\varrho}^{+} = \hat{\varrho}$$

(7.4.8.3)

and positively defined.

These conditions are satisfied by Lindblad's equation [6]

$$\mathrm{i}\,\hbar\,\partial\hat{\varrho}/\partial t = [\,\hat{H}\;\hat{\varrho}\,] + \mathrm{i}\,\hbar\,\hat{D}\,,$$

(7.4.8.4)

where $\hat{H} = \hat{H}(\,t, q)$ is the operator of the total mechanical energy of the system S, $\hat{D}$ is the so-called **dissipative operator** describing the dissipative processes occurring in the system. This operator is equal to

$$\hat{D} = \sum_{jk} C_{jk}\{\,2\,\hat{a}_j\;\hat{\varrho}\;\hat{a}_k^{+} - \hat{a}_k^{+}\hat{a}_j\;\hat{\varrho} - \hat{\varrho}\;\hat{a}_k^{+}\hat{a}_j\,\}\,,$$

(7.4.8.5)

where C_{jk} is some numbers, $\hat{a}_j$ is an arbitrary operator.

7.4.9. Density Matrix

The density matrix $\varrho_{nn'}(t)$ is related to the statistical operator by the formula

$$\varrho_{nn'}(t) = \int \psi_n^{*}(\,t, q)\;\hat{\varrho}(\,t, q)\;\psi_{n'}(t, q)\;\mathrm{d}q\,.$$

(7.4.9.1)

where the functions $\psi_n(t, q)$ can be found from the Schrodinger equation

$$\mathrm{i}\,\hbar\,\partial\psi/\partial t = \hat{H}\psi\,.$$

(7.4.9.2)

The formula (9.1) specifies the density matrix $\varrho_{nn'}(t)$ in some n -representation.

The diagonal element $\varrho_{nn}(t)$ of the density matrix is the probability $w_n(t)$ that the system S is in the state n:

$$\varrho_{nn}(t) = w_n(t) \,. \tag{7.4.9.3}$$

Now the three conditions will look like this. The density matrix at time t must be normalized

$$\sum_n \varrho_{nn}(t) = 1 \,. \tag{7.4.9.4}$$

self-adjoint

$$\varrho^*_{nn'}(t) = \varrho_{n'n}(t) \tag{7.4.9.5}$$

and positive.

The equation for the density matrix was derived by the author of this article from the Liouville-von Neumann equation in [7]. This equation is analogous to Lindblad's equation and has the form

$$\mathrm{i}\,\hbar\,\partial\varrho_{nn'}/\partial t = \sum_m \left(H_{nm}\varrho_{mn'} - \varrho_{nm}H_{mn'}\right) + \mathrm{i}\,\hbar\,D_{nn'} \,, \tag{7.4.9.6}$$

where $H_{nn'}$ are the matrix elements of the Hamiltonian $\widehat{H}$ of the system, $D_{nn'}$ is the dissipative matrix, which is

$$D_{nn'} = \sum_{m\,m'} \gamma_{nm,m'n'}\,\varrho_{mm'} - 1/2 \sum_m \left(\gamma_{nm}\,\varrho_{mn'} + \varrho_{nm}\,\gamma_{mn'}\right), \tag{7.4.9.7}$$

In this formula $\gamma_{nm,m'n'}$ there is some matrix,

$$\gamma_{nn'} = \sum_m \gamma_{mn',nm} \,.$$

(7.4.9.8)

Comparing formula (7.4.8.5) with formula (9.7), establishing that

$$\gamma_{nm,m'n'} = 2 \sum_{jk} C_{jk}\, a_{nm,j}\, a^{+}_{m'n',k} \,,$$

(7.4.9.9)

where $a_{nm,j}$ are the matrix elements of the operator $\hat{a}_j$.

Consider a system where the density matrix ϱnm at an arbitrary time *t* is in the diagonal state:

$$\varrho_{nm} = w_n\, \delta_{nm} \,,$$

(7.4.9.10)

Then equation (7.4.9.6) takes the form

$$\mathrm{i}\,\hbar\, \delta_{nn'}\, \partial w_n/\partial t = \sum_m \left(H_{nm}\, w_m\, \delta_{mn'} - w_n\, \delta_{nm}\, H_{mn'} \right) + \mathrm{i}\,\hbar\, D_{nn'} \,,$$

Sum by *m* and put $n' = n$, we have

$$\partial w_n/\partial t = D_{nn} \,.$$

(7.4.9.11)

Now we find the diagonal elements of the dissipative matrix D_{nn}. Substitute the matrix (9.10) into the formula (9.7), we will have

$$D_{nn'} = \sum_{mm'} \gamma_{nm,m'n'}\, w_m\, \delta_{mm'} - 1/2 \sum_m \left(\gamma_{nm}\, w_m\, \delta_{mn'} + w_n\, \delta_{nm}\, \gamma_{mn'} \right) =$$

$$= \sum_m \gamma_{nm,mn'}\, w_m - 1/2\, \gamma_{nn'} \left(w_{n'} + w_n \right).$$

Thus, we get

$$D_{nn} = \sum_m \gamma_{nm,mn}\, w_m - \gamma_{nn}\, w_n \,.$$

Substituting this expression into equation (7.4.9.11) taking into account (7.4.9.8), we will have a kinetic equation

$$\partial w_n/\partial t = \sum_m \left(p_{nm}\, w_m - p_{mn}\, w_n \right),$$

(7.4.9.12)

where

$$p_{nm} = \gamma_{nm,mn} = (2\,\pi/\hbar) \sum_{NM} |\, v_{nN,mM}\, |^2\, W_M \delta(\, \varepsilon_n - \varepsilon_m + E_N - E_M\,)$$

(7.4.9.13)

there is a probability of transition of the system per unit of time from the state *m* to the state *n*,

$$W_N = \nu \exp\left(-\beta\, E_N \right)$$

there is a probability that the environment surrounding the system is in an equilibrium state with quantum numbers *N*, and E_N is its energy in this state, *ν* is the normalization factor, $\beta = 1/(k_{\mathrm{B}}T)$ is the inverse temperature; $v_{nN,mM}$ are the matrix elements of the interaction of the system with its environment.Formula (7.4.9.13) is **Fermi's Golden rule**.

7.4.10. Lindblad Equation Dissipative Operator

Let $\hat{\varrho}$ be a statistical operator describing the motion of a particle of mass m in a stochastic medium. We write the Lindblad equation (7.4.8.4) with some dissipative operator

$$\mathrm{i}\,\hbar\, \partial\hat{\varrho}/\partial t = [\,\hat{H}\,\hat{\varrho}\,] +$$

$$+\,\mathrm{i}\, D/\hbar\, \{\, 2\,\hat{a}\,\hat{\varrho}\,\hat{a}^{+} - [\,\hat{a}^{+}\hat{a}\,, \hat{\varrho}\,]_{+}\,\} + \mathrm{i}\,\gamma/\hbar\, \{\, 2\,\hat{b}\,\hat{\varrho}\,\hat{b}^{+} - [\,\hat{b}^{+}\hat{b}\,, \hat{\varrho}\,]_{+}\,\},$$

(7.4.10.1)

where

$$\hat{H} = \hat{\mathbf{p}}^2/(2\,m) + \hat{U}$$

(7.4.10.2)

is the Hamiltonian of particles; $\hat{\boldsymbol{r}}$, $\hat{\mathbf{p}}$and $\hat{U}$ are the coordinate, momentum, and potential energy operators, respectively,

$$\boldsymbol{F} = -\nabla U$$

there is an external force acting on the particle; $\beta = 1/(k_{\mathrm{B}}T)$; D is the diffusion coefficient; $\alpha = \beta\,\gamma$ is the coefficient of friction;

$$\hat{\boldsymbol{a}} = \hat{\mathbf{p}} + \mathrm{i}\,\hbar\,\beta\,\hat{\mathbf{F}}/4\,, \qquad \hat{\boldsymbol{b}} = \hat{\boldsymbol{r}} + \mathrm{i}\,\hbar\,\beta\,\hat{\mathbf{p}}/(4\,m)\,. \tag{7.4.10.3}$$

Substituting the operators (7.4.10.3) into equation (7.4.10.1), after simple transformations, we obtain the following equation [8]:

$$\mathrm{i}\,\hbar\,\partial\hat{\varrho}/\partial t = [\,\hat{H}\,\hat{\varrho}\,] -$$
$$-\,\mathrm{i}\,D/\hbar\,\{\,[\,\hat{\mathbf{p}}\,[\,\hat{\mathbf{p}}\,\hat{\varrho}\,]\,] + \mathrm{i}\,\hbar\,\beta/2\,[\,\hat{\mathbf{p}}\,[\,\hat{\boldsymbol{F}}\,\hat{\varrho}\,]_+\,] + (\hbar\,\beta/4)^2\,[\,\hat{\boldsymbol{F}}\,[\,\hat{\boldsymbol{F}}\,\hat{\varrho}\,]\,]\,\} -$$
$$-\,\mathrm{i}\,\gamma/\hbar\,\{\,[\,\hat{\boldsymbol{r}}\,[\,\hat{\boldsymbol{r}}\,\hat{\varrho}\,]\,] + \mathrm{i}\,\hbar\,\beta/(2\,\mathrm{m})\,[\,\hat{\boldsymbol{r}}\,[\,\hat{\mathbf{p}}\,\hat{\varrho}\,]_+\,] + \{\,\hbar\,\beta/(4\,m)\,\}^2\,[\,\hat{\mathbf{p}}\,[\,\hat{\mathbf{p}}\,\hat{\varrho}\,]\,]\,\}. \tag{7.4.10.4}$$

In this equation, there are terms with β^2. To get rid of them, add two more terms to the dissipative operator:

$$\hat{\boldsymbol{a}}_1 = -\,\mathrm{i}\,\hbar\,\beta\,\hat{\mathbf{F}}/4\,, \qquad \hat{\boldsymbol{b}}_1 = -\,\mathrm{i}\,\hbar\,\beta\,\hat{\mathbf{p}}/(4\,m)\,. \tag{7.4.10.5}$$

Then we have the equation

$$\mathrm{i}\,\hbar\,\partial\hat{\varrho}/\partial t = [\,\hat{H}\,\hat{\varrho}\,] -$$
$$-\,\mathrm{i}\,D/\hbar\,\{\,[\,\hat{\mathbf{p}}\,[\,\hat{\mathbf{p}}\,\hat{\varrho}\,]\,] + \mathrm{i}\,\hbar\,\beta/2\,[\,\hat{\mathbf{p}}\,[\,\hat{\boldsymbol{F}}\,\hat{\varrho}\,]_+\,]\,\} -$$
$$-\,\mathrm{i}\,\gamma/\hbar\,\{\,[\,\hat{\boldsymbol{r}}\,[\,\hat{\boldsymbol{r}}\,\hat{\varrho}\,]\,] + \mathrm{i}\,\hbar\,\beta/(2\,\mathrm{m})\,[\,\hat{\boldsymbol{r}}\,[\,\hat{\mathbf{p}}\,\hat{\varrho}\,]_+\,]\,\}. \tag{7.4.10.6}$$

7.4.11. Equations for Density Matrix

In the coordinate representation:

$$\hat{\boldsymbol{r}} = \boldsymbol{r}, \qquad \hat{\boldsymbol{p}} = -\,\mathrm{i}\,\hbar\,\nabla\,,$$

equation (7.4.10.6) takes the form

$$\begin{aligned} \mathrm{i}\,\hbar\,\partial\varrho/\partial t = &-\hbar^2/(2\,m)\,(\,\nabla^2 - \nabla'^2)\,\varrho + (\,U - U'\,)\,\varrho\, - \\ &-\mathrm{i}\,\hbar\,D\,\{\,(\,\nabla + \nabla'\,)^2 - \beta/2\,(\,\nabla + \nabla'\,)\,(\,\boldsymbol{F} + \boldsymbol{F}'\,)\,\}\,\varrho\, - \\ &-\mathrm{i}\,\gamma/\hbar\,\{\,(\,\boldsymbol{r} - \boldsymbol{r}'\,)^2 + \hbar^2\,\beta/(2\,\mathrm{m})\,(\,\boldsymbol{r} - \boldsymbol{r}'\,)\,(\,\nabla - \nabla'\,)\,\}\,\varrho\,, \end{aligned}$$

(7.4.11.1)

where $\varrho = \varrho(\,t,\,\boldsymbol{r},\,\boldsymbol{r}'\,)$. Equation (7.4.11.1) preserves all properties of the density matrix ϱ at all times.

7.4.12. Wigner Equation

To better understand the meaning of equation (11.1), we write Wigner's equation

$$w(t,\boldsymbol{r},\boldsymbol{p}) = 1/(2\,\pi\,\hbar)^3 \int \varrho(t,\,\boldsymbol{r} + \boldsymbol{r}'/2,\,\boldsymbol{r} - \boldsymbol{r}'/2)\,\exp(-\,\mathrm{i}\,\boldsymbol{p}\,\boldsymbol{r}'/\hbar)\,\mathrm{d}\boldsymbol{r}',$$

(7.4.12.1)

which is the quantum analogue of the classical distribution function. Using (7.4.12.1), we find that the desired equation is as follows:

$$\partial w/\partial t = -\,\boldsymbol{p}/m\,\nabla w - \boldsymbol{F}\,\nabla_p w + D\,\nabla\,(\,\nabla - \beta\,\boldsymbol{F}\,)\,w + \gamma\,\{\,\nabla_p\,(\,\nabla_p + \beta/m\,\boldsymbol{p})\,\}\,w.$$

(7.4.12.2)

This equation is analogous to the Fokker – Planck equation for a Brownian particle.

7.4.13. Wigner Equations for Light Diode

Two forces act on the particles (holes h and electrons e) in the diode. One force $\boldsymbol{F}_V$ is created by the applied voltage V, the other force $\boldsymbol{F}_{pn}$ acts on the particles from the *p-n* junction. We assume that the force $\boldsymbol{F}_V$ is constant on the diode segment $[\,-\,l/2,\,l/2\,]$ (see Fig. **8**):

$$\boldsymbol{F}_V = F_V \text{ iby} - l/2 < x < /2 ,$$

(7.4.13.1)

where F_V = const > 0.

The force $\boldsymbol{F}_{pn}$ acts only on the segment $[\,-\,\delta/2,\ \delta/2\,]$ of the length δ of the *p-n* transition. Thus, this force will be positive on the segment $[-\,\delta/2,\ 0\,]$, and negative on the segment $[\,0,\ \delta/2\,]$.

Lets assume that

$$\boldsymbol{F}_{pn} = \begin{cases} F_{pn}\,\mathbf{i} \text{ at } -\delta/2 < x < 0, \\ -F_{pn}\,\mathbf{i} \text{ at } 0 < x < \delta/2 . \end{cases}$$

(7.4.13.2)

где F_{pn} = const > 0.

Now write down the equation of motion for the holes *h*:

$$\partial w^{(+)}/\partial t = -\,\mathbf{p}/m\ \nabla w^{(+)} - \boldsymbol{F}\,\nabla_{\mathbf{p}} w^{(+)} + D\,\nabla\,(\,\nabla - \beta\,\boldsymbol{F}\,)\,w^{(+)} +$$

$$+\,\gamma\,\{\,\nabla_{\mathbf{p}}\,(\,\nabla_{\mathbf{p}} + \,\beta/m\ \mathbf{p})\,\}\,w^{(+)} - \eta\,w^{(+)}\,w^{(-)}\,\delta(x)\,,$$

(7.4.13.3)

where η = const > 0,

$$\delta(x) = \begin{cases} 1 \text{ at } x = 0, \\ 0 \text{ at } x \neq 0 . \end{cases}$$

The last term in this equation describes the relaxation of holes and electrons.

The equation of motion of electrons has the form

$$\partial w^{(-)}/\partial t = -\,\mathbf{p}/m\ \nabla w^{(-)} + \boldsymbol{F}\,\nabla_{\mathbf{p}} w^{(-)} + D\,\nabla\,(\,\nabla + \beta\,\boldsymbol{F}\,)\,w^{(-)} +$$

$$+\,\gamma\,\{\,\nabla_{\mathbf{p}}\,(\,\nabla_{\mathbf{p}} + \,\beta/m\ \mathbf{p})\,\}\,w^{(-)} - \eta\,w^{(+)}\,w^{(-)}\,\delta(x)\,.$$

(7.4.13.4)

The following article will be devoted to the solution of these equations.

7.5. THEORY OF BALL LIGHTNING

7.5.1. Introduction

Ball lightning is a very rare, mysterious and inexplicable natural phenomenon. It appears as a luminous ball that moves relatively slowly and randomly in the air. What can be said about its properties? First, it is formed from ordinary air molecules under the action of a very high electric voltage that breaks them into small parts. Secondly, ball lightning exists for a relatively long time and then explodes. Third, it seems that the temperature inside the balloon is kept high enough.

To describe the ball lightning, the Lindblad equation [1] was used for the statistical operator $\hat{\varrho}$, in which there is a dissipative operator $\widehat{D}$. To obtainthis operator, you need to know the dissipative processes that operate inside the ball lightning. The author of this article proposed **two dissipative operators, which are called diffusion operators $\widehat{a}_1$ and attenuation operators $\widehat{a}_2$** [2].

Since ball lightning exists for a relatively long time, it means that the forces of attraction act in it. Lets assume that the temperature inside the ball is so high that all the atoms in the ball are ionized. So that inside the ball, there are only electrons and nuclei, which we will consider all the same. Electrons are several thousand times lighter than nuclei. Therefore, electrons move faster than nuclei. On this basis, to facilitate the solution of the problem, we assume that the electrons are evenly distributed within the volume of the ball lightning. Based on this assumption, the **force of attraction that acts on the nucleus from the side of the electrons** is found. If we neglect the repulsive forces of the nuclei from each other, we can write the Lindblad equation for the statistical operator of the nuclei. In this case, it is relatively easy to find the average values of the quantities that characterize the movement of the nuclei [11, 12].

7.5.2. Interaction of Electrons with Nuclei

Electrons are several thousand times lighter than nuclei. Therefore, we assume that they **evenly** fill the volume of the ball lightning. Then the probability $w^{(e)}$ will satisfy the relation

$$w^{(e)}\, 4\,\pi\, R^3/3 = N\, Z \tag{7.5.2.1}$$

where R is the radius of the ball lightning, N is the number of nuclei, and Z is the number of electrons in the atom.

If the nucleus is in the center of a ball lightning, then this nucleus will not be affected by the force from the electrons. Put the origin of the radius vector $\boldsymbol{r}$ in the center of the ball.

The larger the modulus of the vector $\boldsymbol{r}$, *i.e.*, the farther the nucleus is from the center of the ball, and the more strongly the nucleus is attracted by electrons to this center. **Approximately** you can write

$$\boldsymbol{F} = -\,2\,Z\,\mathrm{e}^2\,w^{(\mathrm{e})}\,\boldsymbol{r}\,.$$

(7.5.2.2)

In this case, the potential will be equal to

$$U(r) = Z\,\mathrm{e}^2\,w^{(\mathrm{e})}\,r^2.$$

(7.5.2.3)

and the Hamiltonian of the kernel will have the form

$$\widehat{H} = \widehat{\mathbf{p}}^2/(2\,M) + Z\,\mathrm{e}^2\,w^{(\mathrm{e})}\,\widehat{\boldsymbol{r}}^2\,,$$

(7.5.2.4)

where $\widehat{\boldsymbol{r}}$ and $\widehat{\mathbf{p}}$ are the kernel coordinate and momentum operators, and M is the kernel mass.

7.5.3. Lindblad Equation

In quantum physics, the most powerful tool for describing various quantum phenomena is the statistical operator $\hat{\varrho}$. For this operator, Lindblad wrote a phenomenological equation [1]:

$$\mathrm{i}\,\hbar\,\partial\hat{\varrho}/\partial t = [\,\widehat{H}\;\hat{\varrho}\,] + \mathrm{i}\,\hbar\,\widehat{D}\,,$$

(7.5.3.1)

where $\widehat{H}$ is a Hamiltonian, we will call the operator $\widehat{D}$ **dissipative**:

$$\widehat{D} = \sum_{jk} C_{jk} \{ 2\, \hat{a}_j\, \hat{\varrho}\, \hat{a}_k^+ - \hat{a}_k^+ \hat{a}_j\, \hat{\varrho} - \hat{\varrho}\, \hat{a}_k^+ \hat{a}_j \} ,$$

(7.5.3.2)

C_{jk} is some matrix, and $\hat{a}_j$ is an arbitrary operator that still needs to be found.

7.5.4. Dissipative Diffusion and Attenuation Operators

In [2], the author of this article proposed two operators

$$\hat{\boldsymbol{a}}_1 = \hat{\boldsymbol{p}} + \mathrm{i}\, \hbar\, \beta\, \hat{\boldsymbol{F}}/4 , \qquad \hat{\boldsymbol{a}}_2 = \hat{\boldsymbol{r}} + \mathrm{i}\, \hbar\, \beta\, \hat{\boldsymbol{p}}/(4\, M) ,$$

(7.5.4.1)

where $\hat{\boldsymbol{F}}$ is the force operator; M is the mass of the particle,

$$\beta = 1/(k_{\mathrm{B}} T)$$

(7.5.4.2)

– return temperature. These operators are called **dissipative diffusion and attenuation operators**. Substituting these operators into the Linblad equation which leads it to the form

$$\mathrm{i}\, \hbar\, \partial\hat{\varrho}/\partial t = [\, \widehat{H}\, \hat{\varrho}\,] + \mathrm{i}\, D/\hbar\, \{ [\, \hat{\boldsymbol{p}}\, [\, \hat{\boldsymbol{p}}\, \hat{\varrho}\,]\,] + \mathrm{i}\, \hbar\, \beta/2\, [\, \hat{\boldsymbol{p}}\, [\, \hat{\boldsymbol{F}}\, \hat{\varrho}\,]_+\,] \} +$$

$$+ \mathrm{i}\, \gamma/\hbar\, \{ [\, \hat{\boldsymbol{r}}\, [\, \hat{\boldsymbol{r}}\, \hat{\varrho}\,]\,] + \mathrm{i}\, \hbar\, \beta/(2\, M)\, [\, \hat{\boldsymbol{r}}\, [\, \hat{\boldsymbol{p}}\, \hat{\varrho}\,]_+\,] \} ,$$

(7.5.4.3)

here D and γ are the diffusion and attenuation coefficients. In this equation, the terms that hold β^2.

7.5.5. Equation for Statistical Operator of Atomic Nuclei

Now substitute the formulas (7.5.2.2) – (7.5.2.4) in equation (7.5.4.3). We will have an equation that approximates the motion of nuclei in a ball lightning

$$\mathrm{i}\, \hbar\, \partial\hat{\varrho}/\partial t = [\, \hat{\boldsymbol{p}}^2/(2\, M) + Z\, \mathrm{e}^2\, w^{(\mathrm{e})}\, \hat{\boldsymbol{r}}^2, \hat{\varrho}\,] +$$

$$+ \mathrm{i}\, D/\hbar\, \{ [\, \hat{\boldsymbol{p}}\, [\, \hat{\boldsymbol{p}}\, \hat{\varrho}\,]\,] - Z\, \mathrm{e}^2\, w^{(\mathrm{e})}\, \mathrm{i}\, \hbar\, \beta\, [\, \hat{\boldsymbol{p}}\, [\, \hat{\boldsymbol{r}}\, \hat{\varrho}\,]_+\,] \} +$$

$$+\mathrm{i}\,\gamma/\hbar\,\{[\hat{\boldsymbol r}[\hat{\boldsymbol r}\,\hat{\varrho}\,]]+\mathrm{i}\,\hbar\,\beta/(2\,M)[\hat{\boldsymbol r}[\hat{\boldsymbol p}\,\hat{\varrho}\,]_+]\}\,.$$

(7.5.5.1)

The statistical operator must satisfy the normalization condition

$$\mathrm{Tr}\,\hat{\varrho}=N. \tag{7.5.5.2}$$

In equation (7.5.5.1), there is no term that is responsible for repelling the nuclei from each other. This equation will be applied in this article for finding the average values of$\langle\hat{\boldsymbol r}\rangle$, $\langle\hat{\boldsymbol p}\rangle$, $\langle\hat{\boldsymbol r}^2\rangle$, $\langle\hat{\boldsymbol p}^2\rangle$and $\langle\,\hat{\boldsymbol r}\,\hat{\boldsymbol p}+\,\hat{\boldsymbol p}\,\hat{\boldsymbol r}\,\rangle$, that characterize the motion of nuclei. Finding these values turned out to be possible by the presence of terms in equation (7.5.5.1), each of which contains the statistical operator $\hat{\varrho}$ only in the first degree.

To slightly facilitate the solution of the problem, enter the notation

$$g=Z\,\mathrm{e}^2\,w^{(\mathrm{e})}.$$

(7.5.5.3)

Now equation (7.5.5.1) takes the form

$$\mathrm{i}\,\hbar\,\partial\hat{\varrho}/\partial t=[\,\hat{\boldsymbol p}^2/(2\,M)+g\,\hat{\boldsymbol r}^2,\hat{\varrho}\,]+$$

$$+\,\mathrm{i}\,D/\hbar\,\{[\hat{\boldsymbol p}[\hat{\boldsymbol p}\,\hat{\varrho}\,]]-\mathrm{i}\,\hbar\,\beta\,g[\hat{\boldsymbol p}[\hat{\boldsymbol r}\,\hat{\varrho}\,]_+]\}+$$

$$+\mathrm{i}\,\gamma/\hbar\,\{[\hat{\boldsymbol r}[\hat{\boldsymbol r}\,\hat{\varrho}\,]]+\mathrm{i}\,\hbar\,\beta/(2\,M)[\hat{\boldsymbol r}[\hat{\boldsymbol p}\,\hat{\varrho}\,]_+]\}\,.$$

(7.5.5.4)

7.5.6. Average Values of Atomic Nucleus Coordinates

The average coordinate value of$\langle\hat{\boldsymbol r}\rangle$ of the kernel, by definition, is equal to:

$$\langle\hat{\boldsymbol r}\rangle=\mathrm{Tr}\,(\,\hat{\boldsymbol r}\,\hat{\varrho}\,)\,,$$

(7.5.6.1)

Find the change in the value $\langle\hat{\boldsymbol r}\rangle$ over time. It is obvious that

$$\partial\langle\hat{\mathbf{r}}\rangle/\partial t = \mathrm{Tr}\,(\hat{\mathbf{r}}\ \partial\hat{\varrho}/\partial t)\,. \tag{7.5.6.2}$$

Substituting equation (7.5.5.4) here gives [3]

$$\partial\langle\hat{\mathbf{r}}\rangle/\partial t = -\mathrm{i}/\hbar\ \mathrm{Tr}\,(\,\hat{\mathbf{r}}\,\{[\,\hat{\mathbf{p}}^2/(2\,M) + g\,\hat{\mathbf{r}}^2, \hat{\varrho}\,] + \\ + \mathrm{i}\,D/\hbar\ \{\,[\,\hat{\mathbf{p}}\,[\,\hat{\mathbf{p}}\,\hat{\varrho}\,]\,] - \mathrm{i}\,\hbar\,\beta\,g\,[\,\hat{\mathbf{p}}\,[\,\hat{\mathbf{r}}\,\hat{\varrho}\,]_+\,]\,\} + \\ + \mathrm{i}\,\gamma/\hbar\ \{\,[\,\hat{\mathbf{r}}\,[\,\hat{\mathbf{r}}\,\hat{\varrho}\,]\,] + \mathrm{i}\,\hbar\,\beta/(2\,M)\,[\,\hat{\mathbf{r}}\,[\,\hat{\mathbf{p}}\,\hat{\varrho}\,]_+\,]\,\}\,\})\,. \tag{7.5.6.3}$$

1.Calculate the following value

$$\mathrm{Tr}\,(\,\hat{\mathbf{r}}\,[\,\hat{\mathbf{p}}^2\,\hat{\varrho}\,]\,) = \mathrm{Tr}\,\{\,\hat{\mathbf{r}}\,(\,\hat{\mathbf{p}}^2\hat{\varrho} - \hat{\varrho}\,\hat{\mathbf{p}}^2)\,\} = \mathrm{Tr}\,(\,\hat{\mathbf{r}}\,\hat{\mathbf{p}}^2\hat{\varrho} - \hat{\mathbf{r}}\,\hat{\varrho}\,\hat{\mathbf{p}}^2)\,.$$

Now we will use the formula

$$\mathrm{Tr}\,(\,\hat{a}\,\hat{b}\,) = \mathrm{Tr}\,(\,\hat{b}\,\hat{a}\,)\,. \tag{7.5.6.4}$$

Get

$$\mathrm{Tr}\,(\,\hat{\mathbf{r}}\,[\,\hat{\mathbf{p}}^2\,\hat{\varrho}\,]\,) = \mathrm{Tr}\,(\,\hat{\mathbf{r}}\,\hat{\mathbf{p}}^2\hat{\varrho} - \hat{\mathbf{p}}^2\hat{\mathbf{r}}\,\hat{\varrho}\,) = \mathrm{Tr}\,\{\,(\,\hat{\mathbf{r}}\,\hat{\mathbf{p}}\,)\,\hat{\mathbf{p}}\,\hat{\varrho} - \hat{\mathbf{p}}\,(\,\hat{\mathbf{p}}\,\hat{\mathbf{r}}\,)\hat{\varrho}\,\}\,.$$

Lets use the formula

$$\hat{\mathbf{r}}\,\hat{\mathbf{p}} - \hat{\mathbf{p}}\,\hat{\mathbf{r}} = \mathrm{i}\,\hbar\,\delta_{kl}. \tag{7.5.6.5}$$

We will have

$$\mathrm{Tr}\,(\,\hat{\mathbf{r}}\,[\,\hat{\mathbf{p}}^2\,\hat{\varrho}\,]\,) = \mathrm{Tr}\,\{\,(\,\mathrm{i}\,\hbar\,\delta_{kl} + \hat{\mathbf{p}}\,\hat{\mathbf{r}}\,)\,\hat{\mathbf{p}}\,\hat{\varrho} - \hat{\mathbf{p}}\,(\,\hat{\mathbf{r}}\,\hat{\mathbf{p}} - \mathrm{i}\,\hbar\,\delta_{kl}\,)\,\hat{\varrho}\,\} = \\ = 2\,\mathrm{i}\,\hbar\,\delta_{kl}\mathrm{Tr}\,(\,\hat{\mathbf{p}}\,\hat{\varrho}\,) = 2\,\mathrm{i}\,\hbar\,\langle\mathbf{p}\rangle\,. \tag{7.5.6.6}$$

2. Consider the following term

$$\mathrm{Tr}\,(\hat{\boldsymbol{r}}\,[\,\hat{\boldsymbol{r}}^2\,\hat{\varrho}\,]\,) = \mathrm{Tr}\,(\,\hat{\boldsymbol{r}}^3\,\hat{\varrho} - \hat{\boldsymbol{r}}\,\hat{\varrho}\,\hat{\boldsymbol{r}}^2\,)\,.$$

According to the formula (7.5.6.4), we will have

$$\mathrm{Tr}\,(\,\hat{\boldsymbol{r}}\,[\,\hat{\boldsymbol{r}}^2\,\hat{\varrho}\,]\,) = 0\,. \tag{7.5.6.7}$$

3. Now calculate the term

$$\mathrm{Tr}\,(\,\hat{\boldsymbol{r}}\,[\,\hat{\boldsymbol{p}}\,[\,\hat{\boldsymbol{p}}\,\hat{\varrho}\,]\,]\,) = \mathrm{Tr}\,\{\,\hat{\boldsymbol{r}}\,[\,\hat{\boldsymbol{p}}\,(\,\hat{\boldsymbol{p}}\,\hat{\varrho} - \hat{\varrho}\,\hat{\boldsymbol{p}}\,)\,]\,\} = \mathrm{Tr}\,\{\,\hat{\boldsymbol{r}}\,(\,\hat{\boldsymbol{p}}\,(\,\hat{\boldsymbol{p}}\,\hat{\varrho} - \hat{\varrho}\,\hat{\boldsymbol{p}}\,) - (\,\hat{\boldsymbol{p}}\,\hat{\varrho} - \hat{\varrho}\,\hat{\boldsymbol{p}}\,)\hat{\boldsymbol{p}}\,)\,\} = \mathrm{Tr}\,\{\,\hat{\boldsymbol{r}}\,(\,\hat{\boldsymbol{p}}^2\,\hat{\varrho} - 2\,\hat{\boldsymbol{p}}\,\hat{\varrho}\,\hat{\boldsymbol{p}} + \hat{\varrho}\,\hat{\boldsymbol{p}}^2\,)\,\}.$$

Applying the formula (7.5.6.4) to the second term, we get

$$\mathrm{Tr}\,(\,\hat{\boldsymbol{r}}\,[\,\hat{\boldsymbol{p}}\,[\,\hat{\boldsymbol{p}}\,\hat{\varrho}\,]\,]\,) = 0\,. \tag{7.5.6.8}$$

4. The following member

$$\mathrm{Tr}\,(\,\hat{\boldsymbol{r}}\,[\,\hat{\boldsymbol{p}}\,[\,\hat{\boldsymbol{r}}\,\hat{\varrho}\,]_+\,]\,) = \mathrm{Tr}\,\{\,\hat{\boldsymbol{r}}\,[\,\hat{\boldsymbol{p}}\,(\,\hat{\boldsymbol{r}}\,\hat{\varrho} + \hat{\varrho}\,\hat{\boldsymbol{r}}\,)\,]\,\} =$$

$$= \mathrm{Tr}\,\{\,\hat{\boldsymbol{r}}\,(\,\hat{\boldsymbol{p}}\,\hat{\boldsymbol{r}}\,\hat{\varrho} + \hat{\boldsymbol{p}}\,\hat{\varrho}\,\hat{\boldsymbol{r}} - \hat{\boldsymbol{r}}\,\hat{\varrho}\,\hat{\boldsymbol{p}} - \hat{\varrho}\,\hat{\boldsymbol{r}}\,\hat{\boldsymbol{p}}\,)\,\}\,.$$

Lets apply the formula (7.5.6.4), we will have

$$\mathrm{Tr}\,(\,\hat{\boldsymbol{r}}\,[\,\hat{\boldsymbol{p}}\,[\,\hat{\boldsymbol{r}}\,\hat{\varrho}\,]_+\,]\,) = \mathrm{Tr}\,\{\,\hat{\boldsymbol{r}}\,(\,\hat{\boldsymbol{p}}\,\hat{\boldsymbol{r}}\,\hat{\varrho} + \hat{\boldsymbol{r}}\,\hat{\boldsymbol{p}}\,\hat{\varrho} - \hat{\boldsymbol{p}}\,\hat{\boldsymbol{r}}\,\hat{\varrho} - \hat{\boldsymbol{r}}\,\hat{\boldsymbol{p}}\,\hat{\varrho}\,)\,\}\,.$$

It turns out that

$$\mathrm{Tr}\,(\,\hat{\boldsymbol{r}}\,[\,\hat{\boldsymbol{p}}\,[\,\hat{\boldsymbol{r}}\,\hat{\varrho}\,]_+\,]\,) = 0\,. \tag{7.5.6.9}$$

5. The penultimate member

$$\mathrm{Tr}\,(\,\hat{\boldsymbol{r}}\,[\,\hat{\boldsymbol{r}}\,[\,\hat{\boldsymbol{r}}\,\hat{\varrho}\,]\,]\,) = \mathrm{Tr}\,(\,\hat{\boldsymbol{r}}\,[\,\hat{\boldsymbol{r}}\,(\,\hat{\boldsymbol{r}}\,\hat{\varrho} - \hat{\varrho}\,\hat{\boldsymbol{r}}\,)\,]\,) = \mathrm{Tr}\,\{\,\hat{\boldsymbol{r}}\,(\,\hat{\boldsymbol{r}}^2\hat{\varrho} - 2\,\hat{\boldsymbol{r}}\,\hat{\varrho}\,\hat{\boldsymbol{r}} + \hat{\varrho}\,\hat{\boldsymbol{r}}^2\,)\}.$$

Applying the formula (7.5.6.4), we will have

$$\mathrm{Tr}\,(\hat{\boldsymbol{r}}\,[\,\hat{\boldsymbol{r}}\,[\,\hat{\boldsymbol{r}}\,\hat{\varrho}\,]\,]\,) = 0\,. \tag{7.5.6.10}$$

6. Finally, the last member

$$\mathrm{Tr}\,(\hat{\boldsymbol{r}}\,[\,\hat{\boldsymbol{r}}\,[\,\hat{\mathbf{p}}\,\hat{\varrho}\,]_{+}\,]\,) = \mathrm{Tr}\,(\hat{\boldsymbol{r}}\,[\,\hat{\boldsymbol{r}}\,(\hat{\mathbf{p}}\,\hat{\varrho} + \hat{\varrho}\,\hat{\mathbf{p}}\,)\,]\,) =$$
$$= \mathrm{Tr}\,(\,\hat{\boldsymbol{r}}^2\,(\hat{\mathbf{p}}\,\hat{\varrho} + \hat{\varrho}\,\hat{\mathbf{p}}\,) - \hat{\boldsymbol{r}}\,(\hat{\mathbf{p}}\,\hat{\varrho} + \hat{\varrho}\,\hat{\mathbf{p}}\,)\,\hat{\boldsymbol{r}}\,) =$$
$$= \mathrm{Tr}\,(\,\hat{\boldsymbol{r}}^2\,\hat{\mathbf{p}}\,\hat{\varrho} + \hat{\boldsymbol{r}}^2\,\hat{\varrho}\,\hat{\mathbf{p}} - \hat{\boldsymbol{r}}\,\hat{\mathbf{p}}\,\hat{\varrho}\,\hat{\boldsymbol{r}} - \hat{\boldsymbol{r}}\,\hat{\varrho}\,\hat{\mathbf{p}}\,\hat{\boldsymbol{r}}\,) =$$
$$= \mathrm{Tr}\,(\,\hat{\boldsymbol{r}}^2\,\hat{\mathbf{p}}\,\hat{\varrho} + \hat{\mathbf{p}}\,\hat{\boldsymbol{r}}^2\,\hat{\varrho} - \hat{\boldsymbol{r}}^2\,\hat{\mathbf{p}}\,\hat{\varrho} - \hat{\mathbf{p}}\,\hat{\boldsymbol{r}}^2\,\hat{\varrho}\,) = 0\,. \tag{7.5.6.11}$$

Substituting formulas (7.5.6.6) – (7.5.6.11) in formula (7.5.6.3), we get

$$\partial\langle\hat{\boldsymbol{r}}\rangle/\partial t = \langle\hat{\mathbf{p}}\rangle/M\,. \tag{7.5.6.12}$$

7.5.7. Average Value of Atomic Nucleus Pulse

Find the change in the average nucleus pulse

$$\langle\hat{\mathbf{p}}\rangle = \mathrm{Tr}\,(\,\hat{\mathbf{p}}\,\hat{\varrho}\,) \tag{7.5.7.1}$$

value over time:

$$\partial\langle\hat{\mathbf{p}}\rangle/\partial t = \mathrm{Tr}\,(\,\hat{\mathbf{p}}\;\partial\hat{\varrho}/\partial t). \tag{7.5.7.2}$$

Substituting equation (7.5.5.4) into this formula gives

$$\partial\langle\hat{\boldsymbol{p}}\rangle/\partial t = -\mathrm{i}/\hbar\,\mathrm{Tr}\,(\,\hat{\boldsymbol{p}}\,\{[\,\hat{\boldsymbol{p}}^2/(2\,M) + g\,\hat{\boldsymbol{r}}^2, \hat{\varrho}\,] +$$

$$+\,\mathrm{i}\,D/\hbar\,\{\,[\,\hat{\boldsymbol{p}}\,[\,\hat{\boldsymbol{p}}\,\hat{\varrho}\,]\,] - \mathrm{i}\,\hbar\,\beta\,g\,[\,\hat{\boldsymbol{p}}\,[\,\hat{\boldsymbol{r}}\,\hat{\varrho}\,]_+\,]\,\} +$$

$$+\mathrm{i}\,\gamma/\hbar\,\{\,[\,\hat{\boldsymbol{r}}\,[\,\hat{\boldsymbol{r}}\,\hat{\varrho}\,]\,] + \mathrm{i}\,\hbar\,\beta/(2\,M)\,[\,\hat{\boldsymbol{r}}\,[\,\hat{\boldsymbol{p}}\,\hat{\varrho}\,]_+\,]\,\}\,\})\,.$$

(7.5.7.3)

1.Calculate the following value

$$\mathrm{Tr}\,(\,\hat{\boldsymbol{p}}\,[\,\hat{\boldsymbol{p}}^2\,\hat{\varrho}\,]\,) = 0\,.$$

(7.5.7.4)

This trace is zero according to the formula (7.5.6.7).

2. Consider the following term

$$\mathrm{Tr}\,(\,\hat{\boldsymbol{p}}\,[\,\hat{\boldsymbol{r}}^2\,\hat{\varrho}\,]\,) = \mathrm{Tr}\,(\,\hat{\boldsymbol{p}}\,\hat{\boldsymbol{r}}^2\,\hat{\varrho} - \hat{\boldsymbol{p}}\,\hat{\varrho}\,\hat{\boldsymbol{r}}^2\,)\,.$$

According to the formula (7.5.6.4), we will have

$$\mathrm{Tr}\,(\,\hat{\boldsymbol{p}}\,[\,\hat{\boldsymbol{r}}^2\,\hat{\varrho}\,]\,) = \mathrm{Tr}\,(\,\hat{\boldsymbol{p}}\,\hat{\boldsymbol{r}}^2\,\hat{\varrho} - \hat{\boldsymbol{r}}^2\,\hat{\boldsymbol{p}}\,\hat{\varrho})\,.$$

Now applying the formula (7.5.6.5), we get

$$\mathrm{Tr}\,(\,\hat{\boldsymbol{p}}\,[\,\hat{\boldsymbol{r}}^2\,\hat{\varrho}\,]\,) = \mathrm{Tr}\,\{\,(\,\hat{\boldsymbol{r}}\,\hat{\boldsymbol{p}} - \mathrm{i}\,\hbar\,\delta_{kl}\,)\,\hat{\boldsymbol{r}}\,\hat{\varrho} - \hat{\boldsymbol{r}}\,(\,\hat{\boldsymbol{p}}\,\hat{\boldsymbol{r}} + \mathrm{i}\,\hbar\,\delta_{kl}\,)\,\hat{\varrho}\} =$$

$$= -\,2\,\mathrm{i}\,\hbar\,\delta_{kl}\,\mathrm{Tr}\,(\,\hat{\boldsymbol{r}}\,\hat{\varrho}\,) = -\,2\,\mathrm{i}\,\hbar\,\langle\boldsymbol{r}\rangle\,.$$

(7.5.7.5)

3. Now calculate the term

$$\mathrm{Tr}\,(\hat{\boldsymbol{p}}\,[\,\hat{\boldsymbol{p}}\,[\,\hat{\boldsymbol{p}}\,\hat{\varrho}\,]\,]\,) = 0\,.$$

(7.5.7.6)

This term will be zero according to the formula (7.5.6.4).

4. The following member

$$\mathrm{Tr}\,(\hat{\boldsymbol{p}}\,[\,\hat{\boldsymbol{p}}\,[\,\hat{\boldsymbol{r}}\,\hat{\varrho}\,]_+\,]\,) = \mathrm{Tr}\,\{\,\hat{\boldsymbol{p}}\,[\,\hat{\boldsymbol{p}}\,(\,\hat{\boldsymbol{r}}\,\hat{\varrho} + \hat{\varrho}\,\hat{\boldsymbol{r}}\,)\,]\,\} =$$

$$= \mathrm{Tr}\,\{\,\hat{\boldsymbol{p}}\,(\,\hat{\boldsymbol{p}}\,\hat{\boldsymbol{r}}\,\hat{\varrho} + \hat{\boldsymbol{p}}\,\hat{\varrho}\,\hat{\boldsymbol{r}} - \hat{\boldsymbol{r}}\,\hat{\varrho}\,\hat{\boldsymbol{p}} - \hat{\varrho}\,\hat{\boldsymbol{r}}\,\hat{\boldsymbol{p}}\,)\,\}\,.$$

Lets apply the formula (7.5.6.4), we will have

$$\mathrm{Tr}\,(\hat{\boldsymbol{p}}\,[\,\hat{\boldsymbol{p}}\,[\,\hat{\boldsymbol{r}}\,\hat{\varrho}\,]_+\,]\,) = \mathrm{Tr}\,\{\,\hat{\boldsymbol{p}}\,(\,\hat{\boldsymbol{p}}\,\hat{\boldsymbol{r}}\,\hat{\varrho} + \hat{\boldsymbol{r}}\,\hat{\boldsymbol{p}}\,\hat{\varrho} - \hat{\boldsymbol{p}}\,\hat{\boldsymbol{r}}\,\hat{\varrho} - \hat{\boldsymbol{r}}\,\hat{\boldsymbol{p}}\,\hat{\varrho}\,)\,\}\,.$$

It turns out that

$$\mathrm{Tr}\,(\hat{\boldsymbol{p}}\,[\,\hat{\boldsymbol{p}}\,[\,\hat{\boldsymbol{r}}\,\hat{\varrho}\,]_+\,]\,) = 0. \tag{7.5.7.7}$$

5. The penultimate member

$$\mathrm{Tr}\,(\hat{\boldsymbol{p}}\,[\,\hat{\boldsymbol{r}}\,[\,\hat{\boldsymbol{r}}\,\hat{\varrho}\,]\,]\,) = \mathrm{Tr}\,(\hat{\boldsymbol{p}}\,[\,\hat{\boldsymbol{r}}\,(\,\hat{\boldsymbol{r}}\,\hat{\varrho} - \hat{\varrho}\,\hat{\boldsymbol{r}}\,)\,]\,) = \mathrm{Tr}\,\{\,\hat{\boldsymbol{p}}\,(\,\hat{\boldsymbol{r}}^2\hat{\varrho} - 2\,\hat{\boldsymbol{r}}\,\hat{\varrho}\,\hat{\boldsymbol{r}} + \hat{\varrho}\,\hat{\boldsymbol{r}}^2\,)\,\}\,.$$

Applying the formula (7.5.6.4), we will have

$$\mathrm{Tr}\,(\hat{\boldsymbol{p}}\,[\,\hat{\boldsymbol{r}}\,[\,\hat{\boldsymbol{r}}\,\hat{\varrho}\,]\,]\,) = 0\,. \tag{7.5.7.8}$$

6. Finally, the last member

$$\mathrm{Tr}\,(\hat{\boldsymbol{p}}\,[\,\hat{\boldsymbol{r}}\,[\,\hat{\boldsymbol{p}}\,\hat{\varrho}\,]_+\,]\,) = \mathrm{Tr}\,(\,\hat{\boldsymbol{p}}\,[\,\hat{\boldsymbol{r}}\,(\,\hat{\boldsymbol{p}}\,\hat{\varrho} + \hat{\varrho}\,\hat{\boldsymbol{p}}\,)\,]\,) =$$

$$= \mathrm{Tr}\,(\,\hat{\boldsymbol{p}}\,\hat{\boldsymbol{r}}\,(\,\hat{\boldsymbol{p}}\,\hat{\varrho} + \hat{\varrho}\,\hat{\boldsymbol{p}}\,) - \hat{\boldsymbol{p}}\,(\,\hat{\boldsymbol{p}}\,\hat{\varrho} + \hat{\varrho}\,\hat{\boldsymbol{p}}\,)\,\hat{\boldsymbol{r}}\,) =$$

$$= \mathrm{Tr}\,(\,\hat{\boldsymbol{p}}\,\hat{\boldsymbol{r}}\,\hat{\boldsymbol{p}}\,\hat{\varrho} + \hat{\boldsymbol{p}}\,\hat{\boldsymbol{r}}\,\hat{\varrho}\,\hat{\boldsymbol{p}} - \hat{\boldsymbol{p}}\,\hat{\boldsymbol{p}}\,\hat{\varrho}\,\hat{\boldsymbol{r}} - \hat{\boldsymbol{p}}\,\hat{\varrho}\,\hat{\boldsymbol{p}}\,\hat{\boldsymbol{r}}\,) =$$

$$= \mathrm{Tr}\,(\,\hat{\boldsymbol{p}}\,\hat{\boldsymbol{r}}\,\hat{\boldsymbol{p}}\,\hat{\varrho} + \hat{\boldsymbol{p}}\,\hat{\boldsymbol{p}}\,\hat{\boldsymbol{r}}\,\hat{\varrho} - \hat{\boldsymbol{r}}\,\hat{\boldsymbol{p}}\,\hat{\boldsymbol{p}}\,\hat{\varrho} - \hat{\boldsymbol{p}}\,\hat{\boldsymbol{r}}\,\hat{\boldsymbol{p}}\,\hat{\varrho}) =$$

$$\mathrm{Tr}\,(\,\hat{\boldsymbol{p}}\,(\,\hat{\boldsymbol{p}}\,\hat{\boldsymbol{r}}\,)\,\hat{\varrho} - (\,\hat{\boldsymbol{r}}\,\hat{\boldsymbol{p}}\,)\,\hat{\boldsymbol{p}}\,\hat{\varrho})\,.$$

Applying the formula (7.5.6.5)

$$\mathrm{Tr}\,(\hat{\mathbf{p}}\,[\,\hat{\mathbf{r}}\,[\hat{\mathbf{p}}\,\hat{\varrho}\,]_+\,]\,) =$$

$$= \mathrm{Tr}\,(\,\hat{\mathbf{p}}\,(\,\hat{\mathbf{r}}\,\hat{\mathbf{p}} - \mathrm{i}\,\hbar\,\delta_{kl}\,)\,\hat{\varrho} - (\,\hat{\mathbf{p}}\,\hat{\mathbf{r}} + \mathrm{i}\,\hbar\,\delta_{kl}\,)\,\hat{\mathbf{p}}\,\hat{\varrho}\,) =$$

$$= -\,2\,\mathrm{i}\,\hbar\,\mathrm{Tr}\,(\,\hat{\mathbf{p}}\,\hat{\varrho}\,) = -\,2\,\mathrm{i}\,\hbar\,\langle \mathbf{p}\rangle\,.$$

(7.5.7.9)

Substituting formulas (7.5.7.4) – (7.5.7.9) in formula (7.5.7.3), we get

$$\partial\langle\hat{\mathbf{p}}\rangle/\partial t = -\,2\,g\,\langle\hat{\mathbf{r}}\rangle - 2\,\gamma\,\beta/M\,\langle\hat{\mathbf{p}}\rangle\,.$$

(7.5.7.10)

7.5.8. Equation of Motion of Average Value of Nucleus Coordinates

Following arethe obtained equations (7.5.6.12) and (7.5.7.10) for the average values of the coordinate and momentum of the nucleus

$$\begin{cases} \partial\langle\hat{\mathbf{r}}\rangle/\partial t = \langle\hat{\mathbf{p}}\rangle/M\,, \\ \partial\langle\hat{\mathbf{p}}\rangle/\partial t = -\,2\,g\,\langle\hat{\mathbf{r}}\rangle - \,2\,\gamma\,\beta/M\,\langle\hat{\mathbf{p}}\rangle\,. \end{cases}$$

(7.5.8.1)

These equations are obtained exactly from the Lindblad equation (7.5.5.4) for the statistical operator.

Excluding the average value of the pulse, we will have

$$M\,\partial^2\,\langle\hat{\mathbf{r}}\rangle/\partial t^2 = \,-\,2\,\gamma\,\beta\,\partial\langle\hat{\mathbf{r}}\rangle/\partial t - 2\,g\,\langle\hat{\mathbf{r}}\rangle\,.$$

(7.5.8.2)

Lets rewrite this equation using substitution

$$\bar{x} = \langle\hat{\mathbf{r}}\rangle\,.$$

We will have

$$\ddot{\bar{x}} + 2\,p\,\dot{\bar{x}} + q\,\bar{x} = 0\,,$$

(7.5.8.3)

where

$$p = \beta \gamma / M , \qquad q = 2 g / M .$$

We will look for a solution to equation (7.5.8.3) in the form

$$\bar{x}(t) = e^{k t} .$$

(7.5.8.4)

Substituting this function into the equation gives

$$k^2 + 2 p k + q = 0 .$$

(7.5.8.5)

This equation has a solution

$$k_{12} = - p \pm \sqrt{p^2 - q} .$$

(7.5.8.6)

It can be seen that $k_{12} < 0$, so for equation (7.5.8.3), the condition holds

$$\langle \hat{\boldsymbol{r}} \rangle_{t \to \infty} = 0 .$$

(7.5.8.7)

7.5.9. Equation for Average Value $\langle \hat{\boldsymbol{r}}^2 \rangle$

The average value of the square of the coordinate $\langle \hat{\boldsymbol{r}}^2 \rangle$ of the nucleus is by definition equal to:

$$\langle \hat{\boldsymbol{r}}^2 \rangle = \mathrm{Tr} (\hat{\boldsymbol{r}}^2 \hat{\varrho}) .$$

(7.5.9.1)

Find the change in the value $\langle \hat{\boldsymbol{r}}^2 \rangle$ over time. It is obvious that

$$\partial\langle\hat{\boldsymbol{r}}^2\rangle/\partial t = \mathrm{Tr}\,(\,\hat{\boldsymbol{r}}^2\ \partial\hat{\varrho}/\partial t\,)\,.$$

(7.5.9.2)

Substituting equation (7.5.5.4) here gives [4]

$$\partial\langle\hat{\boldsymbol{r}}^2\rangle/\partial t = -\,\mathrm{i}/\hbar\ \mathrm{Tr}\,(\,\hat{\boldsymbol{r}}^2\ \{[\,\hat{\boldsymbol{p}}^2/(2\,M) + g\,\hat{\boldsymbol{r}}^2, \hat{\varrho}\,] +$$
$$+\,\mathrm{i}\,D/\hbar\ \{\,[\,\hat{\boldsymbol{p}}\,[\,\hat{\boldsymbol{p}}\,\hat{\varrho}\,]\,] - \mathrm{i}\,\hbar\,\beta\,g\,[\,\hat{\boldsymbol{p}}\,[\,\hat{\boldsymbol{r}}\,\hat{\varrho}\,]_+\,]\,\} +$$
$$+\mathrm{i}\,\gamma/\hbar\ \{\,[\,\hat{\boldsymbol{r}}\,[\,\hat{\boldsymbol{r}}\,\hat{\varrho}\,]\,] + \mathrm{i}\,\hbar\,\beta/(2\,M)\,[\,\hat{\boldsymbol{r}}\,[\,\hat{\boldsymbol{p}}\,\hat{\varrho}\,]_+\,]\}\}\,)\,.$$

(7.5.9.3)

1.Calculate the following value

$$\mathrm{Tr}\,(\,\hat{\boldsymbol{r}}^2\ [\,\hat{\boldsymbol{p}}^2\,\hat{\varrho}\,]\,) = \mathrm{Tr}\,\{\,\hat{\boldsymbol{r}}^2\,(\,\hat{\boldsymbol{p}}^2\hat{\varrho} - \hat{\varrho}\,\hat{\boldsymbol{p}}^2)\,\} = \mathrm{Tr}\,(\,\hat{\boldsymbol{r}}^2\,\hat{\boldsymbol{p}}^2\hat{\varrho} - \hat{\boldsymbol{r}}^2\,\hat{\varrho}\,\hat{\boldsymbol{p}}^2)\,.$$

Now using the formula (7.5.6.4), we get

$$\mathrm{Tr}\,(\hat{\boldsymbol{r}}^2\ [\,\hat{\boldsymbol{p}}^2\,\hat{\varrho}\,]) = \mathrm{Tr}\,(\,\hat{\boldsymbol{r}}^2\,\hat{\boldsymbol{p}}^2\hat{\varrho} - \hat{\boldsymbol{p}}^2\,\hat{\boldsymbol{r}}^2\,\hat{\varrho}\,) = \mathrm{Tr}\,\{\,\hat{\boldsymbol{r}}\,(\,\hat{\boldsymbol{r}}\,\hat{\boldsymbol{p}}\,)\,\hat{\boldsymbol{p}}\,\hat{\varrho} - \hat{\boldsymbol{p}}\,(\,\hat{\boldsymbol{p}}\,\hat{\boldsymbol{r}}\,)\,\hat{\boldsymbol{r}}\,\hat{\varrho}\,\}\,.$$

Lets use the formula (7.5.6.5), we will have

$$\mathrm{Tr}\,(\,\hat{\boldsymbol{r}}\,[\,\hat{\boldsymbol{p}}^2\,\hat{\varrho}\,]\,) = \mathrm{Tr}\,\{\,\hat{\boldsymbol{r}}\,(\,\mathrm{i}\,\hbar\,\delta_{kl} + \hat{\boldsymbol{p}}\,\hat{\boldsymbol{r}}\,)\,\hat{\boldsymbol{p}}\,\hat{\varrho} - \hat{\boldsymbol{p}}\,(\,\hat{\boldsymbol{r}}\,\hat{\boldsymbol{p}} - \mathrm{i}\,\hbar\,\delta_{kl}\,)\,\hat{\boldsymbol{r}}\,\hat{\varrho}\,\} =$$
$$= \mathrm{Tr}\,\{\,\mathrm{i}\,\hbar\,\delta_{kl}\,\hat{\boldsymbol{r}}\,\hat{\boldsymbol{p}}\,\hat{\varrho} + \hat{\boldsymbol{r}}\,\hat{\boldsymbol{p}}\,\hat{\boldsymbol{r}}\,\hat{\boldsymbol{p}}\,\hat{\varrho} - \hat{\boldsymbol{p}}\,\hat{\boldsymbol{r}}\,\hat{\boldsymbol{p}}\,\hat{\boldsymbol{r}}\,\hat{\varrho} + \mathrm{i}\,\hbar\,\delta_{kl}\,\hat{\boldsymbol{p}}\,\hat{\boldsymbol{r}}\,\hat{\varrho}\,\} =$$
$$= \mathrm{i}\,\hbar\,\delta_{kl}\mathrm{Tr}\,(\,\hat{\boldsymbol{r}}\,\hat{\boldsymbol{p}}\,\hat{\varrho} + \hat{\boldsymbol{p}}\,\hat{\boldsymbol{r}}\,\hat{\varrho}\,) + \mathrm{Tr}\{\,(\hat{\boldsymbol{r}}\,\hat{\boldsymbol{p}})\,\hat{\boldsymbol{r}}\,\hat{\boldsymbol{p}}\,\hat{\varrho} - \hat{\boldsymbol{p}}\,\hat{\boldsymbol{r}}\,(\hat{\boldsymbol{p}}\,\hat{\boldsymbol{r}})\,\hat{\varrho}\,\} =$$
$$= \mathrm{i}\,\hbar\,\langle\,\hat{\boldsymbol{r}}\,\hat{\boldsymbol{p}} + \hat{\boldsymbol{p}}\,\hat{\boldsymbol{r}}\,\rangle + \mathrm{Tr}\{\,(\,\mathrm{i}\,\hbar\,\delta_{kl} + \hat{\boldsymbol{p}}\,\hat{\boldsymbol{r}}\,)\,\hat{\boldsymbol{r}}\,\hat{\boldsymbol{p}}\,\hat{\varrho} - \hat{\boldsymbol{p}}\,\hat{\boldsymbol{r}}\,(\,\hat{\boldsymbol{r}}\,\hat{\boldsymbol{p}}\ - \mathrm{i}\,\hbar\,\delta_{kl}\,)\,\hat{\varrho}\,\} =$$
$$= \mathrm{i}\,\hbar\,\langle\,\hat{\boldsymbol{r}}\,\hat{\boldsymbol{p}} + \hat{\boldsymbol{p}}\,\hat{\boldsymbol{r}}\,\rangle + \mathrm{Tr}\{\,\mathrm{i}\,\hbar\,\hat{\boldsymbol{r}}\,\hat{\boldsymbol{p}}\,\hat{\varrho} + \mathrm{i}\,\hbar\,\hat{\boldsymbol{p}}\,\hat{\boldsymbol{r}}\,\hat{\varrho}\,\} = 2\,\mathrm{i}\,\hbar\,\langle\,\hat{\boldsymbol{r}}\,\hat{\boldsymbol{p}} + \hat{\boldsymbol{p}}\,\hat{\boldsymbol{r}}\,\rangle\,.$$

(7.5.9.4)

2. Consider the following term

$$\mathrm{Tr}\,(\,\hat{\boldsymbol{r}}^2\ [\,\hat{\boldsymbol{r}}^2\,\hat{\varrho}\,]\,) = \mathrm{Tr}\,(\,\hat{\boldsymbol{r}}^4\,\hat{\varrho} - \hat{\boldsymbol{r}}^2\,\hat{\varrho}\,\hat{\boldsymbol{r}}^2\,)\,.$$

According to the formula (7.5.6.4), we will have

$$\mathrm{Tr}\,(\hat{\boldsymbol{r}}^2\,[\,\hat{\boldsymbol{r}}^2\,\hat{\varrho}\,]\,) = 0\,.$$

(7.5.9.5)

3. Now calculate the term

$$\mathrm{Tr}\,(\hat{\boldsymbol{r}}^2\,[\,\hat{\boldsymbol{p}}\,[\,\hat{\boldsymbol{p}}\,\hat{\varrho}\,]\,]\,) = \mathrm{Tr}\,\{\,\hat{\boldsymbol{r}}^2\,[\,\hat{\boldsymbol{p}}\,(\,\hat{\boldsymbol{p}}\,\hat{\varrho} - \hat{\varrho}\,\hat{\boldsymbol{p}}\,)\,]\,\} =$$

$$= \mathrm{Tr}\,\{\,\hat{\boldsymbol{r}}^2\,(\,\hat{\boldsymbol{p}}\,(\,\hat{\boldsymbol{p}}\,\hat{\varrho} - \hat{\varrho}\,\hat{\boldsymbol{p}}\,) - (\,\hat{\boldsymbol{p}}\,\hat{\varrho} - \hat{\varrho}\,\hat{\boldsymbol{p}}\,)\hat{\boldsymbol{p}}\,)\,\} =$$

$$= \mathrm{Tr}\,\{\,\hat{\boldsymbol{r}}^2\,(\,\hat{\boldsymbol{p}}^2\,\hat{\varrho} - 2\,\hat{\boldsymbol{p}}\,\hat{\varrho}\,\hat{\boldsymbol{p}} + \hat{\varrho}\,\hat{\boldsymbol{p}}^2\,)\,\} =$$

$$= \mathrm{Tr}\,\{\,\hat{\boldsymbol{r}}^2\,\hat{\boldsymbol{p}}^2\,\hat{\varrho} - 2\,\hat{\boldsymbol{r}}^2\,\hat{\boldsymbol{p}}\,\hat{\varrho}\,\hat{\boldsymbol{p}} + \hat{\boldsymbol{r}}^2\,\hat{\varrho}\,\hat{\boldsymbol{p}}^2\,\}\,.$$

Applying the formula (7.5.6.4), we get

$$\mathrm{Tr}\,(\hat{\boldsymbol{r}}^2\,[\,\hat{\boldsymbol{p}}\,[\,\hat{\boldsymbol{p}}\,\hat{\varrho}\,]\,]\,) = \mathrm{Tr}\,\{\,\hat{\boldsymbol{r}}^2\,\hat{\boldsymbol{p}}^2\,\hat{\varrho} - 2\,\hat{\boldsymbol{p}}\,\hat{\boldsymbol{r}}^2\,\hat{\boldsymbol{p}}\,\hat{\varrho} + \hat{\boldsymbol{p}}^2\,\hat{\boldsymbol{r}}^2\,\hat{\varrho}\,\}\,.$$

Now applying the formula (7.5.6.5), we will have

$$\mathrm{Tr}\,(\hat{\boldsymbol{r}}^2\,[\,\hat{\boldsymbol{p}}\,[\,\hat{\boldsymbol{p}}\,\hat{\varrho}\,]\,]\,) = \mathrm{Tr}\,\{\,\hat{\boldsymbol{r}}\,(\,\hat{\boldsymbol{r}}\,\hat{\boldsymbol{p}}\,)\,\hat{\boldsymbol{p}}\,\hat{\varrho} - 2\,\hat{\boldsymbol{p}}\,\hat{\boldsymbol{r}}\,(\,\hat{\boldsymbol{r}}\,\hat{\boldsymbol{p}}\,)\,\hat{\varrho} + \hat{\boldsymbol{p}}\,(\,\hat{\boldsymbol{p}}\,\hat{\boldsymbol{r}}\,)\,\hat{\boldsymbol{r}}\,\hat{\varrho}\} =$$

$$= \mathrm{Tr}\,\{\,\hat{\boldsymbol{r}}\,(\,\hat{\boldsymbol{p}}\,\hat{\boldsymbol{r}} + \mathrm{i}\,\hbar\,\delta_{kl})\,\hat{\boldsymbol{p}}\,\hat{\varrho} - 2\,\hat{\boldsymbol{p}}\,\hat{\boldsymbol{r}}\,(\,\hat{\boldsymbol{p}}\,\hat{\boldsymbol{r}} + \mathrm{i}\,\hbar\,\delta_{kl})\,\hat{\varrho} + \hat{\boldsymbol{p}}\,(\,\hat{\boldsymbol{r}}\,\hat{\boldsymbol{p}}\ -\ \mathrm{i}\,\hbar\,\delta_{kl})\,\hat{\boldsymbol{r}}\,\hat{\varrho}\}$$

$$= \mathrm{Tr}\,\{\,\hat{\boldsymbol{r}}\,\hat{\boldsymbol{p}}\,\hat{\boldsymbol{r}}\,\hat{\boldsymbol{p}}\,\hat{\varrho} + \mathrm{i}\,\hbar\,\delta_{kl}\,\hat{\boldsymbol{r}}\,\hat{\boldsymbol{p}}\,\hat{\varrho} - 2\,\hat{\boldsymbol{p}}\,\hat{\boldsymbol{r}}\,\hat{\boldsymbol{p}}\,\hat{\boldsymbol{r}}\,\hat{\varrho} - 2\,\mathrm{i}\,\hbar\,\delta_{kl}\,\hat{\boldsymbol{p}}\,\hat{\boldsymbol{r}}\,\hat{\varrho} + \hat{\boldsymbol{p}}\,\hat{\boldsymbol{r}}\,\hat{\boldsymbol{p}}\,\hat{\boldsymbol{r}}\,\hat{\varrho}\ - \mathrm{i}\,\hbar\,\delta_{kl}\hat{\boldsymbol{p}}\,\hat{\boldsymbol{r}}\,\hat{\varrho}\} =$$

$$= \mathrm{Tr}\,\{\,\hat{\boldsymbol{r}}\,\hat{\boldsymbol{p}}\,(\,\hat{\boldsymbol{r}}\,\hat{\boldsymbol{p}}\,)\,\hat{\varrho} + \mathrm{i}\,\hbar\,\delta_{kl}\,\hat{\boldsymbol{r}}\,\hat{\boldsymbol{p}}\,\hat{\varrho} - (\,\hat{\boldsymbol{p}}\,\hat{\boldsymbol{r}}\,)\,\hat{\boldsymbol{p}}\,\hat{\boldsymbol{r}}\,\hat{\varrho} - 3\,\mathrm{i}\,\hbar\,\delta_{kl}\,\hat{\boldsymbol{p}}\,\hat{\boldsymbol{r}}\,\hat{\varrho}\,\} =$$

$$= \mathrm{Tr}\,\{\,\hat{\boldsymbol{r}}\,\hat{\boldsymbol{p}}\,(\,\hat{\boldsymbol{p}}\,\hat{\boldsymbol{r}} + \mathrm{i}\,\hbar\,\delta_{kl})\,\hat{\varrho} + \mathrm{i}\,\hbar\,\delta_{kl}\,\hat{\boldsymbol{r}}\,\hat{\boldsymbol{p}}\,\hat{\varrho} - (\,\hat{\boldsymbol{r}}\,\hat{\boldsymbol{p}}\ -\ \mathrm{i}\,\hbar\,\delta_{kl}\,)\,\hat{\boldsymbol{p}}\,\hat{\boldsymbol{r}}\,\hat{\varrho} - 3\,\mathrm{i}\,\hbar\,\delta_{kl}\ \hat{\boldsymbol{p}}\,\hat{\boldsymbol{r}}\,\hat{\varrho}\,\} =$$

$$= \mathrm{Tr}\,\{\,\hat{\boldsymbol{r}}\,\hat{\boldsymbol{p}}^2\,\hat{\boldsymbol{r}}\,\hat{\varrho} + \mathrm{i}\,\hbar\,\delta_{kl}\,\hat{\boldsymbol{r}}\,\hat{\boldsymbol{p}}\,\hat{\varrho} + \mathrm{i}\,\hbar\,\delta_{kl}\,\hat{\boldsymbol{r}}\,\hat{\boldsymbol{p}}\,\hat{\varrho} - \hat{\boldsymbol{r}}\,\hat{\boldsymbol{p}}^2\,\hat{\boldsymbol{r}}\,\hat{\varrho} + \mathrm{i}\,\hbar\,\delta_{kl}\,\hat{\boldsymbol{p}}\,\hat{\boldsymbol{r}}\,\hat{\varrho} - 3\ \mathrm{i}\,\hbar\,\delta_{kl}\,\hat{\boldsymbol{p}}\,\hat{\boldsymbol{r}}\,\hat{\varrho}\,\} =$$

$$= 2\,\mathrm{i}\,\hbar\,\delta_{kl}\,\mathrm{Tr}\,\{\,(\,\hat{\boldsymbol{r}}\,\hat{\boldsymbol{p}} - \hat{\boldsymbol{p}}\,\hat{\boldsymbol{r}}\,)\,\hat{\varrho}\,\} = 2\,\mathrm{i}\,\hbar\,\delta_{kl}\,\mathrm{Tr}\,\{\,\mathrm{i}\,\hbar\,\delta_{kl}\,\hat{\varrho}\,\} =$$

$$= -\,2\,\hbar^2\,\mathrm{Tr}\,\hat{\varrho} = -\,2\,\hbar^2\,N\,.$$

(7.5.9.6)

4. The following member

$$\mathrm{Tr}\,(\hat{\boldsymbol{r}}^2\,[\,\hat{\boldsymbol{p}}\,[\,\hat{\boldsymbol{r}}\,\hat{\varrho}\,]_+\,]\,) = \mathrm{Tr}\,\{\,\hat{\boldsymbol{r}}^2\,[\,\hat{\boldsymbol{p}}\,(\,\hat{\boldsymbol{r}}\,\hat{\varrho} + \hat{\varrho}\,\hat{\boldsymbol{r}}\,)\,]\,\} =$$

$$\mathrm{Tr}\,\{\,\hat{\boldsymbol{r}}^2\,(\,\hat{\boldsymbol{p}}\,(\,\hat{\boldsymbol{r}}\,\hat{\varrho} + \hat{\varrho}\,\hat{\boldsymbol{r}}\,) - (\,\hat{\boldsymbol{r}}\,\hat{\varrho} + \hat{\varrho}\,\hat{\boldsymbol{r}}\,)\,\hat{\boldsymbol{p}}\,\} =$$

$$= \mathrm{Tr}\,\{\,\hat{\boldsymbol{r}}^2\,(\,\hat{\boldsymbol{p}}\,\hat{\boldsymbol{r}}\,\hat{\varrho} + \hat{\boldsymbol{p}}\,\hat{\varrho}\,\hat{\boldsymbol{r}} - \hat{\boldsymbol{r}}\,\hat{\varrho}\,\hat{\boldsymbol{p}} - \hat{\varrho}\,\hat{\boldsymbol{r}}\,\hat{\boldsymbol{p}}\,)\,\} =$$

$$= \mathrm{Tr}\,\{\,\hat{\boldsymbol{r}}^2\,\hat{\boldsymbol{p}}\,\hat{\boldsymbol{r}}\,\hat{\varrho} + \hat{\boldsymbol{r}}^2\,\hat{\boldsymbol{p}}\,\hat{\varrho}\,\hat{\boldsymbol{r}} - \hat{\boldsymbol{r}}^2\,\hat{\boldsymbol{r}}\,\hat{\varrho}\,\hat{\boldsymbol{p}} - \hat{\boldsymbol{r}}^2\,\hat{\varrho}\,\hat{\boldsymbol{r}}\,\hat{\boldsymbol{p}}\,)\,\}\,.$$

Lets apply the formula (7.5.6.4), we will have

$$\mathrm{Tr}\,(\hat{\boldsymbol{r}}^2\,[\,\hat{\boldsymbol{p}}\,[\,\hat{\boldsymbol{r}}\,\hat{\varrho}\,]_+\,]\,) =$$

$$= \mathrm{Tr}\,\{\,\hat{\boldsymbol{r}}^2\,\hat{\boldsymbol{p}}\,\hat{\boldsymbol{r}}\,\hat{\varrho} + \hat{\boldsymbol{r}}\,\hat{\boldsymbol{r}}^2\,\hat{\boldsymbol{p}}\,\hat{\varrho} - \hat{\boldsymbol{p}}\,\hat{\boldsymbol{r}}^2\,\hat{\boldsymbol{r}}\,\hat{\varrho} - \hat{\boldsymbol{r}}\,\hat{\boldsymbol{p}}\,\hat{\boldsymbol{r}}^2\,\hat{\varrho}\,\} =$$

$$= \mathrm{Tr}\,\{\,\hat{\boldsymbol{r}}^2\,\hat{\boldsymbol{p}}\,\hat{\boldsymbol{r}}\,\hat{\varrho} + \hat{\boldsymbol{r}}^2\,(\,\hat{\boldsymbol{r}}\,\hat{\boldsymbol{p}}\,)\,\hat{\varrho} - (\,\hat{\boldsymbol{p}}\,\hat{\boldsymbol{r}}\,)\,\hat{\boldsymbol{r}}^2\,\hat{\varrho} - \hat{\boldsymbol{r}}\,(\,\hat{\boldsymbol{p}}\,\hat{\boldsymbol{r}}\,)\,\hat{\boldsymbol{r}}\,\hat{\varrho}\,\}\,.$$

Now applying the formula (7.5.6.5), we will have

$$\mathrm{Tr}\,(\hat{\boldsymbol{r}}^2\,[\,\hat{\boldsymbol{p}}\,[\,\hat{\boldsymbol{r}}\,\hat{\varrho}\,]_+\,]\,) =$$

$$= \mathrm{Tr}\,\{\,\hat{\boldsymbol{r}}^2\,\hat{\boldsymbol{p}}\,\hat{\boldsymbol{r}}\,\hat{\varrho} + \hat{\boldsymbol{r}}^2\,(\,\hat{\boldsymbol{r}}\,\hat{\boldsymbol{p}}\,)\,\hat{\varrho} - (\,\hat{\boldsymbol{p}}\,\hat{\boldsymbol{r}}\,)\,\hat{\boldsymbol{r}}^2\,\hat{\varrho} - \hat{\boldsymbol{r}}\,(\,\hat{\boldsymbol{r}}\,\hat{\boldsymbol{p}} - \mathrm{i}\,\hbar\,\delta_{kl})\,\hat{\boldsymbol{r}}\,\hat{\varrho}\,\} =$$

$$= \mathrm{Tr}\,\{\,\hat{\boldsymbol{r}}^2\,(\,\hat{\boldsymbol{r}}\,\hat{\boldsymbol{p}}\,)\,\hat{\varrho} - (\,\hat{\boldsymbol{p}}\,\hat{\boldsymbol{r}}\,)\,\hat{\boldsymbol{r}}^2\,\hat{\varrho} + \mathrm{i}\,\hbar\,\delta_{kl}\,\hat{\boldsymbol{r}}^2\,\hat{\varrho}\,\} =$$

$$= \mathrm{Tr}\,\{\,\hat{\boldsymbol{r}}^2\,(\,\hat{\boldsymbol{p}}\,\hat{\boldsymbol{r}} + \mathrm{i}\,\hbar\,\delta_{kl})\,\hat{\varrho} - (\,\hat{\boldsymbol{r}}\,\hat{\boldsymbol{p}} - \mathrm{i}\,\hbar\,\delta_{kl}\,)\,\hat{\boldsymbol{r}}^2\,\hat{\varrho} + \mathrm{i}\,\hbar\,\delta_{kl}\,\hat{\boldsymbol{r}}^2\,\hat{\varrho}\,\} =$$

$$= \mathrm{Tr}\,\{\,\hat{\boldsymbol{r}}^2\,\hat{\boldsymbol{p}}\,\hat{\boldsymbol{r}}\,\hat{\varrho} - \hat{\boldsymbol{r}}\,\hat{\boldsymbol{p}}\,\hat{\boldsymbol{r}}^2\hat{\varrho} + 3\,\mathrm{i}\,\hbar\,\delta_{kl}\,\hat{\boldsymbol{r}}^2\,\hat{\varrho}\,\} =$$

$$= \mathrm{Tr}\,\{\,\hat{\boldsymbol{r}}^2\,\hat{\boldsymbol{p}}\,\hat{\boldsymbol{r}}\,\hat{\varrho} - \hat{\boldsymbol{r}}\,(\,\hat{\boldsymbol{p}}\,\hat{\boldsymbol{r}}\,)\,\hat{\boldsymbol{r}}\,\hat{\varrho} + 3\,\mathrm{i}\,\hbar\,\delta_{kl}\,\hat{\boldsymbol{r}}^2\,\hat{\varrho}\,\} =$$

$$= \mathrm{Tr}\,\{\,\hat{\boldsymbol{r}}^2\,\hat{\boldsymbol{p}}\,\hat{\boldsymbol{r}}\,\hat{\varrho} - \hat{\boldsymbol{r}}\,(\,\hat{\boldsymbol{r}}\,\hat{\boldsymbol{p}}\, - \,\mathrm{i}\,\hbar\,\delta_{kl})\,\hat{\boldsymbol{r}}\,\hat{\varrho} + 3\,\mathrm{i}\,\hbar\,\delta_{kl}\,\hat{\boldsymbol{r}}^2\,\hat{\varrho}\,\} =$$

$$= \mathrm{Tr}\,\{\,\mathrm{i}\,\hbar\,\delta_{kl}\,\hat{\boldsymbol{r}}^2\,\hat{\varrho} + 3\,\mathrm{i}\,\hbar\,\delta_{kl}\,\hat{\boldsymbol{r}}^2\,\hat{\varrho}\,\} = 4\,\mathrm{i}\,\hbar\,\mathrm{Tr}\,(\,\hat{\boldsymbol{r}}^2\,\hat{\varrho}\,) = 4\,\mathrm{i}\,\hbar\,\langle\hat{\boldsymbol{r}}^2\rangle\,.$$

(7.5.9.7)

5. The penultimate member

$$\mathrm{Tr}\,(\,\hat{\boldsymbol{r}}^2\,[\,\hat{\boldsymbol{r}}\,[\,\hat{\boldsymbol{r}}\,\hat{\varrho}\,]\,]\,) = \mathrm{Tr}\,(\,\hat{\boldsymbol{r}}^2\,[\,\hat{\boldsymbol{r}}\,(\,\hat{\boldsymbol{r}}\,\hat{\varrho} - \hat{\varrho}\,\hat{\boldsymbol{r}}\,)\,]\,) = \mathrm{Tr}\,\{\,\hat{\boldsymbol{r}}^2\,(\,\hat{\boldsymbol{r}}^2\hat{\varrho} - 2\,\hat{\boldsymbol{r}}\,\hat{\varrho}\,\hat{\boldsymbol{r}} + \hat{\varrho}\,\hat{\boldsymbol{r}}^2\,)\}.$$

Applying the formula (7.5.6.4), we will have

$$\mathrm{Tr}\,(\hat{\boldsymbol{r}}^2\,[\,\hat{\boldsymbol{r}}\,[\,\hat{\boldsymbol{r}}\,\hat{\varrho}\,]\,]\,) = 0\,.$$

(7.5.9.8)

6. Finally, the last member

$$\mathrm{Tr}\,(\hat{\boldsymbol{r}}^2\,[\,[\,\hat{\boldsymbol{r}}\,[\,\hat{\boldsymbol{p}}\,\hat{\varrho}\,]_+\,]\,) = \mathrm{Tr}\,(\hat{\boldsymbol{r}}^2\,[\,\hat{\boldsymbol{r}}\,(\,\hat{\boldsymbol{p}}\,\hat{\varrho} + \hat{\varrho}\,\hat{\boldsymbol{p}}\,)\,]\,) =$$

$$= \mathrm{Tr}\,(\,\hat{\boldsymbol{r}}^3\,(\,\hat{\boldsymbol{p}}\,\hat{\varrho} + \hat{\varrho}\,\hat{\boldsymbol{p}}\,) - \hat{\boldsymbol{r}}^2\,(\,\hat{\boldsymbol{p}}\,\hat{\varrho} + \hat{\varrho}\,\hat{\boldsymbol{p}}\,)\,\hat{\boldsymbol{r}}\,) =$$

$$= \mathrm{Tr}\,(\,\hat{\boldsymbol{r}}^3\,\hat{\boldsymbol{p}}\,\hat{\varrho} + \hat{\boldsymbol{r}}^3\,\hat{\varrho}\,\hat{\boldsymbol{p}} - \hat{\boldsymbol{r}}^2\,\hat{\boldsymbol{p}}\,\hat{\varrho}\,\hat{\boldsymbol{r}} - \hat{\boldsymbol{r}}^2\,\hat{\varrho}\,\hat{\boldsymbol{p}}\,\hat{\boldsymbol{r}}\,) =$$

$$= \mathrm{Tr}\,(\,\hat{\boldsymbol{r}}^3\,\hat{\boldsymbol{p}}\,\hat{\varrho} + \hat{\boldsymbol{p}}\,\hat{\boldsymbol{r}}^3\,\hat{\varrho} - \hat{\boldsymbol{r}}^3\,\hat{\boldsymbol{p}}\,\hat{\varrho} -- \hat{\boldsymbol{p}}\,\hat{\boldsymbol{r}}^3\,\hat{\varrho}\,) = 0\,.$$

(7.5.9.9)

Substituting formulas (7.5.9.4) – (7.5.9.9) in formula (7.5.9.3), we get

$$\partial\langle\hat{\boldsymbol{r}}^2\rangle/\partial t = 1/M\,\langle\,\hat{\boldsymbol{r}}\,\hat{\boldsymbol{p}} + \hat{\boldsymbol{p}}\,\hat{\boldsymbol{r}}\,\rangle - 2\,D\,N + 4\,D\,g\,\beta\,\langle\hat{\boldsymbol{r}}^2\rangle.$$

(7.5.9.10)

7.5.10. Equation for Average Value of $\langle\hat{\boldsymbol{p}}^2\rangle$

The average value of the square of the pulse $\langle\hat{\boldsymbol{p}}^2\rangle$ of the nucleus is defined as

$$\langle\hat{\boldsymbol{p}}^2\rangle = \mathrm{Tr}\,(\,\hat{\boldsymbol{p}}^2\,\hat{\varrho}\,)\,.$$

(7.5.10.1)

Find the change over time of the average value of the square of the pulse $\langle\hat{\boldsymbol{p}}^2\rangle$ of the nucleus

$$\partial\langle\hat{\boldsymbol{p}}^2\rangle/\partial t = \mathrm{Tr}\,(\,\hat{\boldsymbol{p}}\,\partial\hat{\varrho}/\partial t).$$

(7.5.10.2)

Substituting equation (7.5.5.4) into this formula gives

$$\partial\langle\hat{\boldsymbol{p}}^2\rangle/\partial t = -\,\mathrm{i}/\hbar\,\mathrm{Tr}\,(\,\hat{\boldsymbol{p}}^2\,\{[\,\hat{\boldsymbol{p}}^2/(2\,M) + g\,\hat{\boldsymbol{r}}^2, \hat{\varrho}\,] +$$

$$+\,\mathrm{i}\,D/\hbar\,\{\,[\,\hat{\boldsymbol{p}}\,[\,\hat{\boldsymbol{p}}\,\hat{\varrho}\,]\,]-\mathrm{i}\,\hbar\,\beta\,g\,[\,\hat{\boldsymbol{p}}\,[\,\hat{\boldsymbol{r}}\,\hat{\varrho}\,]_+\,]\,\}\,+$$

$$+\mathrm{i}\,\gamma/\hbar\,\{\,[\,\hat{\boldsymbol{r}}\,[\,\hat{\boldsymbol{r}}\,\hat{\varrho}\,]\,]+\mathrm{i}\,\hbar\,\beta/(2\,M)\,[\,\hat{\boldsymbol{r}}\,[\,\hat{\boldsymbol{p}}\,\hat{\varrho}\,]_+\,]\}\,\}\,)\,.$$

(7.5.10.3)

1.Calculate the following value

$$\mathrm{Tr}\,(\,\hat{\boldsymbol{p}}^2\,[\,\hat{\boldsymbol{p}}^2\,\hat{\varrho}\,]\,)=0\,.$$

(7.5.10.4)

This trace is equal to zero according to the formula (7.5.9.5).

2. Consider the following term

$$\mathrm{Tr}\,(\,\hat{\boldsymbol{p}}^2\,[\,\hat{\boldsymbol{r}}^2\,\hat{\varrho}\,]\,)=\mathrm{Tr}\,(\,\hat{\boldsymbol{p}}^2\,\hat{\boldsymbol{r}}^2\,\hat{\varrho}-\hat{\boldsymbol{p}}^2\,\hat{\varrho}\,\hat{\boldsymbol{r}}^2\,)\,.$$

According to the formula (7.5.6.4), we will have

$$\mathrm{Tr}\,(\,\hat{\boldsymbol{p}}^2\,[\,\hat{\boldsymbol{r}}^2\,\hat{\varrho}\,]\,)=\mathrm{Tr}\,(\,\hat{\boldsymbol{p}}^2\,\hat{\boldsymbol{r}}^2\,\hat{\varrho}-\hat{\boldsymbol{r}}^2\,\hat{\boldsymbol{p}}^2\,\hat{\varrho})=\mathrm{Tr}\,\{\,\hat{\boldsymbol{p}}\,(\,\hat{\boldsymbol{p}}\,\hat{\boldsymbol{r}}\,)\,\hat{\boldsymbol{r}}\,\hat{\varrho}-\hat{\boldsymbol{r}}\,(\,\hat{\boldsymbol{r}}\,\hat{\boldsymbol{p}}\,)\,\hat{\boldsymbol{p}}\,\hat{\varrho}\}\,.$$

Now applying the formula (7.5.6.5), we get

$$\mathrm{Tr}\,(\,\hat{\boldsymbol{p}}^2\,[\,\hat{\boldsymbol{r}}^2\,\hat{\varrho}\,]\,)=\mathrm{Tr}\,\{\,\hat{\boldsymbol{p}}\,(\,\hat{\boldsymbol{r}}\,\hat{\boldsymbol{p}}-\mathrm{i}\,\hbar\,\delta_{kl}\,)\,\hat{\boldsymbol{r}}\,\hat{\varrho}-\hat{\boldsymbol{r}}\,(\,\hat{\boldsymbol{p}}\,\hat{\boldsymbol{r}}+\mathrm{i}\,\hbar\,\delta_{kl}\,)\,\hat{\boldsymbol{p}}\,\hat{\varrho}\}=$$

$$=\mathrm{Tr}\,\{-\,\mathrm{i}\,\hbar\,(\,\hat{\boldsymbol{r}}\,\hat{\boldsymbol{p}}+\hat{\boldsymbol{p}}\,\hat{\boldsymbol{r}}\,)\,\hat{\varrho}+(\,\hat{\boldsymbol{p}}\,\hat{\boldsymbol{r}}\,)\,\hat{\boldsymbol{p}}\,\hat{\boldsymbol{r}}\,\hat{\varrho}-\hat{\boldsymbol{r}}\,\hat{\boldsymbol{p}}\,(\,\hat{\boldsymbol{r}}\,\hat{\boldsymbol{p}}\,)\,\hat{\varrho}\}=$$

$$=\mathrm{Tr}\,\{-\,\mathrm{i}\,\hbar\,(\,\hat{\boldsymbol{r}}\,\hat{\boldsymbol{p}}+\hat{\boldsymbol{p}}\,\hat{\boldsymbol{r}}\,)\,\hat{\varrho}+(\,\hat{\boldsymbol{r}}\,\hat{\boldsymbol{p}}-\mathrm{i}\,\hbar\,)\,\hat{\boldsymbol{p}}\,\hat{\boldsymbol{r}}\,\hat{\varrho}-\hat{\boldsymbol{r}}\,\hat{\boldsymbol{p}}\,(\,\hat{\boldsymbol{p}}\,\hat{\boldsymbol{r}}+\mathrm{i}\,\hbar\,)\,\hat{\varrho}\}=$$

$$=\mathrm{Tr}\,\{-\,2\,\mathrm{i}\,\hbar\,(\,\hat{\boldsymbol{r}}\,\hat{\boldsymbol{p}}+\hat{\boldsymbol{p}}\,\hat{\boldsymbol{r}}\,)\,\hat{\varrho}\,\}=-\,2\,\mathrm{i}\,\hbar\,\langle\,\hat{\boldsymbol{r}}\,\hat{\boldsymbol{p}}+\hat{\boldsymbol{p}}\,\hat{\boldsymbol{r}}\,\rangle\,.$$

(7.5.10.5)

3. Now calculate the term

$$\mathrm{Tr}\,(\,\hat{\boldsymbol{p}}^2\,[\,\hat{\boldsymbol{p}}\,[\,\hat{\boldsymbol{p}}\,\hat{\varrho}\,]\,]\,)=0\,.$$

(7.5.10.6)

This term will be zero according to the formula (7.5.9.5).

4. The following member

$$\mathrm{Tr}\,(\hat{\boldsymbol{p}}^2\,[\,\hat{\boldsymbol{p}}\,[\,\hat{\boldsymbol{r}}\,\hat{\varrho}\,]_+\,]\,) = \mathrm{Tr}\,\{\,\hat{\boldsymbol{p}}^2\,[\,\hat{\boldsymbol{p}}\,(\,\hat{\boldsymbol{r}}\,\hat{\varrho} + \hat{\varrho}\,\hat{\boldsymbol{r}}\,)\,]\,\} =$$

$$= \mathrm{Tr}\,\{\,\hat{\boldsymbol{p}}^2\,(\,\hat{\boldsymbol{p}}\,\hat{\boldsymbol{r}}\,\hat{\varrho} + \hat{\boldsymbol{p}}\,\hat{\varrho}\,\hat{\boldsymbol{r}} - \hat{\boldsymbol{r}}\,\hat{\varrho}\,\hat{\boldsymbol{p}} - \hat{\varrho}\,\hat{\boldsymbol{r}}\,\hat{\boldsymbol{p}}\,)\,\} =$$

$$= \mathrm{Tr}\,\{\,\hat{\boldsymbol{p}}^2\,\hat{\boldsymbol{p}}\,\hat{\boldsymbol{r}}\,\hat{\varrho} + \hat{\boldsymbol{p}}^2\,\hat{\boldsymbol{p}}\,\hat{\varrho}\,\hat{\boldsymbol{r}} - \hat{\boldsymbol{p}}^2\,\hat{\boldsymbol{r}}\,\hat{\varrho}\,\hat{\boldsymbol{p}} - \hat{\boldsymbol{p}}^2\,\hat{\varrho}\,\hat{\boldsymbol{r}}\,\hat{\boldsymbol{p}}\,\}\,.$$

Lets apply the formula (7.5.6.4). We will have

$$\mathrm{Tr}\,(\hat{\boldsymbol{p}}^2\,[\,\hat{\boldsymbol{p}}\,[\,\hat{\boldsymbol{r}}\,\hat{\varrho}\,]_+\,]\,) =$$

$$= \mathrm{Tr}\,\{\,\hat{\boldsymbol{p}}^2\,\hat{\boldsymbol{p}}\,\hat{\boldsymbol{r}}\,\hat{\varrho} + \hat{\boldsymbol{r}}\,\hat{\boldsymbol{p}}^2\,\hat{\boldsymbol{p}}\,\hat{\varrho} - \hat{\boldsymbol{p}}\,\hat{\boldsymbol{p}}^2\,\hat{\boldsymbol{r}}\,\hat{\varrho} - \hat{\boldsymbol{r}}\,\hat{\boldsymbol{p}}\,\hat{\boldsymbol{p}}^2\,\hat{\varrho}\} = 0\,. \tag{7.5.10.7}$$

5. The penultimate member

$$\mathrm{Tr}\,(\hat{\boldsymbol{p}}^2\,[\,\hat{\boldsymbol{r}}\,[\,\hat{\boldsymbol{r}}\,\hat{\varrho}\,]\,]\,) =$$

$$= \mathrm{Tr}\,(\hat{\boldsymbol{p}}^2\,[\,\hat{\boldsymbol{r}}\,(\,\hat{\boldsymbol{r}}\,\hat{\varrho} - \hat{\varrho}\,\hat{\boldsymbol{r}}\,)\,]\,) = \mathrm{Tr}\,\{\,\hat{\boldsymbol{p}}^2\,(\,\hat{\boldsymbol{r}}^2\,\hat{\varrho} - 2\,\hat{\boldsymbol{r}}\,\hat{\varrho}\,\hat{\boldsymbol{r}} + \hat{\varrho}\,\hat{\boldsymbol{r}}^2\,)\,\} =$$

$$= \mathrm{Tr}\,\{\,\hat{\boldsymbol{p}}^2\,\hat{\boldsymbol{r}}^2\,\hat{\varrho} - 2\,\hat{\boldsymbol{p}}^2\,\hat{\boldsymbol{r}}\,\hat{\varrho}\,\hat{\boldsymbol{r}} + \hat{\boldsymbol{p}}^2\,\hat{\varrho}\,\hat{\boldsymbol{r}}^2\,\}\,.$$

Applying the formula (7.5.6.4), we will have

$$\mathrm{Tr}\,(\hat{\boldsymbol{p}}^2\,[\,\hat{\boldsymbol{r}}\,[\,\hat{\boldsymbol{r}}\,\hat{\varrho}\,]\,]\,) = \mathrm{Tr}\,\{\,\hat{\boldsymbol{p}}^2\,\hat{\boldsymbol{r}}^2\,\hat{\varrho} - 2\,\hat{\boldsymbol{r}}\,\hat{\boldsymbol{p}}^2\,\hat{\boldsymbol{r}}\,\hat{\varrho} + \hat{\boldsymbol{r}}^2\,\hat{\boldsymbol{p}}^2\,\hat{\varrho}\,\}\,.$$

Now we will apply the formula (7.5.6.5):

$$\mathrm{Tr}\,(\hat{\boldsymbol{p}}^2\,[\,\hat{\boldsymbol{r}}\,[\,\hat{\boldsymbol{r}}\,\hat{\varrho}\,]\,]\,) =$$

$$= \mathrm{Tr}\,\{\,\hat{\boldsymbol{p}}\,(\,\hat{\boldsymbol{p}}\,\hat{\boldsymbol{r}}\,)\,\hat{\boldsymbol{r}}\,\hat{\varrho} - 2\,\hat{\boldsymbol{r}}\,\hat{\boldsymbol{p}}^2\,\hat{\boldsymbol{r}}\,\hat{\varrho} + \hat{\boldsymbol{r}}\,(\,\hat{\boldsymbol{r}}\,\hat{\boldsymbol{p}}\,)\,\hat{\boldsymbol{p}}\,\hat{\varrho}\,\} =$$

$$= \mathrm{Tr}\,\{\,\hat{\boldsymbol{p}}\,\hat{\boldsymbol{r}}\,\hat{\boldsymbol{p}}\,\hat{\boldsymbol{r}}\,\hat{\varrho} - \mathrm{i}\,\hbar\,\delta_{kl}\hat{\boldsymbol{p}}\,\hat{\boldsymbol{r}}\,\hat{\varrho} - 2\,\hat{\boldsymbol{r}}\,\hat{\boldsymbol{p}}^2\,\hat{\boldsymbol{r}}\,\hat{\varrho} + \hat{\boldsymbol{r}}\,\hat{\boldsymbol{p}}\,\hat{\boldsymbol{r}}\hat{\boldsymbol{p}}\,\hat{\varrho} + \mathrm{i}\,\hbar\,\delta_{kl}\,\hat{\boldsymbol{r}}\,\hat{\boldsymbol{p}}\,\hat{\varrho}\,\} =$$

$$= \mathrm{Tr}\,\{\,(\,\hat{\boldsymbol{r}}\,\hat{\boldsymbol{p}} - \mathrm{i}\,\hbar\,\delta_{kl}\,)\,\hat{\boldsymbol{p}}\,\hat{\boldsymbol{r}}\,\hat{\varrho} - \mathrm{i}\,\hbar\,\delta_{kl}\hat{\boldsymbol{p}}\,\hat{\boldsymbol{r}}\,\hat{\varrho} - 2\,\hat{\boldsymbol{r}}\,\hat{\boldsymbol{p}}^2\,\hat{\boldsymbol{r}}\,\hat{\varrho} + \hat{\boldsymbol{r}}\,\hat{\boldsymbol{p}}\,(\,\hat{\boldsymbol{p}}\,\hat{\boldsymbol{r}} + \mathrm{i}\,\hbar\,\delta_{kl}\,)$$
$$\hat{\varrho} + \mathrm{i}\,\hbar\,\delta_{kl}\,\hat{\boldsymbol{r}}\,\hat{\boldsymbol{p}}\,\hat{\varrho}\,\} =$$

$$= 2\,\mathrm{i}\,\hbar\,\mathrm{Tr}\,\{\,(\,\hat{\boldsymbol{r}}\,\hat{\boldsymbol{p}} - \hat{\boldsymbol{p}}\,\hat{\boldsymbol{r}}\,)\,\hat{\varrho}\,\} = -\,2\,\hbar^2\,\mathrm{Tr}\,\hat{\varrho} = -\,2\,\hbar^2\,N\,. \tag{7.5.10.8}$$

6. Finally, the last member

$$\mathrm{Tr}\,(\hat{\boldsymbol{p}}^2\,[\,\hat{\boldsymbol{r}}\,[\,\hat{\boldsymbol{p}}\,\hat{\varrho}\,]_+\,]\,) = \mathrm{Tr}\,\{\,\hat{\boldsymbol{p}}^2\,[\,\hat{\boldsymbol{r}}\,(\,\hat{\boldsymbol{p}}\,\hat{\varrho} + \hat{\varrho}\,\hat{\boldsymbol{p}}\,)\,]\,\} =$$

$$= \mathrm{Tr}\,\{\,\hat{\boldsymbol{p}}^2\,\hat{\boldsymbol{r}}\,(\,\hat{\boldsymbol{p}}\,\hat{\varrho} + \hat{\varrho}\,\hat{\boldsymbol{p}}\,) - \hat{\boldsymbol{p}}^2\,(\,\hat{\boldsymbol{p}}\,\hat{\varrho} + \hat{\varrho}\,\hat{\boldsymbol{p}}\,)\,\hat{\boldsymbol{r}}\,\} =$$

$$= \mathrm{Tr}\,(\,\hat{\boldsymbol{p}}^2\,\hat{\boldsymbol{r}}\,\hat{\boldsymbol{p}}\,\hat{\varrho} + \hat{\boldsymbol{p}}^2\,\hat{\boldsymbol{r}}\,\hat{\varrho}\,\hat{\boldsymbol{p}} - \hat{\boldsymbol{p}}^2\,\hat{\boldsymbol{p}}\,\hat{\varrho}\,\hat{\boldsymbol{r}} - \hat{\boldsymbol{p}}^2\,\hat{\varrho}\,\hat{\boldsymbol{p}}\,\hat{\boldsymbol{r}}\,) =$$

$$= \mathrm{Tr}\,(\,\hat{\boldsymbol{p}}^2\,\hat{\boldsymbol{r}}\,\hat{\boldsymbol{p}}\,\hat{\varrho} + \hat{\boldsymbol{p}}\,\hat{\boldsymbol{p}}^2\,\hat{\boldsymbol{r}}\,\hat{\varrho} - \hat{\boldsymbol{r}}\,\hat{\boldsymbol{p}}^2\,\hat{\boldsymbol{p}}\,\hat{\varrho} - \hat{\boldsymbol{p}}\,\hat{\boldsymbol{r}}\,\hat{\boldsymbol{p}}^2\,\hat{\varrho}) =$$

$$= \mathrm{Tr}\,\{\,\hat{\boldsymbol{p}}^2\,\hat{\boldsymbol{r}}\,\hat{\boldsymbol{p}}\,\hat{\varrho} + \hat{\boldsymbol{p}}^2(\,\hat{\boldsymbol{p}}\,\hat{\boldsymbol{r}}\,)\,\hat{\varrho} - (\hat{\boldsymbol{r}}\,\hat{\boldsymbol{p}}\,)\,\hat{\boldsymbol{p}}^2\,\hat{\varrho} - \hat{\boldsymbol{p}}\,\hat{\boldsymbol{r}}\,\hat{\boldsymbol{p}}^2\,\hat{\varrho}\,\}\,.$$

Applying the formula (7.5.6.5), we get

$$\mathrm{Tr}\,(\hat{\boldsymbol{p}}^2\,[\,\hat{\boldsymbol{r}}\,[\,\hat{\boldsymbol{p}}\,\hat{\varrho}\,]_+\,]\,) =$$

$$= \mathrm{Tr}\,\{\,\hat{\boldsymbol{p}}^2\,\hat{\boldsymbol{r}}\,\hat{\boldsymbol{p}}\,\hat{\varrho} + \hat{\boldsymbol{p}}^2(\,\hat{\boldsymbol{r}}\,\hat{\boldsymbol{p}}\, - \,\mathrm{i}\,\hbar\,)\,\hat{\varrho} - (\,\hat{\boldsymbol{p}}\,\hat{\boldsymbol{r}}\, + \,\mathrm{i}\,\hbar\,)\,\hat{\boldsymbol{p}}^2\,\hat{\varrho} - \hat{\boldsymbol{p}}\,\hat{\boldsymbol{r}}\,\hat{\boldsymbol{p}}^2\,\hat{\varrho}\,\} =$$

$$= 2\,\mathrm{Tr}\,\{\,\hat{\boldsymbol{p}}^2\,\hat{\boldsymbol{r}}\,\hat{\boldsymbol{p}}\,\hat{\varrho} - \mathrm{i}\,\hbar\,\hat{\boldsymbol{p}}^2\,\hat{\varrho} - \hat{\boldsymbol{p}}\,\hat{\boldsymbol{r}}\,\hat{\boldsymbol{p}}^2\,\hat{\varrho}\,\} =$$

$$= 2\,\mathrm{Tr}\,\{\,\hat{\boldsymbol{p}}\,(\,\hat{\boldsymbol{p}}\,\hat{\boldsymbol{r}}\,)\,\hat{\boldsymbol{p}}\,\hat{\varrho} - \mathrm{i}\,\hbar\,\hat{\boldsymbol{p}}^2\,\hat{\varrho} - \hat{\boldsymbol{p}}\,\hat{\boldsymbol{r}}\,\hat{\boldsymbol{p}}^2\,\hat{\varrho}\,\} =$$

$$= 2\,\mathrm{Tr}\,\{\,\hat{\boldsymbol{p}}\,(\,\hat{\boldsymbol{r}}\,\hat{\boldsymbol{p}}\, - \mathrm{i}\,\hbar\,)\,\hat{\boldsymbol{p}}\,\hat{\varrho} - \mathrm{i}\,\hbar\,\hat{\boldsymbol{p}}^2\,\hat{\varrho} - \hat{\boldsymbol{p}}\,\hat{\boldsymbol{r}}\,\hat{\boldsymbol{p}}^2\,\hat{\varrho}\,\} =$$

$$= -\,4\,\mathrm{i}\,\hbar\,\mathrm{Tr}\,(\,\hat{\boldsymbol{p}}^2\,\hat{\varrho}\,) = -\,4\,\mathrm{i}\,\hbar\,\langle\hat{\boldsymbol{p}}^2\rangle\,.$$

(7.5.10.9)

Substituting the formulas (7.5.10.4) – (7.5.10.9) in the formula (7.5.10.3), we get

$$\partial\langle\hat{\boldsymbol{p}}^2\rangle/\partial t = -\,2\,g\,\langle\,\hat{\boldsymbol{r}}\,\hat{\boldsymbol{p}} + \hat{\boldsymbol{p}}\,\hat{\boldsymbol{r}}\,\rangle - 2\,\gamma\,N + 2\,\gamma\,\beta/M\,\langle\hat{\boldsymbol{p}}^2\rangle\,.$$

(7.5.10.10)

7.5.11. Equation for Average Value of Operator $\hat{\boldsymbol{r}}\,\hat{\boldsymbol{p}} + \hat{\boldsymbol{p}}\,\hat{\boldsymbol{r}}$

The average value $\langle\,\hat{\boldsymbol{r}}\,\hat{\boldsymbol{p}} + \hat{\boldsymbol{p}}\,\hat{\boldsymbol{r}}\,\rangle$ of the operator $\hat{\boldsymbol{r}}\,\hat{\boldsymbol{p}} + \hat{\boldsymbol{p}}\,\hat{\boldsymbol{r}}$ by definition is given by the formula

$$\langle\,\hat{\boldsymbol{r}}\,\hat{\boldsymbol{p}} + \hat{\boldsymbol{p}}\,\hat{\boldsymbol{r}}\,\rangle = \mathrm{Tr}\,\{\,(\,\hat{\boldsymbol{r}}\,\hat{\boldsymbol{p}} + \hat{\boldsymbol{p}}\,\hat{\boldsymbol{r}}\,)\,\hat{\varrho}\,\}\,.$$

(7.5.11.1)

Find the change in the value $\langle \hat{\boldsymbol{r}}\,\hat{\boldsymbol{p}} + \hat{\boldsymbol{p}}\,\hat{\boldsymbol{r}} \rangle$ over time. It is obvious that

$$\partial \langle \hat{\boldsymbol{r}}\,\hat{\boldsymbol{p}} + \hat{\boldsymbol{p}}\,\hat{\boldsymbol{r}} \rangle / \partial t = \mathrm{Tr}\,\{\,(\,\hat{\boldsymbol{r}}\,\hat{\boldsymbol{p}} + \hat{\boldsymbol{p}}\,\hat{\boldsymbol{r}}\,)\,\partial \hat{\varrho}/\partial t\,\}\,.$$

(7.5.11.2)

Substituting equation (7.5.5.4) here gives

$$\partial \langle \hat{\boldsymbol{r}}\,\hat{\boldsymbol{p}} + \hat{\boldsymbol{p}}\,\hat{\boldsymbol{r}} \rangle / \partial t = -\,\mathrm{i}/\hbar\,\mathrm{Tr}\,\{\,(\,\hat{\boldsymbol{r}}\,\hat{\boldsymbol{p}} + \hat{\boldsymbol{p}}\,\hat{\boldsymbol{r}}\,)\,\{[\,\hat{\boldsymbol{p}}^2/(2\,M) + g\,\hat{\boldsymbol{r}}^2, \hat{\varrho}\,] +$$

$$+\,\mathrm{i}\,D/\hbar\,\{\,[\,\hat{\boldsymbol{p}}\,[\,\hat{\boldsymbol{p}}\,\hat{\varrho}\,]\,] - \mathrm{i}\,\hbar\,\beta\,g\,[\,\hat{\boldsymbol{p}}\,[\,\hat{\boldsymbol{r}}\,\hat{\varrho}\,]_+\,]\,\} +$$

$$+\mathrm{i}\,\gamma/\hbar\,\{\,[\,\hat{\boldsymbol{r}}\,[\,\hat{\boldsymbol{r}}\,\hat{\varrho}\,]\,] + \mathrm{i}\,\hbar\,\beta/(2\,M)\,[\,\hat{\boldsymbol{r}}\,[\,\hat{\boldsymbol{p}}\,\hat{\varrho}\,]_+\,]\}\}\,.$$

(7.5.11.3)

1. Calculate the following value

$$\mathrm{Tr}\,\{\,(\,\hat{\boldsymbol{r}}\,\hat{\boldsymbol{p}} + \hat{\boldsymbol{p}}\,\hat{\boldsymbol{r}}\,)\,[\,\hat{\boldsymbol{p}}^2\,\hat{\varrho}\,]\,) = \mathrm{Tr}\,\{\,(\,\hat{\boldsymbol{r}}\,\hat{\boldsymbol{p}} + \hat{\boldsymbol{p}}\,\hat{\boldsymbol{r}}\,)\,(\,\hat{\boldsymbol{p}}^2\hat{\varrho} - \hat{\varrho}\,\hat{\boldsymbol{p}}^2)\,\} =$$

$$= \mathrm{Tr}\,(\,\hat{\boldsymbol{r}}\,\hat{\boldsymbol{p}}\,\hat{\boldsymbol{p}}^2\hat{\varrho} - \hat{\boldsymbol{r}}\,\hat{\boldsymbol{p}}\,\hat{\varrho}\,\hat{\boldsymbol{p}}^2 + \hat{\boldsymbol{p}}\,\hat{\boldsymbol{r}}\,\hat{\boldsymbol{p}}^2\hat{\varrho} - \hat{\boldsymbol{p}}\,\hat{\boldsymbol{r}}\,\hat{\varrho}\,\hat{\boldsymbol{p}}^2\,).$$

Lets use the formula (7.5.6.4), we get

$$\mathrm{Tr}\,\{\,(\,\hat{\boldsymbol{r}}\,\hat{\boldsymbol{p}} + \hat{\boldsymbol{p}}\,\hat{\boldsymbol{r}}\,)\,[\,\hat{\boldsymbol{p}}^2\,\hat{\varrho}\,]\,\} =$$

$$= \mathrm{Tr}\,(\,\hat{\boldsymbol{r}}\,\hat{\boldsymbol{p}}\,\hat{\boldsymbol{p}}^2\hat{\varrho} - \hat{\boldsymbol{p}}^2\,\hat{\boldsymbol{r}}\,\hat{\boldsymbol{p}}\,\hat{\varrho} + \hat{\boldsymbol{p}}\,\hat{\boldsymbol{r}}\,\hat{\boldsymbol{p}}^2\hat{\varrho} - \hat{\boldsymbol{p}}^2\,\hat{\boldsymbol{p}}\,\hat{\boldsymbol{r}}\,\hat{\varrho}\,).$$

By using the formula (6.5), we will have

$$\mathrm{Tr}\,\{\,(\,\hat{\boldsymbol{r}}\,\hat{\boldsymbol{p}} + \hat{\boldsymbol{p}}\,\hat{\boldsymbol{r}}\,)\,[\,\hat{\boldsymbol{p}}^2\,\hat{\varrho}\,]\,\} =$$

$$= \mathrm{Tr}\,\{\,(\,\hat{\boldsymbol{r}}\,\hat{\boldsymbol{p}}\,)\,\hat{\boldsymbol{p}}^2\hat{\varrho} - \hat{\boldsymbol{p}}^2\,\hat{\boldsymbol{r}}\,\hat{\boldsymbol{p}}\,\hat{\varrho} + \hat{\boldsymbol{p}}\,(\,\hat{\boldsymbol{r}}\,\hat{\boldsymbol{p}}\,)\,\hat{\boldsymbol{p}}\,\hat{\varrho} - \hat{\boldsymbol{p}}^2\,(\,\hat{\boldsymbol{p}}\,\hat{\boldsymbol{r}}\,)\,\hat{\varrho}\,\} =$$

$$= \mathrm{Tr}\,\{\,(\,\hat{\boldsymbol{p}}\,\hat{\boldsymbol{r}} + \mathrm{i}\,\hbar\,)\,\hat{\boldsymbol{p}}^2\hat{\varrho} - \hat{\boldsymbol{p}}^2\,\hat{\boldsymbol{r}}\,\hat{\boldsymbol{p}}\,\hat{\varrho} + \hat{\boldsymbol{p}}\,(\,\hat{\boldsymbol{p}}\,\hat{\boldsymbol{r}} + \mathrm{i}\,\hbar\,)\,\hat{\boldsymbol{p}}\,\hat{\varrho} - \hat{\boldsymbol{p}}^2\,(\,\hat{\boldsymbol{p}}\,\hat{\boldsymbol{r}}\,)\,\hat{\varrho}\,\} =$$

$$= \mathrm{Tr}\,\{\,\hat{\boldsymbol{p}}\,\hat{\boldsymbol{r}}\,\hat{\boldsymbol{p}}^2\hat{\varrho} + \mathrm{i}\,\hbar\,\hat{\boldsymbol{p}}^2\hat{\varrho} - \hat{\boldsymbol{p}}^2\,\hat{\boldsymbol{r}}\,\hat{\boldsymbol{p}}\,\hat{\varrho} + \hat{\boldsymbol{p}}^2\,\hat{\boldsymbol{r}}\,\hat{\boldsymbol{p}}\,\hat{\varrho} + \mathrm{i}\,\hbar\,\hat{\boldsymbol{p}}^2\,\hat{\varrho} - \hat{\boldsymbol{p}}^2\,(\,\hat{\boldsymbol{p}}\,\hat{\boldsymbol{r}}\,)\,\hat{\varrho}\,\} =$$

$$= \mathrm{Tr}\,\{\,\hat{\boldsymbol{p}}\,(\,\hat{\boldsymbol{r}}\,\hat{\boldsymbol{p}}\,)\,\hat{\boldsymbol{p}}\,\hat{\varrho} + 2\,\mathrm{i}\,\hbar\,\hat{\boldsymbol{p}}^2\hat{\varrho} - \hat{\boldsymbol{p}}^2\,(\,\hat{\boldsymbol{p}}\,\hat{\boldsymbol{r}}\,)\,\hat{\varrho}\,\} =$$

$$= \mathrm{Tr}\,\{\,\hat{\boldsymbol{p}}\,(\,\hat{\boldsymbol{p}}\,\hat{\boldsymbol{r}} + \mathrm{i}\,\hbar\,)\,\hat{\boldsymbol{p}}\,\hat{\varrho} + 2\,\mathrm{i}\,\hbar\,\hat{\boldsymbol{p}}^2\hat{\varrho} - \hat{\boldsymbol{p}}^2\,(\,\hat{\boldsymbol{r}}\,\hat{\boldsymbol{p}} - \mathrm{i}\,\hbar\,)\,\hat{\varrho}\,\} =$$

$$= \mathrm{Tr}\,\{\, \hat{\boldsymbol{p}}^2\, \hat{\boldsymbol{r}}\, \hat{\boldsymbol{p}}\, \hat{\varrho} + \mathrm{i}\,\hbar\, \hat{\boldsymbol{p}}^2\, \hat{\varrho} + 2\,\mathrm{i}\,\hbar\, \hat{\boldsymbol{p}}^2 \hat{\varrho} - \hat{\boldsymbol{p}}^2\, \hat{\boldsymbol{r}}\, \hat{\boldsymbol{p}}\, \hat{\varrho} + \mathrm{i}\,\hbar\, \hat{\boldsymbol{p}}^2\, \hat{\varrho}\,\} =$$

$$= 4\,\mathrm{i}\,\hbar\, \mathrm{Tr}\,\{\, \hat{\boldsymbol{p}}^2 \hat{\varrho}\,\} = 4\,\mathrm{i}\,\hbar\, \langle \hat{\boldsymbol{p}}^2 \rangle\,.$$

(7.5.11.4)

2. Consider the following term

$$\mathrm{Tr}\,\{\, (\, \hat{\boldsymbol{r}}\, \hat{\boldsymbol{p}} + \hat{\boldsymbol{p}}\, \hat{\boldsymbol{r}}\,)\, [\, \hat{\boldsymbol{r}}^2\, \hat{\varrho}\,]\,\} = \mathrm{Tr}\,\{\, (\, \hat{\boldsymbol{r}}\, \hat{\boldsymbol{p}} + \hat{\boldsymbol{p}}\, \hat{\boldsymbol{r}}\,)\, (\, \hat{\boldsymbol{r}}^2\, \hat{\varrho} - \hat{\varrho}\, \hat{\boldsymbol{r}}^2\,)\,\} =$$

$$= \mathrm{Tr}\,\{\, \hat{\boldsymbol{r}}\, \hat{\boldsymbol{p}}\, \hat{\boldsymbol{r}}^2\, \hat{\varrho} + \hat{\boldsymbol{p}}\, \hat{\boldsymbol{r}}\, \hat{\boldsymbol{r}}^2\, \hat{\varrho} - \hat{\boldsymbol{r}}\, \hat{\boldsymbol{p}}\, \hat{\varrho}\, \hat{\boldsymbol{r}}^2 - \hat{\boldsymbol{p}}\, \hat{\boldsymbol{r}}\, \hat{\varrho}\, \hat{\boldsymbol{r}}^2\,\} =$$

$$= \mathrm{Tr}\,\{\, \hat{\boldsymbol{r}}\, \hat{\boldsymbol{p}}\, \hat{\boldsymbol{r}}^2\, \hat{\varrho} + \hat{\boldsymbol{p}}\, \hat{\boldsymbol{r}}\, \hat{\boldsymbol{r}}^2\, \hat{\varrho} - \hat{\boldsymbol{r}}^2\, \hat{\boldsymbol{r}}\, \hat{\boldsymbol{p}}\, \hat{\varrho} - \hat{\boldsymbol{r}}^2\, \hat{\boldsymbol{p}}\, \hat{\boldsymbol{r}}\, \hat{\varrho}\,\} =$$

$$= \mathrm{Tr}\,\{\, \hat{\boldsymbol{r}}\, \hat{\boldsymbol{p}}\, \hat{\boldsymbol{r}}^2\, \hat{\varrho} + (\, \hat{\boldsymbol{p}}\, \hat{\boldsymbol{r}}\,)\, \hat{\boldsymbol{r}}^2\, \hat{\varrho} - \hat{\boldsymbol{r}}^2\, (\, \hat{\boldsymbol{r}}\, \hat{\boldsymbol{p}}\,)\, \hat{\varrho} - \hat{\boldsymbol{r}}^2\, \hat{\boldsymbol{p}}\, \hat{\boldsymbol{r}}\, \hat{\varrho}\,\}\,.$$

According to the formula (6.5), we will have

$$\mathrm{Tr}\,\{\, (\, \hat{\boldsymbol{r}}\, \hat{\boldsymbol{p}} + \hat{\boldsymbol{p}}\, \hat{\boldsymbol{r}}\,)\, [\, \hat{\boldsymbol{r}}^2\, \hat{\varrho}\,]\,\} =$$

$$= \mathrm{Tr}\,\{\, \hat{\boldsymbol{r}}\, \hat{\boldsymbol{p}}\, \hat{\boldsymbol{r}}^2\, \hat{\varrho} + (\, \hat{\boldsymbol{r}}\, \hat{\boldsymbol{p}} - \mathrm{i}\,\hbar\,)\, \hat{\boldsymbol{r}}^2\, \hat{\varrho} - \hat{\boldsymbol{r}}^2\, (\, \hat{\boldsymbol{p}}\, \hat{\boldsymbol{r}} + \mathrm{i}\,\hbar\,)\, \hat{\varrho} - \hat{\boldsymbol{r}}^2\, \hat{\boldsymbol{p}}\, \hat{\boldsymbol{r}}\, \hat{\varrho}\,\} =$$

$$= 2\,\mathrm{Tr}\,\{\, \hat{\boldsymbol{r}}\, \hat{\boldsymbol{p}}\, \hat{\boldsymbol{r}}^2\, \hat{\varrho} - \mathrm{i}\,\hbar\, \hat{\boldsymbol{r}}^2\, \hat{\varrho} - \hat{\boldsymbol{r}}^2\, \hat{\boldsymbol{p}}\, \hat{\boldsymbol{r}}\, \hat{\varrho}\,\} =$$

$$= 2\,\mathrm{Tr}\,\{\, \hat{\boldsymbol{r}}\, (\, \hat{\boldsymbol{p}}\, \hat{\boldsymbol{r}}\,)\, \hat{\boldsymbol{r}}\, \hat{\varrho} - \mathrm{i}\,\hbar\, \hat{\boldsymbol{r}}^2\, \hat{\varrho} - \hat{\boldsymbol{r}}^2\, \hat{\boldsymbol{p}}\, \hat{\boldsymbol{r}}\, \hat{\varrho}\,\} =$$

$$= 2\,\mathrm{Tr}\,\{\, \hat{\boldsymbol{r}}\, (\, \hat{\boldsymbol{r}}\, \hat{\boldsymbol{p}} - \mathrm{i}\,\hbar\,)\, \hat{\boldsymbol{r}}\, \hat{\varrho} - \mathrm{i}\,\hbar\, \hat{\boldsymbol{r}}^2\, \hat{\varrho} - \hat{\boldsymbol{r}}^2\, \hat{\boldsymbol{p}}\, \hat{\boldsymbol{r}}\, \hat{\varrho}\,\} =$$

$$= -\,4\,\mathrm{i}\,\hbar\, \mathrm{Tr}\,\{\, \hat{\boldsymbol{r}}^2\, \hat{\varrho}\,\} = -\,4\,\mathrm{i}\,\hbar\, \langle \hat{\boldsymbol{r}}^2 \rangle\,.$$

(7.5.11.5)

3. Now calculate the term

$$\mathrm{Tr}\,\{\, (\, \hat{\boldsymbol{r}}\, \hat{\boldsymbol{p}} + \hat{\boldsymbol{p}}\, \hat{\boldsymbol{r}}\,)\, [\, \hat{\boldsymbol{p}}\, [\, \hat{\boldsymbol{p}}\, \hat{\varrho}\,]\,]\,\} =$$

$$= \mathrm{Tr}\,\{\, (\, \hat{\boldsymbol{r}}\, \hat{\boldsymbol{p}} + \hat{\boldsymbol{p}}\, \hat{\boldsymbol{r}}\,)\, (\, \hat{\boldsymbol{p}}^2\, \hat{\varrho} - 2\, \hat{\boldsymbol{p}}\, \hat{\varrho}\, \hat{\boldsymbol{p}} + \hat{\varrho}\, \hat{\boldsymbol{p}}^2\,)\,\} =$$

$$= \mathrm{Tr}\,\{\, (\, \hat{\boldsymbol{r}}\, \hat{\boldsymbol{p}} + \hat{\boldsymbol{p}}\, \hat{\boldsymbol{r}}\,)\, \hat{\boldsymbol{p}}^2\, \hat{\varrho} - 2\, (\, \hat{\boldsymbol{r}}\, \hat{\boldsymbol{p}} + \hat{\boldsymbol{p}}\, \hat{\boldsymbol{r}}\,)\, \hat{\boldsymbol{p}}\, \hat{\varrho}\, \hat{\boldsymbol{p}} + (\, \hat{\boldsymbol{r}}\, \hat{\boldsymbol{p}} + \hat{\boldsymbol{p}}\, \hat{\boldsymbol{r}}\,)\, \hat{\varrho}\, \hat{\boldsymbol{p}}^2\,)\,\} =$$

$$= \mathrm{Tr}\,\{\, \hat{\boldsymbol{r}}\, \hat{\boldsymbol{p}}\, \hat{\boldsymbol{p}}^2\, \hat{\varrho} + \hat{\boldsymbol{p}}\, \hat{\boldsymbol{r}}\, \hat{\boldsymbol{p}}^2\, \hat{\varrho} - 2\, \hat{\boldsymbol{r}}\, \hat{\boldsymbol{p}}\, \hat{\boldsymbol{p}}\, \hat{\varrho}\, \hat{\boldsymbol{p}} - 2\, \hat{\boldsymbol{p}}\, \hat{\boldsymbol{r}}\, \hat{\boldsymbol{p}}\, \hat{\varrho}\, \hat{\boldsymbol{p}} + \hat{\boldsymbol{r}}\, \hat{\boldsymbol{p}}\, \hat{\varrho}\, \hat{\boldsymbol{p}}^2 + \hat{\boldsymbol{p}}\, \hat{\boldsymbol{r}}\, \hat{\varrho}\, \hat{\boldsymbol{p}}^2\,\}\,.$$

Lets use the formula (7.5.6.4), we will have

$$\mathrm{Tr}\,\{\,(\,\hat{\boldsymbol r}\,\hat{\boldsymbol p}\,+\,\hat{\boldsymbol p}\,\hat{\boldsymbol r}\,)\,[\,\hat{\boldsymbol p}\,[\,\hat{\boldsymbol p}\,\hat{\varrho}\,]\,]\,\}=$$

$$=\mathrm{Tr}\,\{\,\hat{\boldsymbol r}\,\hat{\boldsymbol p}\,\hat{\boldsymbol p}^2\,\hat{\varrho}\,+\,\hat{\boldsymbol p}\,\hat{\boldsymbol r}\,\hat{\boldsymbol p}^2\,\hat{\varrho}-2\,\hat{\boldsymbol p}\,\hat{\boldsymbol r}\,\hat{\boldsymbol p}\,\hat{\boldsymbol p}\,\hat{\varrho}-2\,\hat{\boldsymbol p}\,\hat{\boldsymbol p}\,\hat{\boldsymbol r}\,\hat{\boldsymbol p}\,\hat{\varrho}+\hat{\boldsymbol p}^2\,\hat{\boldsymbol r}\,\hat{\boldsymbol p}\,\hat{\varrho}\,+\,\hat{\boldsymbol p}^2\,\hat{\boldsymbol p}\,\hat{\boldsymbol r}\,\hat{\varrho}\,\}\,.$$

Now we will use the formula (7.5.6.5):

$$\mathrm{Tr}\,\{\,(\,\hat{\boldsymbol r}\,\hat{\boldsymbol p}\,+\,\hat{\boldsymbol p}\,\hat{\boldsymbol r}\,)\,[\,\hat{\boldsymbol p}\,[\,\hat{\boldsymbol p}\,\hat{\varrho}\,]\,]\,\}=$$

$$=\mathrm{Tr}\,\{\,(\,\hat{\boldsymbol r}\,\hat{\boldsymbol p}\,)\,\hat{\boldsymbol p}^2\,\hat{\varrho}\,+\,\hat{\boldsymbol p}\,\hat{\boldsymbol r}\,\hat{\boldsymbol p}^2\,\hat{\varrho}-2\,\hat{\boldsymbol p}\,(\,\hat{\boldsymbol r}\,\hat{\boldsymbol p}\,)\,\hat{\boldsymbol p}\,\hat{\varrho}-2\,\hat{\boldsymbol p}\,\hat{\boldsymbol p}\,\hat{\boldsymbol r}\,\hat{\boldsymbol p}\,\hat{\varrho}+\hat{\boldsymbol p}^2\,\hat{\boldsymbol r}\,\hat{\boldsymbol p}\,\hat{\varrho}\,+\,\hat{\boldsymbol p}^2\,(\,\hat{\boldsymbol p}\,\hat{\boldsymbol r}\,)\,\hat{\varrho}\,\}=$$

$$=\mathrm{Tr}\,\{\,(\,\hat{\boldsymbol p}\,\hat{\boldsymbol r}\,+\,\mathrm{i}\,\hbar\,)\,\hat{\boldsymbol p}^2\,\hat{\varrho}\,+\,\hat{\boldsymbol p}\,\hat{\boldsymbol r}\,\hat{\boldsymbol p}^2\,\hat{\varrho}-2\,\hat{\boldsymbol p}\,(\,\hat{\boldsymbol p}\,\hat{\boldsymbol r}\,+\,\mathrm{i}\,\hbar\,)\,\hat{\boldsymbol p}\,\hat{\varrho}-2\,\hat{\boldsymbol p}\,\hat{\boldsymbol p}\,\hat{\boldsymbol r}\,\hat{\boldsymbol p}\,\hat{\varrho}+$$

$$+\,\hat{\boldsymbol p}^2\,\hat{\boldsymbol r}\,\hat{\boldsymbol p}\,\hat{\varrho}\,+\,\hat{\boldsymbol p}^2\,(\,\hat{\boldsymbol r}\,\hat{\boldsymbol p}\,-\mathrm{i}\,\hbar\,)\hat{\varrho}\,\}=$$

$$=2\,\mathrm{Tr}\,\{\,\hat{\boldsymbol p}\,\hat{\boldsymbol r}\,\hat{\boldsymbol p}^2\,\hat{\varrho}-\hat{\boldsymbol p}\,\hat{\boldsymbol p}\,\hat{\boldsymbol r}\,\hat{\boldsymbol p}\,\hat{\varrho}-\mathrm{i}\,\hbar\,\hat{\boldsymbol p}^2\,\hat{\varrho}\,\}=$$

$$=2\,\mathrm{Tr}\,\{\,\hat{\boldsymbol p}\,\hat{\boldsymbol r}\,\hat{\boldsymbol p}^2\,\hat{\varrho}-\hat{\boldsymbol p}\,(\,\hat{\boldsymbol r}\,\hat{\boldsymbol p}\,-\mathrm{i}\,\hbar\,)\,\hat{\boldsymbol p}\,\hat{\varrho}-\mathrm{i}\,\hbar\,\hat{\boldsymbol p}^2\,\hat{\varrho}\,\}=$$

$$=2\,\mathrm{Tr}\,\{\mathrm{i}\,\hbar\,\hat{\boldsymbol p}^2\,\hat{\varrho}-\mathrm{i}\,\hbar\,\hat{\boldsymbol p}^2\,\hat{\varrho}\,\}=0\,.$$

(7.5.11.6)

4. The following member

$$\mathrm{Tr}\,\{\,(\,\hat{\boldsymbol r}\,\hat{\boldsymbol p}\,+\,\hat{\boldsymbol p}\,\hat{\boldsymbol r}\,)\,[\,\hat{\boldsymbol p}\,[\,\hat{\boldsymbol r}\,\hat{\varrho}\,]_+\,]\,\}=$$

$$=\mathrm{Tr}\,\{\,(\,\hat{\boldsymbol r}\,\hat{\boldsymbol p}\,+\,\hat{\boldsymbol p}\,\hat{\boldsymbol r}\,)\,(\,\hat{\boldsymbol p}\,\hat{\boldsymbol r}\,\hat{\varrho}+\hat{\boldsymbol p}\,\hat{\varrho}\,\hat{\boldsymbol r}-\hat{\boldsymbol r}\,\hat{\varrho}\,\hat{\boldsymbol p}-\hat{\varrho}\,\hat{\boldsymbol r}\,\hat{\boldsymbol p}\,)\,\}=$$

$$=\mathrm{Tr}\,\{\,\hat{\boldsymbol r}\,\hat{\boldsymbol p}\,\hat{\boldsymbol p}\,\hat{\boldsymbol r}\,\hat{\varrho}+\hat{\boldsymbol r}\,\hat{\boldsymbol p}\,\hat{\boldsymbol p}\,\hat{\varrho}\,\hat{\boldsymbol r}-\hat{\boldsymbol r}\,\hat{\boldsymbol p}\,\hat{\boldsymbol r}\,\hat{\varrho}\,\hat{\boldsymbol p}\,-\hat{\boldsymbol r}\,\hat{\boldsymbol p}\,\hat{\varrho}\,\hat{\boldsymbol r}\,\hat{\boldsymbol p}+$$

$$+\,\hat{\boldsymbol p}\,\hat{\boldsymbol r}\,\hat{\boldsymbol p}\,\hat{\boldsymbol r}\,\hat{\varrho}+\hat{\boldsymbol p}\,\hat{\boldsymbol r}\,\hat{\boldsymbol p}\,\hat{\varrho}\,\hat{\boldsymbol r}-\hat{\boldsymbol p}\,\hat{\boldsymbol r}\,\hat{\boldsymbol r}\,\hat{\varrho}\,\hat{\boldsymbol p}-\hat{\boldsymbol p}\,\hat{\boldsymbol r}\,\hat{\varrho}\,\hat{\boldsymbol r}\,\hat{\boldsymbol p}\,\}\,.$$

Lets apply the formula (7.5.6.4). We will have

$$\mathrm{Tr}\,\{\,(\,\hat{\boldsymbol r}\,\hat{\boldsymbol p}\,+\,\hat{\boldsymbol p}\,\hat{\boldsymbol r}\,)\,[\,\hat{\boldsymbol p}\,[\,\hat{\boldsymbol r}\,\hat{\varrho}\,]_+\,]\,\}=$$

$$=\mathrm{Tr}\,\{\,\hat{\boldsymbol r}\,\hat{\boldsymbol p}\,\hat{\boldsymbol p}\,\hat{\boldsymbol r}\,\hat{\varrho}+\hat{\boldsymbol r}\,\hat{\boldsymbol r}\,\hat{\boldsymbol p}\,\hat{\boldsymbol p}\,\hat{\varrho}\,-\hat{\boldsymbol p}\,\hat{\boldsymbol r}\,\hat{\boldsymbol p}\,\hat{\boldsymbol r}\,\hat{\varrho}\,-\hat{\boldsymbol r}\,\hat{\boldsymbol p}\,\hat{\boldsymbol r}\,\hat{\boldsymbol p}\,\hat{\varrho}+$$

$$+\hat{p}\hat{r}\hat{p}\hat{r}\hat{\varrho}+\hat{r}\hat{p}\hat{r}\hat{p}\hat{\varrho}-\hat{p}\hat{p}\hat{r}\hat{r}\hat{\varrho}-\hat{r}\hat{p}\hat{p}\hat{r}\hat{\varrho}\} =$$

$$= \mathrm{Tr}\{\hat{r}\hat{r}\hat{p}\hat{p}\hat{\varrho}-\hat{p}\hat{p}\hat{r}\hat{r}\hat{\varrho}\}.$$

Formula (7.5.6.5) gives

$$\mathrm{Tr}\{(\hat{r}\hat{p}+\hat{p}\hat{r})[\hat{p}[\hat{r}\hat{\varrho}]_+]\} = \mathrm{Tr}\{\hat{r}(\hat{r}\hat{p})\hat{p}\hat{\varrho}-\hat{p}(\hat{p}\hat{r})\hat{r}\hat{\varrho}\} =$$

$$= \mathrm{Tr}\{\hat{r}(\hat{p}\hat{r}+\mathrm{i}\hbar)\hat{p}\hat{\varrho}-\hat{p}(\hat{r}\hat{p}-\mathrm{i}\hbar)\hat{r}\hat{\varrho}\} =$$

$$= \mathrm{Tr}\{\hat{r}\hat{p}(\hat{r}\hat{p})\hat{\varrho}-(\hat{p}\hat{r})\hat{p}\hat{r}\hat{\varrho}+\mathrm{i}\hbar(\hat{r}\hat{p}+\hat{p}\hat{r})\hat{\varrho}\} =$$

$$= \mathrm{Tr}\{\hat{r}\hat{p}(\hat{p}\hat{r}+\mathrm{i}\hbar)\hat{\varrho}-(\hat{r}\hat{p}-\mathrm{i}\hbar)\hat{p}\hat{r}\hat{\varrho}+\mathrm{i}\hbar(\hat{r}\hat{p}+\hat{p}\hat{r})\hat{\varrho}\} =$$

$$= 2\mathrm{i}\hbar\,\mathrm{Tr}\{(\hat{r}\hat{p}+\hat{p}\hat{r})\hat{\varrho}\} = 2\mathrm{i}\hbar\langle\hat{r}\hat{p}+\hat{p}\hat{r}\rangle. \quad \textbf{(7.5.11.7)}$$

5. The penultimate member

$$\mathrm{Tr}\{(\hat{r}\hat{p}+\hat{p}\hat{r})[\hat{r}[\hat{r}\hat{\varrho}]]\} = \mathrm{Tr}\{(\hat{r}\hat{p}+\hat{p}\hat{r})(\hat{r}^2\hat{\varrho}-2\hat{r}\hat{\varrho}\hat{r}+\hat{\varrho}\hat{r}^2)\} =$$

$$= \mathrm{Tr}\{\hat{r}\hat{p}\hat{r}^2\hat{\varrho}-2\hat{r}\hat{p}\hat{r}\hat{\varrho}\hat{r}+\hat{r}\hat{p}\hat{\varrho}\hat{r}^2+\hat{p}\hat{r}\hat{r}^2\hat{\varrho}-2\hat{p}\hat{r}\hat{r}\hat{\varrho}\hat{r}+\hat{p}\hat{r}\hat{\varrho}\hat{r}^2\}.$$

The formula (7.5.6.4) leads to the ratio

$$\mathrm{Tr}\{(\hat{r}\hat{p}+\hat{p}\hat{r})[\hat{r}[\hat{r}\hat{\varrho}]]\} =$$

$$= \mathrm{Tr}\{\hat{r}\hat{p}\hat{r}^2\hat{\varrho}-2\hat{r}\hat{r}\hat{p}\hat{r}\hat{\varrho}+\hat{r}^2\hat{r}\hat{p}\hat{\varrho}+\hat{p}\hat{r}\hat{r}^2\hat{\varrho}-2\hat{r}\hat{p}\hat{r}\hat{r}\hat{\varrho}+\hat{r}^2\hat{p}\hat{r}\hat{\varrho}\}.$$

Now the formula (7.5.6.5) gives

$$\mathrm{Tr}\{(\hat{r}\hat{p}+\hat{p}\hat{r})[\hat{r}[\hat{r}\hat{\varrho}]]\} =$$

$$= \mathrm{Tr}\{\hat{r}(\hat{p}\hat{r})\hat{r}\hat{\varrho}-2\hat{r}\hat{r}\hat{p}\hat{r}\hat{\varrho}+\hat{r}^2(\hat{r}\hat{p})\hat{\varrho}+(\hat{p}\hat{r})\hat{r}^2\hat{\varrho}-2\hat{r}\hat{p}\hat{r}\hat{r}\hat{\varrho}+\hat{r}(\hat{r}\hat{p})\hat{r}\hat{\varrho}\} =$$

$$= \mathrm{Tr}\{\hat{r}(\hat{r}\hat{p}-\mathrm{i}\hbar)\hat{r}\hat{\varrho}-2\hat{r}\hat{r}\hat{p}\hat{r}\hat{\varrho}+\hat{r}^2(\hat{p}\hat{r}+\mathrm{i}\hbar)\hat{\varrho}+$$

$$+(\hat{r}\hat{p}-\mathrm{i}\hbar)\hat{r}^2\hat{\varrho}-2\hat{r}\hat{p}\hat{r}\hat{r}\hat{\varrho}+\hat{r}(\hat{p}\hat{r}+\mathrm{i}\hbar)\hat{r}\hat{\varrho}\} =$$

$$=2\,\mathrm{Tr}\,\{-\mathrm{i}\hbar\,\hat{\boldsymbol r}^2\,\hat\varrho+\mathrm{i}\hbar\,\hat{\boldsymbol r}^2\,\hat\varrho\}=0.$$

(7.5.11.8)

6. Finally, the last member

$$\mathrm{Tr}\,\{(\hat{\boldsymbol r}\hat{\boldsymbol p}+\hat{\boldsymbol p}\hat{\boldsymbol r})[[\hat{\boldsymbol r}[\hat{\boldsymbol p}\hat\varrho]_+]\}=$$

$$=\mathrm{Tr}\,\{(\hat{\boldsymbol r}\hat{\boldsymbol p}+\hat{\boldsymbol p}\hat{\boldsymbol r})(\hat{\boldsymbol r}(\hat{\boldsymbol p}\hat\varrho+\hat\varrho\hat{\boldsymbol p})-(\hat{\boldsymbol p}\hat\varrho+\hat\varrho\hat{\boldsymbol p})\hat{\boldsymbol r})\}=$$

$$=\mathrm{Tr}\,\{(\hat{\boldsymbol r}\hat{\boldsymbol p}+\hat{\boldsymbol p}\hat{\boldsymbol r})(\hat{\boldsymbol r}\hat{\boldsymbol p}\hat\varrho+\hat{\boldsymbol r}\hat\varrho\hat{\boldsymbol p}-\hat{\boldsymbol p}\hat\varrho\hat{\boldsymbol r}-\hat\varrho\hat{\boldsymbol p}\hat{\boldsymbol r})\}=$$

$$=\mathrm{Tr}\,\{\hat{\boldsymbol r}\hat{\boldsymbol p}\hat{\boldsymbol r}\hat{\boldsymbol p}\hat\varrho+\hat{\boldsymbol r}\hat{\boldsymbol p}\hat{\boldsymbol r}\hat\varrho\hat{\boldsymbol p}-\hat{\boldsymbol r}\hat{\boldsymbol p}\hat{\boldsymbol p}\hat\varrho\hat{\boldsymbol r}-\hat{\boldsymbol r}\hat{\boldsymbol p}\hat\varrho\hat{\boldsymbol p}\hat{\boldsymbol r}+$$

$$+\hat{\boldsymbol p}\hat{\boldsymbol r}\hat{\boldsymbol r}\hat{\boldsymbol p}\hat\varrho+\hat{\boldsymbol p}\hat{\boldsymbol r}\hat{\boldsymbol r}\hat\varrho\hat{\boldsymbol p}-\hat{\boldsymbol p}\hat{\boldsymbol r}\hat{\boldsymbol p}\hat\varrho\hat{\boldsymbol r}-\hat{\boldsymbol p}\hat{\boldsymbol r}\hat\varrho\hat{\boldsymbol p}\hat{\boldsymbol r}\}=$$

$$=\mathrm{Tr}\,\{\hat{\boldsymbol r}\hat{\boldsymbol p}\hat{\boldsymbol r}\hat{\boldsymbol p}\hat\varrho+\hat{\boldsymbol p}\hat{\boldsymbol r}\hat{\boldsymbol p}\hat{\boldsymbol r}\hat\varrho-\hat{\boldsymbol r}\hat{\boldsymbol r}\hat{\boldsymbol p}\hat{\boldsymbol p}\hat\varrho-\hat{\boldsymbol p}\hat{\boldsymbol r}\hat{\boldsymbol r}\hat{\boldsymbol p}\hat\varrho+$$

$$+\hat{\boldsymbol p}\hat{\boldsymbol r}\hat{\boldsymbol r}\hat{\boldsymbol p}\hat\varrho+\hat{\boldsymbol p}\hat{\boldsymbol p}\hat{\boldsymbol r}\hat{\boldsymbol r}\hat\varrho-\hat{\boldsymbol r}\hat{\boldsymbol p}\hat{\boldsymbol r}\hat{\boldsymbol p}\hat\varrho-\hat{\boldsymbol p}\hat{\boldsymbol r}\hat{\boldsymbol p}\hat{\boldsymbol r}\hat\varrho\}=$$

$$=\mathrm{Tr}\,\{-\hat{\boldsymbol r}(\hat{\boldsymbol p}\hat{\boldsymbol r}+\mathrm{i}\hbar)\hat{\boldsymbol p}\hat\varrho-(\hat{\boldsymbol r}\hat{\boldsymbol p}-\mathrm{i}\hbar)\hat{\boldsymbol r}\hat{\boldsymbol p}\hat\varrho+\hat{\boldsymbol p}\hat{\boldsymbol r}(\hat{\boldsymbol p}\hat{\boldsymbol r}+\mathrm{i}\hbar)\hat\varrho+\hat{\boldsymbol p}(\hat{\boldsymbol r}\hat{\boldsymbol p}-\mathrm{i}\hbar)\hat{\boldsymbol r}\hat\varrho\}=$$

$$=2\,\mathrm{Tr}\,\{-(\hat{\boldsymbol r}\hat{\boldsymbol p})\hat{\boldsymbol r}\hat{\boldsymbol p}\hat\varrho+\hat{\boldsymbol p}\hat{\boldsymbol r}(\hat{\boldsymbol p}\hat{\boldsymbol r})\hat\varrho\}=$$

$$=2\,\mathrm{Tr}\,\{-(\hat{\boldsymbol p}\hat{\boldsymbol r}+\mathrm{i}\hbar)\hat{\boldsymbol r}\hat{\boldsymbol p}\hat\varrho+\hat{\boldsymbol p}\hat{\boldsymbol r}(\hat{\boldsymbol r}\hat{\boldsymbol p}-\mathrm{i}\hbar)\hat\varrho\}=$$

$$=-2\mathrm{i}\hbar\,\mathrm{Tr}\,\{(\hat{\boldsymbol r}\hat{\boldsymbol p}+\hat{\boldsymbol p}\hat{\boldsymbol r})\hat\varrho\}=-2\mathrm{i}\hbar\,\langle\hat{\boldsymbol r}\hat{\boldsymbol p}+\hat{\boldsymbol p}\hat{\boldsymbol r}\rangle.$$

(7.5.11.9)

Substituting formulas (7.5.11.4) – (7.5.11.9) in formula (7.5.11.3), we get

$$\partial\langle\hat{\boldsymbol r}\hat{\boldsymbol p}+\hat{\boldsymbol p}\hat{\boldsymbol r}\rangle/\partial t=\langle\hat{\boldsymbol p}^2\rangle\,2/M-4\,g\,\langle\hat{\boldsymbol r}^2\rangle+$$

$$+2\,D\,g\,\beta\,\langle\hat{\boldsymbol r}\hat{\boldsymbol p}+\hat{\boldsymbol p}\hat{\boldsymbol r}\rangle+\gamma\,\beta/M\,\langle\hat{\boldsymbol r}\hat{\boldsymbol p}+\hat{\boldsymbol p}\hat{\boldsymbol r}\rangle.$$

(7.5.11.10)

7.5.12. Solution of Obtained Equations

Let's put all the obtained equations (7.5.9.10), (7.5.10.10) and (7.5.11.10) together:

$$\begin{cases} \partial\langle\hat{\mathbf{r}}^2\rangle/\partial t = 1/M\ \langle\hat{\mathbf{r}}\,\hat{\mathbf{p}} + \hat{\mathbf{p}}\,\hat{\mathbf{r}}\rangle - 2\,D\,N + 4\,D\,g\,\beta\,\langle\hat{\mathbf{r}}^2\rangle, \\ \partial\langle\hat{\mathbf{p}}^2\rangle/\partial t = -2\,g\,\langle\hat{\mathbf{r}}\,\hat{\mathbf{p}} + \hat{\mathbf{p}}\,\hat{\mathbf{r}}\rangle - 2\,\gamma\,N + 2\,\gamma\,\beta/M\,\langle\hat{\mathbf{p}}^2\rangle, \\ \partial\langle\hat{\mathbf{r}}\,\hat{\mathbf{p}} + \hat{\mathbf{p}}\,\hat{\mathbf{r}}\rangle/\partial t = \langle\hat{\mathbf{p}}^2\rangle\,2/M - 4\,g\,\langle\hat{\mathbf{r}}^2\rangle + \\ +2\,D\,g\,\beta\,\langle\hat{\mathbf{r}}\,\hat{\mathbf{p}} + \hat{\mathbf{p}}\,\hat{\mathbf{r}}\rangle + \gamma\,\beta/M\,\langle\hat{\mathbf{r}}\,\hat{\mathbf{p}} + \hat{\mathbf{p}}\,\hat{\mathbf{r}}\rangle. \end{cases}$$

(7.5.12.1)

These equations are a system of three linear inhomogeneous equations. We introduce the notation:

$$x(t) = \langle\hat{\mathbf{r}}^2\rangle\,,\ y(t) = \langle\hat{\mathbf{p}}^2\rangle\,, \qquad z(t) = \langle\hat{\mathbf{r}}\,\hat{\mathbf{p}} + \hat{\mathbf{p}}\,\hat{\mathbf{r}}\rangle\,.$$

(7.5.12.2)

Now the equations (7.5.12.1) take the form

$$\begin{cases} \partial x/\partial t - 4\,D\,g\,\beta\,x - z/M = -2\,D\,N\,, \\ \partial y/\partial t - 2\,\gamma\,\beta/M\,y + 2\,g\,z = -2\,\gamma\,N\,, \\ \partial z/\partial t + 4\,g\,x - 2/M\,y - (\gamma\,\beta/M + 2\,D\,g\,\beta)\,z = 0\,. \end{cases}$$

(7.5.12.3)

Let's write the system of equations as follows

$$\begin{cases} \partial x/\partial t - a_{11}\,x - a_{13}\,z = -b_1\,, \\ \partial y/\partial t - a_{22}\,y + a_{23}\,z = -b_2\,, \\ \partial z/\partial t + a_{31}\,x - a_{32}\,y - a_{33}\,z = 0\,. \end{cases}$$

(7.5.12.4)

where

$$\begin{cases} a_{11} = 4\,D\,g\,\beta\,, a_{12} = 0\,, a_{13} = 1/M\,; \\ a_{21} = 0\,, a_{22} = 2\,\gamma\,\beta/M\,, a_{23} = 2\,g\,; \\ a_{31} = 4\,g\,, a_{32} = 2/M\,, a_{33} = \gamma\,\beta/M + 2\,D\,g\,\beta\,; \end{cases}$$

(7.5.12.5)

$$b_1 = 2\,D\,N, \qquad b_2 = 2\,\gamma\,N\,,\, b_3 = 0\,.$$

(7.5.12.6)

Since ball lightning is stationary for some time, we write the system (7.5.12.4) as follows

$$\begin{cases} a_{11}\,x + a_{13}\,z = b_1\,, \\ a_{22}\,y - a_{23}\,z = b_2\,, \\ a_{31}\,x - a_{32}\,y - a_{33}\,z = 0\,. \end{cases}$$

(7.5.12.7)

Let's express the variable z from the third equation of this system:

$$z = (\,a_{31}\,x - a_{32}\,y\,)/a_{33}\,.$$

Let's substitute this expression in the first two equations. We get a system with two variables:

$$\begin{cases} a_{11}\,x + a_{13}\,(\,a_{31}\,x - a_{32}\,y\,)/a_{33} = b_1\,, \\ a_{22}\,y - a_{23}\,(\,a_{31}\,x - a_{32}\,y\,)/a_{33} = b_2\,. \end{cases}$$

(7.5.12.8)

Simplifying, we get

$$\begin{cases} (\,a_{11}\,a_{33} + a_{13}\,a_{31}\,)\,x - a_{13}\,a_{32}\,y = a_{33}\,b_1\,, \\ -\,a_{23}\,a_{31}\,x + (\,a_{22}\,a_{33} + a_{23}\,a_{32}\,)\,y = a_{33}\,b_2\,. \end{cases}$$

(7.5.12.9)

Calculate the determinant:

$$\Delta = \left\| \begin{matrix} a_{11}\,a_{33} + a_{13}\,a_{31} & -\,a_{13}\,a_{32} \\ -\,a_{23}\,a_{31} & a_{22}\,a_{33} + a_{23}\,a_{32} \end{matrix} \right\| =$$

$$= (\,a_{11}\,a_{33} + a_{13}\,a_{31}\,)\,(\,a_{22}\,a_{33} + a_{23}\,a_{32}\,) - a_{13}\,a_{32}\,a_{23}\,a_{31} =$$

$$= a_{11}\,a_{33}\,a_{22}\,a_{33} + a_{13}\,a_{31}\,a_{22}\,a_{33} + a_{11}\,a_{33}a_{23}\,a_{32}\,.$$

Substituting coefficients, we will have

$$\Delta = 8/M\, g\ \{ D\, \beta^2 \gamma\, (\gamma\ \beta/M + 2\, D\, g\, \beta) +$$

$$+ \gamma\, \beta/M + 2\, D\, g\, \beta \}\ (\gamma\ \beta/M + 2\, D\, g\, \beta)\,.$$

(7.5.12.10)

Let's calculate two more determinants:

$$\Delta_x = \begin{Vmatrix} a_{33}\, b_1 & - a_{13}\, a_{32} \\ a_{33}\, b_2 & a_{22}\, a_{33} + a_{23}\, a_{32} \end{Vmatrix} =$$

$$= a_{33}\, \{ (a_{22}\, a_{33} + a_{23}\, a_{32})\, b_1 + a_{13}\, a_{32}\, b_2 \},$$

$$\Delta_y = \begin{Vmatrix} a_{11}\, a_{33} + a_{13}\, a_{31} & a_{33}\, b_1 \\ - a_{23}\, a_{31} & a_{33}\, b_2 \end{Vmatrix} =$$

$$= a_{33}\, \{ (a_{11}\, a_{33} + a_{13}\, a_{31}) b_2 + a_{23}\, a_{31}\, b_1 \}\,.$$

Let's substitute coefficients for these determinants, we get

$$\Delta_x = 2\, a_{33} \{ ((\gamma\ \beta/M)^2 + 2\, \gamma\ \beta/M\, D\, g\, \beta + 2/M\, g)\, b_1 + 1/M^2\, b_2 \},$$

$$\Delta_y = 4\, a_{33}\, g \{ 2\, g\, b_1 + (D\, \beta^2\, \gamma\ 1/M + 2\, D^2\, g\, \beta^2 + 1/M) b_2 \}\,.$$

(7.5.12.11)

Now lets find the values

$$x = \Delta_x/\Delta \text{ and } y = \Delta_y/\Delta\,.$$

According to these formulas, we will have

$$x = 1/F\ \{ [(\gamma\ \beta/M)^2 + 2\, \beta^2\, \gamma/M\, D\, g + 2/M\, g]\, b_1 + 1/M^2\, b_2 \},$$

(7.5.12.12)

$$y = 2/F\, g \{ 2\, g\, b_1 + (D\, \beta^2\, \gamma/M + 2\, D^2\, \beta^2\, g + 1/M) b_2 \},$$

(7.5.12.13)

where

$$F = 4/M\,g\,\{\,D\,\beta^2\,\gamma\,(\,\gamma\,\beta/M + 2\,D\,g\,\beta\,) + \gamma\,\beta/M + 2\,D\,g\,\beta\,\}\,.$$

Consider two extreme cases, when either $D = 0$ or $\gamma = 0$. So first let $D = 0$. In this case, the formulas (7.5.12.6) turn into

$$b_1 = 0, \qquad b_2 = 2\,\gamma\,N\,;$$

and the formulas (7.5.12.12) and (7.5.12.13) give

$$\langle \hat{\boldsymbol{r}}^2 \rangle = N/(\,2\,g\,\beta\,), \qquad \langle \hat{\boldsymbol{p}}^2 \rangle = M\,N/\beta\,.$$

(7.5.12.14)

Now consider the case when $\gamma = 0$. In this case

$$b_1 = 2\,D\,N, \qquad b_2 = 0\,.$$

At the same time again

$$\langle \hat{\boldsymbol{r}}^2 \rangle = N/(\,2\,g\,\beta\,), \qquad \langle \hat{\boldsymbol{p}}^2 \rangle = M\,N/\beta\,.$$

(7.5.12.15)

It turned out that these average values coincide with the same values that correspond to the case $D = 0$.

7.5.13. Equation for Atomic Nucleus Density Matrix in Coordinate Representation

Lets write equation (7.5.5.4) in the coordinate representation when the operators $\hat{\boldsymbol{r}}$ and $\hat{\boldsymbol{p}}$ have the following types

$$\hat{\boldsymbol{r}} = \boldsymbol{r}, \qquad \hat{\boldsymbol{p}} = -\mathrm{i}\,\hbar\,\nabla\,.$$

(7.5.13.1)

Equation (7.5.5.4) will look like this

$$\mathrm{i}\,\hbar\,\partial\varrho/\partial t = -\,\hbar^2/(2\,M)\,(\,\nabla^2 - \,\nabla'^2)\,\varrho - g\,(\,r^2 - r'^2\,)\,\varrho +$$

$$+\,\mathrm{i}\,\hbar\,D\,\{\,(\,\nabla + \nabla')^2\,\varrho + \beta\,g\,(\,\nabla + \nabla')\,(\,\boldsymbol{r} + \boldsymbol{r}'\,)\,\varrho\,\} +$$

$$+\,\mathrm{i}\,\gamma/\hbar\,\{\,(\,\boldsymbol{r} - \boldsymbol{r}')^2\,\varrho + \hbar^2\,\beta/(2\,M)\,(\,\boldsymbol{r} - \boldsymbol{r}'\,)\,(\,\nabla - \nabla'\,)\,\varrho\,\} + \ldots,$$

(7.5.13.2)

where $\varrho = \varrho(\,t, \boldsymbol{r}, \boldsymbol{r}'\,)$.

7.5.14. Wigner Equation

Writing down the equation for the Wigner function

$$w(t, \boldsymbol{r}, \boldsymbol{p}) = 1/(2\,\pi\,\hbar)^3 \int \varrho(t, \boldsymbol{r} + \boldsymbol{r}'/2, \boldsymbol{r} - \boldsymbol{r}'/2)\,\exp(\,-\,\mathrm{i}\,\boldsymbol{p}\,\boldsymbol{r}'/\hbar)\,\mathrm{d}\boldsymbol{r}'.$$

(7.5.14.1)

After substituting in equation (7.5.13.2), we get the equation

$$\partial w/\partial t = -\,\boldsymbol{p}/M\,\nabla w + 2\,g\,\boldsymbol{r}\,\nabla_{\boldsymbol{p}} w + D\,\nabla\,(\,\nabla + 2g\,\boldsymbol{r}\,)\,w +$$

$$+\,\gamma\,\{\,\nabla_{\boldsymbol{p}}\,(\,\nabla_{\boldsymbol{p}} + \,\beta/m\,\boldsymbol{p})\,\}\,w + \ldots$$

(7.5.14.2)

7.5.15. Distribution of Atomic Nucleus by Coordinates

The average values (7.5.12.14) and (7.5.12.15) are obtained in two extreme cases, when either the diffusion coefficient $D = 0$ or the attenuation coefficient $\gamma = 0$. In these two cases, the average values of $\langle\hat{\boldsymbol{r}}^2\rangle$ and $\langle\hat{\boldsymbol{p}}^2\rangle$ have the same formulas. This means that the Wigner equation has the form

$$\partial w/\partial t = -\,\boldsymbol{p}/M\,\nabla w + 2\,g\,\boldsymbol{r}\,\nabla_{\boldsymbol{p}} w\,.$$

(7.5.15.1)

The stationary equation will be

$$-\boldsymbol{p}/M\, \nabla w + 2\, g\, \boldsymbol{r}\, \nabla_{\boldsymbol{p}} \mathrm{w} = 0\,.$$

(7.5.15.2)

Using the Boltzmann principle, the solution to this equation is not difficult to find:

$$w(\boldsymbol{r}, \boldsymbol{p}) = N\, B\, B_p \exp \{ -\beta\, (\, 2\, g\, \boldsymbol{r}^2 + \boldsymbol{p}^2/M\,)\, \}\,.$$

(7.5.15.3)

where B and B_p are the normalizing factors. Knowing the function $w(\boldsymbol{r}, \boldsymbol{p})$, the average values of $\langle \hat{\boldsymbol{r}}^2 \rangle$ and $\langle \hat{\boldsymbol{p}}^2 \rangle$ will be found by the formulas

$$\langle \hat{\boldsymbol{r}}^2 \rangle = \int \boldsymbol{r}^2\, w(\boldsymbol{r}, \boldsymbol{p})\, \mathrm{d}\boldsymbol{r}\, \mathrm{d}\boldsymbol{p}\,, \qquad \langle \hat{\boldsymbol{p}}^2 \rangle = \int \boldsymbol{p}^2\, w(\boldsymbol{r}, \boldsymbol{p})\, \mathrm{d}\boldsymbol{r}\, \mathrm{d}\boldsymbol{p}\,.$$

These formulas lead to the same results.

We integrate the function (7.5.15.3) by pulses and find how the cores will be distributed over the volume of ball lightning:

$$w(\boldsymbol{r}) = N\, B \exp(\, -\alpha\, \boldsymbol{r}^2\,).$$

(7.5.15.4)

where

$$\alpha = 2\, \beta\, g\,.$$

The normalizing factor B must satisfy the equality:

$$B \int \exp(\, -\alpha\, \boldsymbol{r}^2\,)\, \mathrm{d}\boldsymbol{r} = 1.$$

(7.5.15.5)

Consider the integral

$$\int \exp(\, -\alpha\, \boldsymbol{r}^2\,)\, \mathrm{d}\boldsymbol{r} = 4\, \pi \int_0^R \exp(\, -\alpha\, r^2\,)\, r^2\, \mathrm{d}r\,.$$

After integration, we will have a function that will depend on R and α. So $B = B(R, \alpha)$.

7.6. THEORY OF TUNNEL TRANSITIONS

7.6.1. Introduction

Consider a simple crystal whose nodes are denoted by vectors $\boldsymbol{R}$. The nodes contain identical atoms, except for one. This atom differs from all other atoms in that the potential energy of an electron near this atom is lower than its energy near other atoms.

If all the atoms in the crystal are exactly the same, then the electron in such a crystal can be found near any atom with the same probability. If the potential energy of the electron near one atom is less than the energy of an electron near other atoms, then when moving through a crystal, the electron is more likely to be located near this atom. With this movement, the electron will make **tunnel transitions** from one potential well to another.

7.6.2. Lindblad Equation and diSsipative Damping Operator

Write down the Lindblad equation

$$\mathrm{i}\,\hbar\,\partial\hat{\varrho}/\partial t = [\,\hat{H}\,\hat{\varrho}\,] + \mathrm{i}\,\gamma/\hbar\,\{\,2\,\hat{a}\,\hat{\varrho}\,\hat{a}^{+} - [\,\hat{a}^{+}\hat{a}\,,\hat{\varrho}\,]_{+}\,\}\,\}\,,$$

(7.6.1)

where

$$\hat{H} = \hat{\boldsymbol{p}}^2/(2\,m) + \hat{U}(\hat{\boldsymbol{r}})$$

(7.6.2)

is the particle Hamiltonian; $\hat{\boldsymbol{r}}$, $\hat{\boldsymbol{p}}$ and $\hat{U}$ are the operators of the coordinates, momentum and potential energy, respectively. The operator $\hat{U}(\hat{\boldsymbol{r}})$ for the electron moving through the crystal with a single atom at the node $\boldsymbol{R} = 0$ will have the form

$$\hat{U}(\hat{\boldsymbol{r}}) = \hat{U}_{0}(\hat{\boldsymbol{r}}) + \sum_{\boldsymbol{R}\neq 0} \hat{U}_1\,(\hat{\boldsymbol{r}} - \boldsymbol{R})\,,$$

(7.6.3)

where $\widehat{U}_0(\hat{\boldsymbol{r}})$ is the potential energy of the electron that is located near the atom at node $\boldsymbol{R} = 0$; $\widehat{U}_1(\hat{\boldsymbol{r}} - \boldsymbol{R})$ is the potential energy of the electron that moves near the atom at node $\boldsymbol{R} \neq 0$. The operator $\hat{\boldsymbol{a}}$ is a dissipative damping operator:

$$\hat{\boldsymbol{a}} = \hat{\boldsymbol{r}} + \mathrm{i}\, \hbar\, \beta\, \hat{\boldsymbol{p}}/(4\, m) \, . \tag{7.6.4}$$

The statistical operator $\hat{\varrho}$ describes the tunneling transitions of an electron from one potential well to another.

7.6.3. The Probability of Increasing the Crystal with Increasing Temperature

In paragraph **7.1** has been found for the density matrix $\varrho(\xi, \xi')$ of the harmonic oscillator and

the probability $w(\xi) = \varrho(\xi, \xi)$, that the distance between two atoms of the oscillator will increase by the value of ξ:

$$w(\xi) = \sqrt{\alpha/\pi}\, \exp\left(-\, \alpha\, \xi^2\right),$$

where

$$\alpha = m\, \omega/\hbar\, \mathrm{th}\left(\beta\, \hbar\, \omega/2\right).$$

The probability $w(\xi)$ was used in paragraph **1.3**.

REFERENCES

[1] B.V. Bondarev, "Lindblad equation for the harmonic oscillator ", *Uncertainty relation depending on temperature, Applied Mathematics*, vol. 8, pp. 1529-1538, 2017.
[2] G. Lindblad, "On the generators of guantum dynamical semigroups", *Commun. Math. Phys.*, vol. 48, pp. 119-130, 1976.
[3] B.V. Bondarev, "Statistical operator in the theory of quantum oscillator. Dissipative decay operator", *Scientific discussion*, vol. 1, no 17, pp. 49-51, 2018.
[4] B.V. Bondarev, "Quantum Markovian master equation for a system of identical particles interacting with a heat reservoir", *Physica A*, vol. 176, no 2, pp. 366-386, 1991.
[5] H.J. Round, "A note on carborundum", *Electrical World*, vol. 49, pp. 309, 1907.
[6] O.V. Losev, "Luminous carborundum detector and detection with crystals", *Telegraphy and Telephony without Wires*, vol. 5, no. 44, pp. 485-494, 1927.
[7] O.V. Losev, "Influence of temperature on luminous carborundum contact. On the application of the quantum theory equation to the detector glow phenomenon", *Telegraphy and Telephony without Wires,* vol. 2, no. 53, pp. 153-161, 1929.

[8] O.V. Losev, "On the application of quantum theory to the phenomena of detector luminescence", *Sat. Physics and Production Leningrad, LPI*, pp. 43-46, 1929.
[9] O.V. Losev, "Electrical conductivity of carborundum and unipolar conductivity of detectors", *Bulletin of Electrical Engineering,* vol. 8, pp. 247-255, 1931.
[10] B.V. Bondarev, "Derivation of the quantum kinetic equation from the equation c – von Neumann", *Theor. Math. Phys*., vol. 100, pp. 33-43, 1994.
[11] B.V. Bondarev, "Ball lightning. Density matrix method", *Scientific discussion*, vol. 1, no 38, pp. 37-44, 2019.
[12] B.V. Bondarev, "Equation for statistical operator in theory of ball lightning", *Scientific discussion*, vol. 1, no 39, pp. 10-16, 2020.

CHAPTER 8

The Beginning of Theoretical Nanophysics

8.1. EQUATION FOR DENSITY MATRIX SYSTEMS OF IDENTICAL PARTICLES

The equations for the statistical operator and the density matrix are considered here for a single particle and a system of identical particles when dissipative forces act on them. From the equation for the density matrix, a kinetic equation can be obtained when the density matrix is diagonal. These equations are the basis for the study of the simplest models of nanophysics [1].

8.1.1. Introduction

In quantum mechanics, the most general description of the system is the statistical operator $\hat{\varrho}$. The statistical operator must be normalized at any time

$$\mathrm{Tr}\,\hat{\varrho} = 1\,, \tag{8.1.1.1}$$

self-adjoint

$$\hat{\varrho}^* = \hat{\varrho} \tag{8.1.1.2}$$

and positively definite. A correct equation describing the evolution of a statistical operator must ensure that these properties are preserved over time.

For the first time the equation for the statistical operator

$$\hat{\varrho} = \hat{\varrho}\,(t, q), \tag{8.1.1.3}$$

where q is the quantum coordinate of the system, which was obtained by Lindblad [1]. This equation has the form

$$\mathrm{i}\,\hbar\,\dot{\hat{\varrho}} = [\,\hat{H}\,\hat{\varrho}\,] + \mathrm{i}\,\hbar\,\hat{D}, \tag{8.1.1.4}$$

where $\hat{H}$ is the Hamiltonian of the system,

$$\hat{D} = \sum{}_{jk}\; C_{jk}\{\,[\,\hat{a}_j\,\hat{\varrho}\,,\,\hat{a}_k^+] + [\,\hat{a}_j\,,\,\hat{\varrho}\,\hat{a}_k^+]\,\}, \tag{8.1.1.5}$$

Boris V. Bondarev

C_{jk} are some numbers, and $\hat{a}_j$ is an arbitrary operator. The operator $\widehat{D}$ is called the dissipative operator. This statement can be written as

$$\widehat{D} = \sum_{jk} C_{jk} \{ 2\, \hat{a}_j\, \hat{\varrho}\, \hat{a}_k^+ - \hat{a}_k^+\, \hat{a}_j\, \hat{\varrho} - \hat{\varrho}\, \hat{a}_k^+\, \hat{a}_j \}. \quad \textbf{(8.1.1.6)}$$

8.1.2. Equation for the Density Matrix of One Particle

The density matrix is related to the operator $\hat{\varrho}\,(t, q)$ formula

$$\varrho_{nn'}\,(t) = \int \varphi_n^*(t, q)\, \hat{\varrho}\,(t, q)\, \varphi_{n'}(t, q)\, dq. \quad \textbf{(8.1.2.1)}$$

This formula specifies the density matrix $\varrho_{nn'}\,(t)$ in the n-representation. Wave function $\varphi_n(t, q)$ can be found from the Schrödinger equation

$$\mathrm{i}\, \hbar\, \dot{\varphi}_n = \widehat{H}\, \varphi_n\,. \quad \textbf{(8.1.2.2)}$$

The equation for the density matrix was derived from the Liouville - von Neumann equation. This equation is analogous to the Lindblad equation and has the form

$$\mathrm{i}\, \hbar\, \dot{\varrho}_{nn'} = \sum_m (\, H_{nm}\, \varrho_{mn'} - \varrho_{nm}\, H_{mn'}\,) + \mathrm{i}\, \hbar\, \{ \sum_{mm'} \gamma_{nm,m'n'}\, \varrho_{mm'} - 1/2 \sum_m (\, \gamma_{nm}\, \varrho_{mn'} + \varrho_{nm}\, \gamma_{mn'}\,) \}\,, \quad \textbf{(8.1.2.3)}$$

where H_{nm} are the matrix elements of the Hamiltonian $\widehat{H}$ system, $\gamma_{nm,m'n'}$ is some matrix,

$$\gamma_{nn'} = \sum_m \gamma_{mn'nm}. \quad \textbf{(8.1.2.4)}$$

The equation (8.1.2.3) can be written as

$$\mathrm{i}\, \hbar\, \dot{\varrho}_{nn'} = \sum_m (\, H_{nm}\, \varrho_{mn'} - \varrho_{nm}\, H_{mn'}\,) + \mathrm{i}\, \hbar\, D_{nn'}, \quad \textbf{(8.1.2.5)}$$

where $D_{nn'}$ is a dissipative matrix, which will now be equal to

$$D_{nn'} = \sum_{mm'} \gamma_{nm,m'n'}\, \varrho_{mm'} - 1/2 \sum_m (\, \gamma_{nm}\, \varrho_{mn'} + \varrho_{nm}\, \gamma_{mn'}\,). \quad \textbf{(8.1.2.6)}$$

Compare this formula with the formula (8.1.1.6), we establish that

$$\gamma_{nm,m'n'} = 2 \sum_{jk} C_{jk}\, a_{nm,j}\, a^{+}_{m'n',k}, \quad \textbf{(8.1.2.7)}$$

where $a_{nm,j}$ are matrix elements of the operator $\hat{a}_j$.

The diagonal element $\varrho_{nn'}$ is the probability w_n that the system is in the state *n*. This value satisfies the normalization condition

$$\sum_n \varrho_{nn} = 1. \quad \textbf{(8.1.2.8)}$$

This formula is similar to formula (8.1.1.1).

In addition, the density matrix ϱ_{nm} satisfies the following condition

$$\varrho^{*}_{nm} = \varrho_{mn}\, . \quad \textbf{(8.1.2.9)}$$

The same condition is subject to Hamiltonian:

$$H^{*}_{nm} = H_{mn}\, . \quad \textbf{(8.1.2.10)}$$

Consider the system where the density matrix ϱ_{nm} at an arbitrary time *t* is in the diagonal state:

$$\varrho_{nm} = w_n\, \delta_{nm}\, , \quad \textbf{(8.1.2.11)}$$

where δ_{nm} is the Kronecker symbol. Then from equation (3.3), we obtain

$$\dot{w}_n = \sum_m (p_{nm}\, w_m - p_{mn}\, w_n), \quad \textbf{(8.1.2.12)}$$

where

$$p_{nm} = \gamma_{nm,mn} = 2\,\pi/\hbar \sum_{NM} |\, v_{nN,mM}\, |^2\, W_M\, \delta(\varepsilon_n - \varepsilon_m + E_N - E_M), \quad \textbf{(8.1.2.13)}$$

there is a probability of transition of the system in a unit of time from the state *m* to the state *n*,

$$W_N = \nu \exp\,(- \beta\, E_N)$$

there is a possibility that the environment is in an equilibrium state with quantum numbers *N*, and E_N is its energy in this state, ν is the normalization factor, $\beta = 1/(k_B\, T)$ is the inverse temperature; $v_{nN,mM}$ are the matrix elements of the system interaction with its environment. Formula (2.13) is the **Golden rule of Fermi**.

8.1.3. The Hierarchy for Statistical Operators

Consider a system consisting of N identical particles. The statistical operator $\hat{\varrho}$ describing the state of such a system can be written as:

$$\hat{\varrho} = \hat{\varrho}\,(\,t, q_1, q_2, \ldots, q_N\,) \equiv \hat{\varrho}\,(\,1, 2, \ldots, N\,)\,. \tag{8.1.3.1}$$

Here, the numbers in parentheses indicate the indexes of the variables that are affected by this operator. The statistical operator, as well as any other operator describing the state of the system of identical particles, must be symmetric:

$$\hat{\varrho}\,(\,\ldots, i, \ldots, j, \ldots\,) = \hat{\varrho}\,(\,\ldots, j, \ldots, i, \ldots\,)\,. \tag{8.1.3.2}$$

We take the following normalization condition for the statistical operator:

$$\mathrm{Tr}_{1\ldots N}\,\hat{\varrho}\,(\,1, 2, \ldots, N) = N!\,. \tag{8.1.3.3}$$

The Hamilton operator for the system of identical particles can be represented as a sum

$$\hat{H}(\,1, 2, \ldots, N) = \Sigma_{i=1}^{N}\,\hat{H}(i) + 1/2\,\Sigma_{i=1, j=1, j\neq i}^{N}\,\hat{H}(i,j), \tag{8.1.3.4}$$

where $\hat{H}(i)$ is a single-particle Hamiltonian, *i.e.*, the energy operator of one particle without taking into account its interaction with other particles; $\hat{H}(i, j)$ is the interaction operator of two particles. A single-particle Hamiltonian may contain dissipative terms. The two-particle Hamiltonian must be symmetric:

$$\hat{H}(i,j) = \hat{H}(j, i)\,. \tag{8.1.3.5}$$

Under this condition, the Hamiltonian (8.1.3.4) will also be symmetric.

Let's temporarily omit the summation operation in the expression (8.1.1.5) and write it as follows:

$$\hat{D} = [\,\hat{a}\,\hat{\varrho}, \hat{a}^{+}] + [\,\hat{a}, \hat{\varrho}\,\hat{a}^{+}]\,. \tag{8.1.3.6}$$

Suppose that the dissipative processes occurring in the system are due to the stochastic interaction of particles with the heat reservoir. And each particle interacts with it independently of other particles. In this case, the multi-particle operator a in expression (3.6) can be represented as a sum:

$$\hat{a} = \hat{a}(1, \dots, N) = \sum_{i=1}^{N} \hat{a}(i), \quad \textbf{(8.1.3.7)}$$

where operator $\hat{a}(i)$ characterizes the effect of the heat reservoir on one of the particles. Substituting this sum into expression (8.1.3.6) gives

$$\widehat{D} = \sum_{i=1}^{N} \sum_{j=1}^{N} \{ [\, \hat{a}(i)\, \hat{\varrho}, \hat{a}^{+}(j) \,] + [\, \hat{a}(i), \hat{\varrho}\, \hat{a}^{+}(j) \,] \} \,. \quad \textbf{(8.1.3.8)}$$

Of practical interest, it is the equation for the single-particle statistical operator $\hat{\varrho}(1)$, which is determined by the ratio

$$\hat{\varrho}(1) = 1/(N-1)!\, \mathrm{Tr}_{2\,\dots\,N}\, \hat{\varrho}(1, 2, \dots, N)\,. \quad \textbf{(8.1.3.9)}$$

From this definition and condition, (8.1.3.3) implies the normalization condition for

$$\mathrm{Tr}_1\, \hat{\varrho}(1) = N\,. \quad \textbf{(8.1.3.10)}$$

We define a two-particle statistical operator $\hat{\varrho}(1, 2)$ as

$$\hat{\varrho}(1, 2) = 1/(N-2)!\, \mathrm{Tr}_{3\,\dots\,N}\, \hat{\varrho}(1, 2, \dots, N)\,. \quad \textbf{(8.1.3.11)}$$

This operator satisfies the normalization condition

$$\mathrm{Tr}_{12}\, \hat{\varrho}(1, 2) = N(N-1)\,.$$

8.1.4. The Equation for Statistical Operators

We assume that equation (8.1.1.4) is true for operator (8.1.3.1). To obtain the equation for the derivative of the one-particle operator $\hat{\varrho}(1)$, we apply to both parts of equation (1.4) the convolution operation $\mathrm{Tr}_{2\,\dots\,N}$. Since

$$\mathrm{Tr}_i\, [\, \widehat{H}(i)\, \hat{\varrho}(1, 2, \dots, N) \,] \equiv 0$$

using definition (8.1.3.9) we obtain

$$\mathrm{Tr}_{2\,\dots\,N} \sum_{i=1}^{N} [\, \widehat{H}(i)\, \hat{\varrho}(1, 2, \dots, N) \,] = (N-1)!\, [\, \widehat{H}(1)\, \hat{\varrho}(1) \,]\,.$$

By virtue of identity

$$\mathrm{Tr}_{ij}\, [\, \widehat{H}(i,j)\, \hat{\varrho}(1, 2, \dots, N) \,] \equiv 0$$

taking into account property (8.1.3.5) and definition (8.1.3.11) we will have

$$\mathrm{Tr}_{2\,\dots\,N}\, 1/2 \sum_{i=1, j=1, j\neq i}^{N} [\,\widehat{H}(i,j)\,\hat{\varrho}(1,2,\dots,N)\,] =$$

$$= 1/2\, \mathrm{Tr}_{2\,\dots\,N} \{ \sum_{i=1, j=2, j\neq i}^{N} [\,\widehat{H}(1,j)\,\hat{\varrho}(1,2,\dots,N)\,] +$$

$$+ \sum_{i=2, j=2, j\neq i}^{N} [\,\widehat{H}(1,j)\,\hat{\varrho}(1,2,\dots,N)\,] \} =$$

$$= 1/2\,(N-1)!\, \mathrm{Tr}_2\, [\widehat{H}(1,2)\,\hat{\varrho}(1,2)\,] .$$

Likewise

$$\mathrm{Tr}_{2\,\dots\,N}\, \widehat{D} = \mathrm{Tr}_{2\,\dots\,N} \{ [\,\hat{a}(1)\,\hat{\varrho}, \hat{a}^{+}(1)\,] + [\,\hat{a}(1), \hat{\varrho}\,\hat{a}^{+}(1)\,] +$$

$$+ \sum_{i=2}^{N} ([\,\hat{a}(i)\,\hat{\varrho}, \hat{a}^{+}(1)\,] + [\,\hat{a}(i), \hat{\varrho}\,\hat{a}^{+}(1)\,] +$$

$$+ [\,\hat{a}(1)\,\hat{\varrho}, \hat{a}^{+}(i)\,] + [\,\hat{a}(1), \hat{\varrho}\,\hat{a}^{+}(i)\,]) +$$

$$+ \sum_{i=2}^{N} \sum_{j=2}^{N} ([\,\hat{a}(i)\,\hat{\varrho}, \hat{a}^{+}(j)\,] + [\,\hat{a}(i), \hat{\varrho}\,\hat{a}^{+}(j)\,]) \} =$$

$$= (N-1)!\, \{ [\,\hat{a}(1)\,\hat{\varrho}(1), \hat{a}^{+}(1)\,] + [\,\hat{a}(1), \hat{\varrho}(1)\,\hat{a}^{+}(1)\,] +$$

$$+ [\,\mathrm{Tr}_2\, \hat{a}(2)\,\hat{\varrho}(1,2), \hat{a}^{+}(1)\,] + [\,\hat{a}(1), \mathrm{Tr}_2\, \hat{\varrho}(1,2)\,\hat{a}^{+}(2)\,] \} ,$$

Since

$$\mathrm{Tr}_i\, [\,\hat{a}(i), \hat{\varrho}\,\hat{a}^{+}(j)\,] \equiv 0\, , \mathrm{Tr}_j\, [\hat{a}(i)\,\hat{\varrho}, \hat{a}^{+}(j)\,] \equiv 0 .$$

Having collected together the expressions, we come to the equation

$$\mathrm{i}\,\hbar\, \dot{\hat{\varrho}}(1) = [\,\widehat{H}(1)\,\hat{\varrho}(1)] + 1/2\, \mathrm{Tr}_2\, [\widehat{H}(1,2)\,\hat{\varrho}(1,2)\,] +$$

$$+ \mathrm{i}\,\hbar\, \{ [\,\hat{a}(1)\,\hat{\varrho}(1), \hat{a}^{+}(1)\,] + [\,\hat{a}(1), \hat{\varrho}(1)\,\hat{a}^{+}(1)\,] +$$

$$+ [\,\mathrm{Tr}_2\, \hat{a}(2)\,\hat{\varrho}(1,2), \hat{a}^{+}(1)\,] + [\,\hat{a}(1), \mathrm{Tr}_2\, \hat{\varrho}(1,2)\,\hat{a}^{+}(2)\,] \} .$$

(8.1.4.1)

8.1.5. The Equation for Density Matrix

We denote the matrix elements of operators $\hat{\varrho}(1)$, $\hat{\varrho}(1, 2)$, $\hat{H}(1)$, $\hat{H}(1, 2)$ and $\hat{a}(1)$ in a certain n-representation as

$$\varrho_{n_1 n_1'} = \varrho_{11'} , \qquad \varrho_{n_1 n_2, n_1' n_2'} = \varrho_{12,1'2'} ,$$

$$H_{n_1 n_1'} = H_{11'} , \qquad H_{n_1 n_2, n_1' n_2'} = H_{12,1'2'} ,$$

$$a_{n_1 n_1'} = a_{11'} ,$$

and write equation (4.1) in matrix form:

$$\mathrm{i}\,\hbar\,\dot{\varrho}_{11'} = \sum\nolimits_{n_2} [\, H_{12}\, \varrho_{21'}\,] + 1/2 \sum\nolimits_{n_2 n_3 n_4} [\, H_{12,34}\, \varrho_{43,21'}] +$$
$$+ \mathrm{i}\,\hbar\, \{ \sum\nolimits_{n_2 n_3} ([\, a_{12}\, \varrho_{23} \, , a^{+}_{31'}] + [\, a_{12} \, , \varrho_{23}\, a^{+}_{31'}]) +$$
$$+ \sum\nolimits_{n_2 n_3 n_4} ([\, a_{34}\, \varrho_{43,12} \, , a^{+}_{21'}] + [\, a_{12} \, , \varrho_{21',34}\, a^{+}_{43}]) \} .$$

(8.1.5.1)

The two-particle density matrix $\varrho_{12,1'2'}$, describing the state of the boson system must be symmetric. Whereas the two-particle density matrix of the fermion system must be antisymmetric. Approximately this matrix can be expressed through a single-particle density matrix as follows:

$$\varrho_{12,1'2'} \cong \varrho_{11'}\, \varrho_{22'} \pm \varrho_{12'}\, \varrho_{21'} , \qquad \textbf{(8.1.5.2)}$$

where the plus sign corresponds to bosons and the minus sign corresponds to fermions. Swap the first two variables and the second two in this formula. Get

$$\varrho_{21,2'1'} \cong \varrho_{22'}\, \varrho_{11'} \pm \varrho_{21'}\, \varrho_{12'} . \qquad \textbf{(8.1.5.3)}$$

It is seen that these two-particle density matrices are symmetric:

$$\varrho_{12,1'2'} = \varrho_{21,2'1'} . \qquad \textbf{(8.1.5.4)}$$

Substitute expression (8.1.5.2) in equation (5.1). We come to the equation for a single-particle density matrix:

$$
\begin{gathered}
\mathrm{i}\,\hbar\,\dot{\varrho}_{11'} = \sum\nolimits_{n_2} [H_{12}\,\varrho_{21'}] + \\
+ 1/2 \sum\nolimits_{n_2 n_3 n_4} [H_{12,34}\,(\varrho_{42}\,\varrho_{31'} \pm \varrho_{41'}\,\varrho_{32})] + \\
+ \mathrm{i}\,\hbar\, \{ \sum\nolimits_{n_2 n_3} ([a_{12}\,\varrho_{23}\,, a^{+}_{31'}] + [a_{12}\,, \varrho_{23}\,a^{+}_{31'}]) + \\
+ \sum\nolimits_{n_2 n_3 n_4} ([a_{34}\,(\varrho_{41}\,\varrho_{32} \pm \varrho_{42}\,\varrho_{31}), a^{+}_{21'}] + \\
+ [a_{12}\,, (\varrho_{23}\,\varrho_{1'4} \pm \varrho_{24}\,\varrho_{1'3})\,a^{+}_{43}])\} .
\end{gathered}
\tag{8.1.5.5}
$$

Two-part matrix $H_{12,1'2'}$ the fermions are anti-symmetric, *i.e.*, it satisfies the ratio of

$$
H_{12,1'2'} = - H_{21,1'2'} = - H_{12,2'1'} = H_{21,2'1'} . \tag{8.1.5.6}
$$

For bosons, this matrix is symmetric:

$$
H_{12,1'2'} = H_{21,1'2'} = H_{12,2'1'} = H_{21,2'1'} . \tag{8.1.5.7}
$$

We transform the summand with a two-particle Hamiltonian:

$$
\begin{gathered}
\sum\nolimits_{n_2 n_3 n_4} ([H_{12,34}\,(\varrho_{42}\,\varrho_{31'} \pm \varrho_{41'}\,\varrho_{32})] = \\
= \sum\nolimits_{n_2 n_3 n_4} ([H_{12,34}\,, \varrho_{42}\,\varrho_{31'}] \pm [H_{12,34}\,, \varrho_{41'}\,\varrho_{32}]) .
\end{gathered}
$$

Swap the indexes n_3 and n_4 in the second part of this equation. We will have

$$
\begin{gathered}
\sum\nolimits_{n_2 n_3 n_4} ([H_{12,34}\,(\varrho_{42}\,\varrho_{31'} \pm \varrho_{41'}\,\varrho_{32})] = \\
= \sum\nolimits_{n_2 n_3 n_4} ([H_{12,34}\,, \varrho_{42}\,\varrho_{31'}] \pm [H_{12,43}\,, \varrho_{31'}\,\varrho_{42}]) = \\
= \sum\nolimits_{n_2 n_3 n_4} [(H_{12,34} \pm H_{12,43}), \varrho_{42}\,\varrho_{31'}] = \\
= 2 \sum\nolimits_{n_2 n_3 n_4} [H_{12,34}\,, \varrho_{42}\,\varrho_{31'}] .
\end{gathered}
$$

This equality is true by virtue of formulas (8.1.5.6) and (8.1.5.7).

Enter the symbol

$$\gamma_{12,31'} = 2\, a_{12}\, a^{+}_{31'} \,, \qquad \textbf{(8.1.5.8)}$$

Write equation (8.1.5.5) with this notation:

$$\begin{aligned} \mathrm{i}\,\hbar\,\dot{\varrho}_{11'} &= \textstyle\sum_{n_2} [\, H_{12}\, \varrho_{21'}\,] + \sum_{n_2 n_3 n_4} [\, H_{12,34}\,,\, \varrho_{42}\, \varrho_{31'}] + \\ &+ \mathrm{i}\,\hbar/2 \textstyle\sum_{n_2 n_3} (2\,\gamma_{12,31'}\, \varrho_{23} - \gamma_{23,12}\, \varrho_{31'} - \gamma_{31',23}\, \varrho_{12}\,) + \\ &+ \textstyle\sum_{n_2 n_3 n_4} \{ (\, \varrho_{14}\, \varrho_{23} \pm \varrho_{24}\, \varrho_{13})\, \gamma_{34,21'} - \\ &- \gamma_{34,12}\, (\, \varrho_{24}\, \varrho_{1'3} \pm \varrho_{1'4}\, \varrho_{23}) + \gamma_{12,43}\, (\, \varrho_{32}\, \varrho_{41'} \pm \varrho_{42}\, \varrho_{31'}) - \\ &- (\, \varrho_{31}\, \varrho_{42} \pm \varrho_{41}\, \varrho_{32})\, \gamma_{21',43}\, \} \,. \end{aligned}$$

(8.1.5.9)

This is the desired equation for the density matrix.

8.1.6. Kinetic Equation

Substitute diagonal matrix (8.1.2.11) in equation (8.1.5.9):

$$\varrho_{12} = w_1\, \delta_{12} \,. \qquad \textbf{(8.1.6.1)}$$

We will have, after we put $n_2 = n_1$, the following equation

$$\begin{aligned} \partial w_1/\partial t &= \textstyle\sum_{n_2} (\, \gamma_{12,21}\, w_2 - \gamma_{21,12}\, w_1) + \\ &+ 1/2 \textstyle\sum_{n_2} w_1\, w_2\, \{\, \gamma_{21,21} \pm \gamma_{12,21} - (\,\gamma_{12,12} \pm \gamma_{21,12}) + \\ &+ \gamma_{12,12} \pm \gamma_{12,21} - (\,\gamma_{21,21} \pm \gamma_{21,12})\, \} \,. \end{aligned}$$

Reducing the terms containing elements of the matrix $\gamma_{12,12}$ and $\gamma_{21,21}$, we obtain the kinetic equation

$$\partial w_1/\partial t = \textstyle\sum_{n_2} \{\, \gamma_{12,21}\, w_2\, (\, 1 \pm w_1) - \gamma_{21,12}\, w_1\, (\, 1 \pm w_2)\, \} \,.$$

(8.1.6.2)

Substitution of formula (8.1.2.13)

$$p_{12} = \gamma_{12,21} \, , \tag{8.1.6.3}$$

results in equation (8.1.6.2) can be written as

$$\partial w_1/\partial t = \sum_{n_2} \{ \, p_{12} \, w_2 \, (\, 1 \pm w_1) - p_{21} \, w_1 \, (\, 1 \pm w_2) \, \} \, . \tag{8.1.6.4}$$

where p_{12} is the probability of transition of a particle per unit time from the state n_2 to the state n_1.

8.1.7. The Energy of the System of Identical Particles Unitary Matrix

We write a known expression for the average energy of the particle system

$$E = \sum_{n_1 n_2} H_{12} \, \varrho_{21} + 1/2 \sum_{\{n\}} H_{12,34} \, \varrho_{34,12} \, , \tag{8.1.7.1}$$

where $\{n\} = n_1, n_2, \; n_3, n_4$.

Let a κ -representation of the density matrix will be diagonal, *i.e.*, have the form

$$\tilde{\varrho}_{\kappa\kappa'} = w_\kappa \, \delta_{\kappa\kappa'}, \tag{8.1.7.2}$$

$$\tilde{\varrho}_{\kappa_1\kappa_2,\kappa_1'\kappa_2'} \equiv \tilde{\varrho}_{12,1'2'} = w_{12}^{(11)} \, (\, \delta_{11'} \, \delta_{22'} \pm \delta_{12'} \, \delta_{21'}) \, . \tag{8.1.7.3}$$

where probabilities w_κ and $w_{12}^{(11)}$ satisfy the conditions

$$\sum_\kappa w_\kappa = N \, , \sum_{\kappa_1\kappa_2 \neq \kappa_1} w_{12}^{(11)} = N \, (\, N - 1) \, . \tag{8.1.7.4}$$

The probability $w_{12}^{(11)}$ determines that both of the conditions of κ_1 and $\kappa_2 \neq \kappa_1$ occupied by particles.

The transition from the n -representation to the κ -representation is carried out using the unitary matrix $U_{n\kappa}$, which is determined by the ratio

$$\sum_n U_{n\kappa} \, U_{n\kappa'}^* = \delta_{\kappa\kappa'} \, . \tag{8.1.7.5}$$

If the unitary matrix is known, then the density matrix in the κ -representation is found by the formula

$$\tilde{\varrho}_{\kappa\kappa'} = \sum_{nn'} U_{n\kappa}\, \varrho_{nn'}\, U^*_{n'\kappa'} \,. \qquad \textbf{(8.1.7.6)}$$

Since expression (8.1.7.1) in any representation have the same form, substituting expressions (8.1.7.2) and (8.1.7.3) in this formula will have

$$E = \sum_{\kappa} \varepsilon_{\kappa}\, w_{\kappa} + 1/2 \sum_{\kappa_1\kappa_2 \neq \kappa_1} \varepsilon_{12}\, w^{(11)}_{12} \,, \qquad \textbf{(8.1.7.7)}$$

where

$$\varepsilon_{\kappa} = \sum_{nn'} U^*_{n\kappa}\, H_{nn'}\, U_{n'\kappa} \qquad \textbf{(8.1.7.8)}$$

is the energy of a particle in a state κ,

$$\varepsilon_{12} = 2 \sum_{\{n\}} U^*_{1\kappa_1}\, U^*_{2\kappa_2}\, H_{12,34}\, U_{3\kappa_1} U_{4\kappa_2} \qquad \textbf{(8.1.7.9)}$$

is the interaction energy of two particles, one of which is in the state κ_1, and the other in the state κ_2.

8.1.8. Variational principle

In thermodynamics, the variational principle was developed, which, along with quantum equation (5.7), leads to the same results. Now, systems of particles in the state of thermodynamic equilibrium are discussed. That's why now, briefly, we will discuss the variation principle.

The thermodynamic potential Ω of the particle system is related to the average internal energy E, entropy S, absolute temperature T, chemical potential μ and the average number of particles N

$$\Omega = E - S\,T - \mu\, N \,. \qquad \textbf{(8.1.8.1)}$$

The function $y = y(x)$ describing the equilibrium state of the system can be found by the Lagrange method. To do this, the thermodynamic potential should be expressed through this function $\Omega = \Omega(y)$, differentiated by this function and equate the derivative with zero:

$$\partial\Omega/\partial y = 0 \,. \qquad \textbf{(8.1.8.2)}$$

As a result, we obtain the equations for the equilibrium function.

With energy, almost everything is clear from the previous section, but it is not clear why we started discussing the unitary matrix. The fact is that entropy can be written only for the density matrix in diagonal form (8.1.7.2). For example, the entropy of the fermion system in the first approximation, when the binary probability

$$w_{12}^{(11)} = w_1 \, w_2 \qquad \textbf{(8.1.8.3)}$$

for $\kappa_1 \neq \kappa_2$, equal to

$$S = - k_B \sum_\kappa \{ (1 - w_\kappa) \ln (1 - w_\kappa) + w_\kappa \ln w_\kappa \} . \qquad \textbf{(8.1.8.4)}$$

In the first approximation, the energy of the fermions is

$$E = \sum_\kappa \varepsilon_\kappa \, w_\kappa + 1/2 \sum_{\kappa_1 \kappa_2 \neq \kappa_1} \varepsilon_{12} \; w_1 \, w_2 \qquad \textbf{(8.1.8.5)}$$

provided that $\varepsilon_{11} = 0$.

Now that we know the thermodynamic potential as a function of w_κ, we will find the derivative of it. As a result, we get

$$w_\kappa = 1/\{ 1 + e^{\beta (\overline{\varepsilon_\kappa} - \mu)} \} , \qquad \textbf{(8.1.8.6)}$$

where

$$\overline{\varepsilon_\kappa} = \varepsilon_\kappa + \sum_{\kappa'} \varepsilon_{\kappa\kappa'} \, w_{\kappa'} \qquad \textbf{(8.1.8.7)}$$

is the average energy of one fermion in the state κ.

8.1.9. Correlation Function

The variational principle allows us to find not only the probability w_κ, but also the binary probability $w_{12}^{(11)}$. To do this, we introduce the so-called correlation function $\xi_{\kappa\kappa'}$ using the ratio

$$w_{\kappa\kappa'}^{(11)} = w_\kappa \, w_{\kappa'} \, (1 - \delta_{\kappa\kappa'}) + \xi_{\kappa\kappa'} . \qquad \textbf{(8.1.9.1)}$$

8.1.10. The Probability of Transition of Particle

The probability of transition p_{12} particles from the state of κ_2 in the state κ_1, according to Fermi rule (2.13) must depend on the difference $\overline{\varepsilon_1} - \overline{\varepsilon_2}$ medium energy particles in these states. As it is known, the thermodynamic equilibrium of a particle system obeys the Boltzmann rule. Thus, the probability of transition of a particle in equilibrium can always be represented as

$$p_{12} = p^{(o)}_{12}\, e^{-\beta(\overline{\varepsilon_1} - \overline{\varepsilon_2})/2}, \qquad \textbf{(8.1.10.1)}$$

where

$$p^{(o)}_{12} = p^{(o)}_{21},$$

$\overline{\varepsilon_i}$ is the average energy of particles in a state of κ_i.

Let the system of particles come to a state of thermodynamic equilibrium. Since the probability w_κ is now time-independent, the right side of equation (8.1.6.4) is zero. Thus, the principle of detailed equilibrium, in this case, is expressed by equality

$$p_{12}\, w_2\,(1 \pm w_1) = p_{21}\, w_1\,(1 \pm w_2) \qquad \textbf{(8.1.10.2)}$$

Substituting expression (8.1.10.1) into this equation, after simple transformations, we obtain

$$(1 \pm w_1)/w_1\, e^{-\beta\overline{\varepsilon_1}} = (1 \pm w_2)/w_2\, e^{-\beta\overline{\varepsilon_2}}, \qquad \textbf{(8.1.10.3)}$$

where the left part depends on κ_1, and the right on κ_2. This is possible only if both parts of equality are equal to the same constant value. We denote this value as $e^{-\beta\mu}$, where μ is the chemical potential. Come to the equation

$$(1 \pm w)/w = e^{\beta(\overline{\varepsilon} - \mu)}, \qquad \textbf{(8.1.10.4)}$$

in which the average energy of the particle $\overline{\varepsilon}$ according to formula (8.1.8.7) is a function depending on the distribution function w. Considering the dependence of the energy $\overline{\varepsilon}$ on the probability w, it is possible to find the distribution of particles by States.

For equilibrium systems of non-interacting particles, equation (8.1.10.4) leads to Fermi – Dirac or Bose – Einstein functions:

$$w = 1/[1 \pm e^{\beta(\varepsilon - \mu)}], \qquad \textbf{(8.1.10.5)}$$

where the energy ε is the eigenvalue of a single-particle Hamiltonian.

8.1.11. Conclusion

Equation (5.9) here gives an opportunity to use it to describe the motion of identical particles in various problems of nanophysics. It should be noted that the presence of a dissipative term in this equation leads this equation to a kinetic equation for the probability.

8.2. NEW QUANTUM SPASER THEORY METHOD OF DENSITY MATRIX

In nanophysics, structures of very small sizes were predicted theoretically and created experimentally - so-called **spasers,** which are analogous to lasers. Earlier, the author of this article developed a laser theory based on the density matrix method. In this article, the new laser theory is applied to the description of the spaser operation.

8.2.1. Introduction

In the last decade, progress in creating structures of very small dimensions has led to the emergence of so-called **metamaterials**, which have actually been around for quite a long time. For example, observing the diverse beauty of butterfly wings, scientists have found that it is not in the paint of butterflies, but in the special structure of the wings. The interaction of light with filamentous formations in the wings of butterflies leads to the fact that only the color that is determined by the distribution of structures on the wings is reflected from the wings. Nature itself has taken care of creating special structures that can be called metamaterials.

This idea was experimentally established in ancient times, when people learned to add metal to glass and achieved the color of the glass. If you add metal to the glass, the glass has a different color for the lumen and reflection. Thus, people made wonderful stained-glass windows in churches and made beautiful dishes.

Later, scientists figured out what was going on with the metal by dissolving the metal in the glass. The metal does not dissolve to the molecular level, but to a particle size of about 50 nm. Visible light has wavelengths ranging from 380 nm to 780 nm. If visible light falls on glass with metal particles whose wavelength is approximately 10 times the size of the particles, the particles will be located almost in a uniform electric field of this wave. The electric field begins to move electrons in the metal particle at certain frequencies. There are resonant vibrations at certain frequencies and these resonant vibrations correspond to amazing colors. By changing the metal that is inserted into the glass, you can change the color of the glass. This is how people made stained glass windows. This phenomenon is called plasmon resonance, and quasiparticles, which are collective oscillations of free electrons, are called plasmons. We will denote resonant plasmons with the letter p (see Fig. **8.2.1**). The energy dissipation of plasmons is observed in various metals. As a rule, silver and gold are used, since these metals have low energy losses.

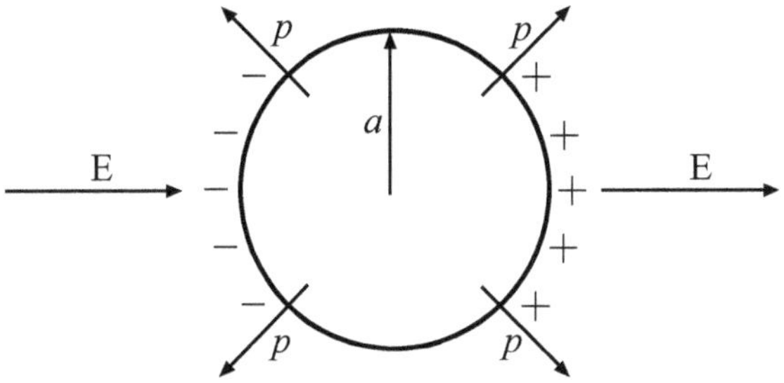

Fig. (8.2.1). The metal particle inside a glass where visible light is applied surface resonant plasmons are excited *p*.

Repeating the above, let's consider a metal ball of radius a inside the glass lattice, which is under the constant influence of an electromagnetic wave of visible light. The electric field penetrates inside the ball and causes electrons to shift inside it. The displaced electrons are affected by the Coulomb return force from the stationary ions of the crystal lattice. The magnitude of this force, together with the effective mass of the electrons, determines the oscillation frequency ω_p, which characterizes the surface plasmons. When ω_p frequency coincides with the frequency of the applied field, a resonance occurs in the system, which leads to the amplification of plasmons.

Now the question arises on how to create an active device that will accept plasmons so that they generate waves of visible light of a certain frequency. The possibility of creating such a device was theoretically predicted in 2003 by D. Bergman and M. Shtokman [2]. By analogy with the laser, the authors suggested calling such devices a spaser (SPASER = Surface Plasmon Amplification by Stimulated Emission of Radiation). The successful experimental implementation of the spaser was first reported in the work of M. Noginov and co-authors [3].

Such a spaser was created by surrounding a metal ball with a dielectric shell with **active atoms** embedded in it (see Fig. **8.2.2**). There are a lot of such devices in the glass plate. In the study [3], the spaser is a Golden ball with a radius of a = 7nm, surrounded by a dielectric shell with a thickness of h = 15nm.

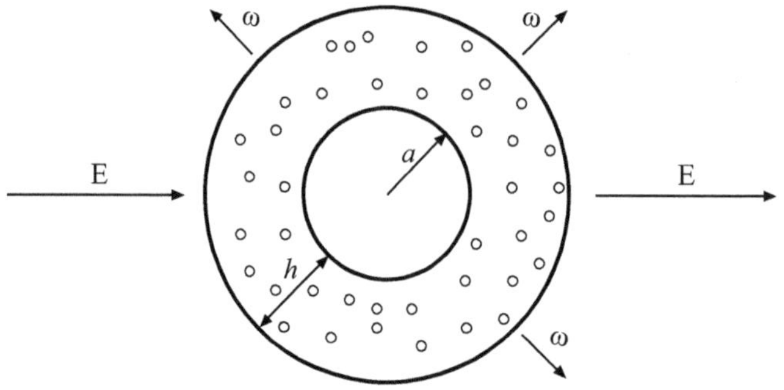

Fig. (8.2.2). A metal particle surrounded by a dielectric with active atoms. Coherent radiation is flying in all directions from the **spaser.**

The role of the visible light wave, which constantly falls on the glass, is to pump the active medium with plasmons. These plasmons excite active atoms, *i.e.*, they cause the atoms to move from the ground state φ_1 to the excited state φ_2. Then the excited atoms go back to the state φ_1 and they emit light with almost the same frequency ω. Fig. (**8.2.3**) shows the four transitions between two energy levels that the active atom makes.

8.2.2. Kinetic Equation

We will consider the spaser theory from the point of view of the semi-classical theory, *i.e.*, we will consider the electric field as a classical field, and for the active medium, we will apply the quantum theory.

In 1964, N. G. Basov, A. M. Prokhorov, and C. H. Townes won the Nobel prize. They were presented this award for fundamental research in the field of quantum electronics. These studies led to the creation of masers and lasers [4, 5].

The most convenient method is the **two-level method** proposed by Basov and Prokhorov in 1955. This method corresponds to the diagram shown in Fig. (**8.2.3**). This figure shows the four transitions between the two energy levels ε_1 and ε_2 that the active atom makes.

Resonant plasmons, excited by visible light, fall on the atoms of the active substance. This process is characterized by the number Q, which is equal to the number of atoms that pass from the state $\varphi_1(\boldsymbol{r})$ to the state $\varphi_2(\boldsymbol{r})$ under the action of pumping, multiplied by the number N_1 of active atoms in the first state: $Q\,N_1$.

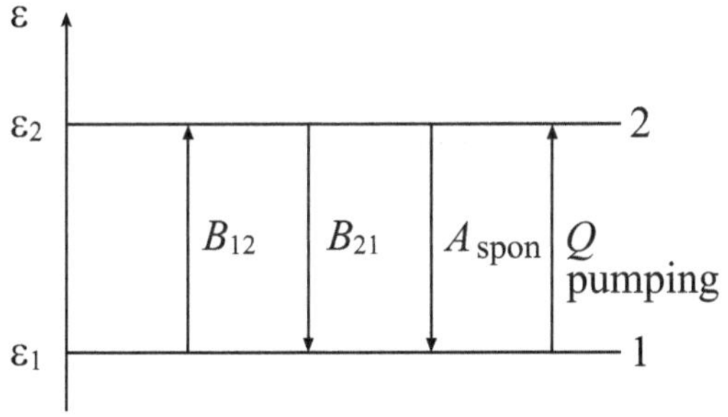

Fig. (8.2.3). Two-level diagram of the interaction of an atom with radiation.

The active transitions of the atom between the states φ_1 and φ_2.

Then the excited atoms make two transitions to the ground state $\varphi_1(\boldsymbol{r})$. From state 2, the atom can go to state 1 with the emission of a photon spontaneously. This transition is characterized by the parameter A, which is the number of atoms making such a transition per unit of time multiplied by the number N_2 excited atoms: $A\ N_2$. But the main thing in the next transition is that it is performed under the action of another photon of frequency ω. After the collision of this photon with an excited atom, it goes into the ground state with the emission of another photon of the same frequency. This transition is called **induced**. The number of atoms that make induced transitions per unit of time will be equal to $B_{21}\ W(\omega)\ N_2$, where B_{21} is a coefficient, the value of $W(\omega)$ is called the spectral density of the radiation energy. What is the main thing here? The most important thing is that after the induced transition from the atom, two **coherent photons** are already flying, *i.e.*, these photons are exactly similar to each other.

Finally, the transition of the active atom from state 1 to state 2 can occur with the absorption of a photon of the frequency ω. The number of such transitions per unit of time will be equal to $B_{12}\ W(\omega)\ N_1$. The frequency ω of the photon is approximately equal to

$$\omega = (\varepsilon_2 - \varepsilon_1)/\hbar\,. \qquad \textbf{(8.2.2.1)}$$

Such transitions occur sequentially with a number of active atoms. Light does not immediately fly out of the dielectric with a length of h, but returns many times from the boundaries of this layer.

The dielectric here plays the role of a **resonator**. So it flies out of it and is observed in the form of a sufficiently strong **coherent radiation**.

The kinetics of these transitions can be described by a phenomenological equation

$$\mathrm{d}N_1/\mathrm{d}t = -[Q + B_{12} W(\omega)] N_1 + [A + B_{21} W(\omega)] N_2 , \quad \textbf{(8.2.2.2)}$$

where

$$N_1 + N_2 = N, \quad \textbf{(8.2.2.3)}$$

$N_1 = N_1(t)$ и $N_2 = N_2(t)$ is the number of atoms in states 1 and 2 at time t, N is the total number of active atoms. It can be shown that

$$B_{12} = B_{21} = B .$$

All these formulas are valid for spaser operation. Only now the active medium is pumped when surface plasmons are absorbed.

As a result of sufficiently intensive pumping Q, saturation is achieved, at which the number N_2 electrons in the state $\varphi_2(\boldsymbol{r})$ becomes equal to the number N_1 electrons in the ground state $\varphi_1(\boldsymbol{r})$ (r). In this case, for a pair of levels ε_1 and ε_2, there is an inversion of populations ($N_2 > N_1$). The results of the work developed in the articles [6, 7] will be described below.

8.2.3. Lindblad Equation

In a study [8], Lindblad phenomenologically wrote an equation for a statistical operator $\hat{\varrho}$:

$$\mathrm{i}\hbar\, \partial\hat{\varrho}/\partial t = [\hat{H}\, \hat{\varrho}\,] + \mathrm{i}\hbar \hat{D} , \quad \textbf{(8.2.3.1)}$$

where $\hat{H}$ is a Hamiltonian of some nonequilibrium system S, the operator $\hat{D}$ can be called a **dissipative operator**. This operator is equal to

$$\hat{D} = \sum_{jk} C_{jk} \{ 2\, \hat{a}_j\, \hat{\varrho}\, \hat{a}_k^{+} - \hat{a}_k^{+} \hat{a}_j\, \hat{\varrho} - \hat{\varrho}\, \hat{a}_k^{+} \hat{a}_j \} , \quad \textbf{(8.2.3.2)}$$

C_{jk} are some numbers, and $\hat{a}_j$ is an arbitrary operator that still needs to be found.

8.2.4. Equation for Density Matrix

In the quantum theory of nonequilibrium processes, the most effective tool is the density matrices $\varrho_{nn'}$, where n is a quantum variable of some system. By definition of the density matrix its diagonal element ϱ_{nn} is equal to the probability w_n of finding the system in the state n:

$$\varrho_{nn} = w_n \,.$$

In the studies [9-11], the equation for the density matrix was derived from the Liouville – von Neumann equation. We considered a small non-equilibrium system S that was in thermal contact with a large equilibrium reservoir R, for a composite system $S + R$, the Liouville – von Neumann equation was considered valid. The equation for the density matrix $\varrho_{nn'}$ of the system S was derived from this equation

$$\mathrm{i}\,\hbar\; \partial\varrho_{nn'}/\partial t = \sum_m \left(H_{nm}\,\varrho_{mn'} - \varrho_{nm}\,H_{mn'}\right) + \mathrm{i}\,\hbar\, D_{nn'}, \qquad \textbf{(8.2.4.1)}$$

where H_{nm} are the matrix elements of the Hamiltonian $\widehat{H}$, $D_{nn'}$_is a dissipative matrix that is equal to

$$D_{nn'} = \sum_{m\,m'} \gamma_{nm,m'n'}\,\varrho_{mm'} - 1/2 \sum_m \left(\gamma_{nm}\,\varrho_{mn'} + \varrho_{nm}\,\gamma_{mn'}\right), \qquad \textbf{(8.2.4.2)}$$

$\gamma_{nm,m'n'}$ – some matrix,

$$\gamma_{nn'} = \sum_m \gamma_{mn',nm} \,. \qquad \textbf{(8.2.4.3)}$$

The matrix $D_{nn'}$ is associated with the operator $\widehat{D}$ by the formula

$$\gamma_{nm,m'n'} = 2\sum_{jk} C_{jk}\, a_{nm,j}\, a^{+}_{m'n',k} \,, \qquad \textbf{(8.2.4.4)}$$

where $a_{nm,j}$ are matrix elements of the operator $\hat{a}_j$. We assume that

$$\sum_n \varrho_{nn} = 1 \,. \qquad \textbf{(8.2.4.5)}$$

8.2.5. Derivation of Quantum Kinetic Equation from Equation for the Density Matrix

Let the density matrix ϱ_{nm} have a diagonal form in some representation

$$\varrho_{nm} = w_n\, \delta_{nm} \,, \qquad \textbf{(8.2.5.1)}$$

where δ_{nm} is the Kronecker symbol. Then the quantum kinetic equation follows from equation (4.1)

$$\partial w_n/\partial t = \sum_m \left(p_{nm}\, w_m - p_{mn}\, w_n\right), \qquad \textbf{(8.2.5.2)}$$

where

$$p_{nm} = \gamma_{nm,mn} = (2\,\pi/\hbar)\sum_N \sum_M |\, v_{nN,mM}\,|^2 W_M \delta(\,\varepsilon_n - \varepsilon_m + E_N - E_M\,) \quad \textbf{(8.2.5.3)}$$

there is a probability that the system *S* will move from the state *m* to the state *n* in a unit of time, $v_{nN,mM}$ – matrix elements of interaction between the system *S* and its environment,

$$W_N = \nu \exp(\,-\,\beta\, E_N)$$

there is a probability that the environment surrounding the system *S* with *R* is in an equilibrium state with quantum numbers *N*, and E_N is its energy in this state, ν is the normalization multiplier,

$$\beta = 1/(k_{\mathrm{B}}\, T)$$

– return temperature. Formula (8.2.5.2) is the **Golden rule of Fermi**.

Using the Fermi formula and the Boltzmann principle, we find the probability of transition

$$p_{nm} = p_{nm}^{(\mathrm{o})}\, \mathrm{e}^{-\,\beta\,(\,\varepsilon_n - \varepsilon_m\,)/2}, \quad \textbf{(8.2.5.4)}$$

where $p_{nm}^{(\mathrm{o})} = p_{mn}^{(\mathrm{o})}$.

8.2.6. Hamiltonian of an Atom with Two Energy Levels

The matrix elements $H_{\alpha\beta}$ of the active atom in the coordinate representation may look like this:

$$H_{\alpha\beta} = \varepsilon_\alpha^{(\mathrm{o})}\, \delta_{\alpha\beta} + V_{\alpha\beta}\,, \quad \textbf{(8.2.6.1)}$$

where $\alpha, \beta = 1, 2$; $\varepsilon_1^{(\mathrm{o})}$ and $\varepsilon_2^{(\mathrm{o})}$ – the energy at the first and second levels, $V_{\alpha\beta}$ – matrix elements of the operator of the dipole interaction of a stationary atom and radiation.

If the intensity of the electric field that acts on the atom is equal to

$$\boldsymbol{E} = \boldsymbol{E}_{\mathrm{o}} \cos \omega\, t\,, \quad \textbf{(8.2.6.2)}$$

where $\boldsymbol{E}_{\text{o}}$ and ω are the amplitude and frequency of the light incident on the atom, then the matrix elements will be equal

$$V_{11} = V_{22} = 0\,,\, V_{12} = V_{21}^{*} = -\,\boldsymbol{d} \cdot \boldsymbol{E}_{\text{o}} \cos \omega\, t = -\,\hbar\, \Omega \cos \omega\, t\,, \tag{8.2.6.3}$$

$\boldsymbol{d}$ – the electric dipole moment of an atom is a vector of the electric field intensity that acts on the atom from the radiation side. Frequency

$$\omega_{\text{o}} = (\,\varepsilon_{2}^{(\text{o})} - \varepsilon_{1}^{(\text{o})}\,)/\hbar \tag{8.2.6.4}$$

there is a frequency of transition between levels when there is no electric field. Frequency

$$\Omega = \boldsymbol{d} \cdot \boldsymbol{E}_{\text{o}}/\hbar\,, \tag{8.2.6.5}$$

called the Rabi frequency.

The matrix elements of the Hamiltonian $\widehat{H}$ are equal

$$H_{11} = \varepsilon_{1}^{(\text{o})}, H_{12} = -\,\hbar\, \Omega\, \cos \omega\, t,\; H_{21} = -\,\hbar\, \Omega\, \cos \omega\, t\;, H_{22} = \varepsilon_{2}^{(\text{o})}; \tag{8.2.6.6}$$

or the matrix $H_{\alpha\beta}$ can be written as follows

$$H_{\alpha\beta} = \begin{Vmatrix} \varepsilon_{1}^{(\text{o})} & -\,\hbar\, \Omega\, \cos \omega\, t \\ -\,\hbar\, \Omega\, \cos \omega\, t & \varepsilon_{2}^{(\text{o})} \end{Vmatrix}.$$

8.3.7. Eigenvalue of Atom's Energy. Unitary Matrix

The eigenvalue of the energy of an atom can simply be found if we know the Hamiltonian $\widetilde{H}_{\kappa\kappa'}$, which is in the diagonal form:

$$\widetilde{H}_{\kappa\kappa'} = \varepsilon_{\kappa}\, \delta_{\kappa\kappa'}, \tag{8.2.7.1}$$

where ε_{κ} is the desired eigenvalue of the energy of the atom.

To go from the Hamiltonian $H_{\alpha\beta}$ to the diagonal Hamiltonian $\widetilde{H}_{\kappa\kappa'}$, you need to find a unitary matrix $U_{\alpha\kappa}$ that satisfies the condition

$$\sum_{\alpha} U_{\alpha\kappa}\, U^{*}_{\alpha\kappa'} = \delta_{\kappa\kappa'} \,. \qquad \textbf{(8.2.7.2)}$$

This transition from the α-representation to the κ-representation is performed using the formula

$$\widetilde{H}_{\kappa\kappa'} = \sum_{\alpha\beta} U_{\alpha\kappa}\; H_{\alpha\beta}\, \boldsymbol{U}^{*}_{\beta\kappa'}. \qquad \textbf{(8.2.7.3)}$$

It is not difficult to check that the unitary matrix will have the form

$$U_{\alpha\kappa} = \left\| \begin{matrix} \cos\vartheta & -\sin\vartheta \\ \sin\vartheta & \cos\vartheta \end{matrix} \right\| , \qquad \textbf{(8.2.7.4)}$$

where the angle ϑ will satisfy the equation

$$\omega_{\text{o}} \cos\vartheta \sin + \Omega\, (\cos^{2}\vartheta - \sin^{2}\vartheta)\, \cos\omega\, t = 0\,. \qquad \textbf{(8.2.7.5)}$$

Frequencies (8.2.6.4) and (8.2.6.5) are such that

$$\Omega/\omega_{\text{o}} \ll 1. \qquad \textbf{(8.2.7.6)}$$

Taking this inequality into account, we will have an approximate solution of equation (7.5)

$$\operatorname{tg}\vartheta \cong -\,\xi(t)\,, \qquad \textbf{(8.2.7.7)}$$

where

$$\xi(t) = \Omega \cos\omega\, t/\omega_{\text{o}}\,. \qquad \textbf{(8.2.7.8)}$$

In this case, the eigenvalues of the energy of the atom will be equal

$$\varepsilon_1 = \varepsilon_1^{(\text{o})} - \hbar\,\omega_{\text{o}}\,\xi^{2}(t)\,,\;\; \varepsilon_2 = \varepsilon_2^{(\text{o})} + \hbar\,\omega_{\text{o}}\,\xi^{2}(t)\,. \qquad \textbf{(8.2.7.9)}$$

The radiation that comes from an atom has frequencies

$$\widetilde{\omega} = (\varepsilon_2 - \varepsilon_1)\,/\,\hbar\,,$$

which is equal to

$$\widetilde{\omega}(t) = \omega_{\text{o}}\,\{\,1 + 2\,\xi^{2}(t)\,\}\,. \qquad \textbf{(8.2.7.10)}$$

8.2.8. Transition of Density Matrix from α-Representation to κ-Representation

We now find the relationship of the density matrix $\varrho_{\alpha\beta}$ to the density matrix $\tilde{\varrho}_{\kappa\kappa'}$ in the κ-representation, which is performed using the formula

$$\varrho_{\alpha\beta} = \sum_{\kappa\kappa'} U^*_{\alpha\kappa}\, \tilde{\varrho}_{\kappa\kappa'}\, U_{\beta\kappa'} . \quad \textbf{(8.2.8.1)}$$

According to this formula, we will have

$$\varrho_{11} = \tilde{\varrho}_{11} + (\tilde{\varrho}_{22} - \tilde{\varrho}_{11})\, \xi^2(t) + 2\,\mathrm{Re}\, \tilde{\varrho}_{12}\, \xi(t) ,$$

$$\varrho_{12} = \tilde{\varrho}_{12} + (\tilde{\varrho}_{22} - \tilde{\varrho}_{11})\, \xi(t) - 2\,\mathrm{Re}\, \tilde{\varrho}_{12}\, \xi^2(t) ,$$

$$\varrho_{21} = \tilde{\varrho}_{21} + (\tilde{\varrho}_{22} - \tilde{\varrho}_{11})\, \xi(t) - 2\,\mathrm{Re}\, \tilde{\varrho}_{12}\, \xi^2(\mathrm{t}) ,$$

$$\varrho_{22} = \tilde{\varrho}_{22} - (\tilde{\varrho}_{22} - \tilde{\varrho}_{11})\, \xi^2(t) - 2\,\mathrm{Re}\, \tilde{\varrho}_{12}\, \xi(t) .$$

(8.2.8.2)

Add the first and fourth equality:

$$\varrho_{11} + \varrho_{22} = \tilde{\varrho}_{11} + \tilde{\varrho}_{22} = 1, \quad \textbf{(8.2.8.3)}$$

where condition (8.2.4.5) for normalizing the density matrix was taken into account. Subtract the third from the second equality:

$$\varrho_{12} - \varrho_{21} = \tilde{\varrho}_{12} - \tilde{\varrho}_{21} . \quad \textbf{(8.2.8.4)}$$

8.2.9. Equations for Density Matrices in Two Representations α and κ

Now you can write two equations for the density matrix. The first equation can be written in the α-representation, and the second – in the κ-representation:

$$\mathrm{i}\,\hbar\, \partial \varrho_{\alpha\alpha'}/\partial t = \sum_\beta \left(H_{\alpha\beta}\, \varrho_{\beta\alpha'} - \varrho_{\alpha\beta}\, H_{\beta\alpha'}\right) + \mathrm{i}\,\hbar\, D_{\alpha\alpha'} , \quad \textbf{(8.2.9.1)}$$

$$\mathrm{i}\,\hbar\, \partial \tilde{\varrho}_{\kappa\kappa'}/\partial t = \sum_\eta \left(\widetilde{H}_{\kappa\eta}\, \tilde{\varrho}_{\eta\kappa'} - \tilde{\varrho}_{\kappa\eta}\, \widetilde{H}_{\eta\kappa'}\right) + \mathrm{i}\,\hbar\, \widetilde{D}_{\kappa\kappa'} . \quad \textbf{(8.2.9.2)}$$

Let's assume that the density matrix $\tilde{\varrho}_{\kappa\kappa'}$ has a diagonal form, *i.e.*

$$\tilde{\varrho}_{\kappa\kappa'} = \widetilde{w}_\kappa\, \delta_{\kappa\kappa'} . \quad \textbf{(8.2.9.3)}$$

Then equation (8.2.9.2) turns into a quantum kinetic equation. Transform equation (8.2.9.2) using formulas (8.2.7.1) and (9.3):

$$\mathrm{i}\,\hbar\,\delta_{\kappa\kappa'}\,\partial\widetilde{w}_{\kappa}/\partial t = (\,\varepsilon_{\kappa}\,\widetilde{w}_{\kappa'} \;-\widetilde{w}_{\kappa}\,\varepsilon_{\kappa'}\,)\,\delta_{\kappa\kappa'} \;+\mathrm{i}\,\hbar\,\widetilde{D}_{\kappa\kappa'}\,. \quad \textbf{(8.2.9.4)}$$

If $\kappa = \kappa'$, then this equation turns into a quantum kinetic equation

$$\partial\widetilde{w}_{\kappa}/\partial t = \widetilde{D}_{\kappa\kappa}\,. \quad \textbf{(8.2.9.5)}$$

But we already know this kinetic equation – this equation (8.2.2.2).

8.2.10. Dissipative Matrix

Find the dissipative matrix $D_{\kappa\kappa'}$, which is included in equation (9.1). Now we need to find out what the matrix $\gamma_{nm,m'n'}$ looks like. The easiest way is to use formula (8.2.4.4), which connects the matrix $D_{\kappa\kappa'}$ with the operator $\widehat{D}$. Only now this formula will look like this

$$\gamma_{nm,m'n'} \;= \textstyle\sum_j\, C_j\, a_{nm,j}\, a^{+}_{m'n',j}\,, \quad \textbf{(8.2.10.1)}$$

where $a_{nm,j}$ are matrix elements of the operator $\hat{a}_j$, C_j are constants. Here $\{\,n, m = \kappa, \kappa'\}$. When we find the matrix $\gamma_{nm,m'n'}$, we find the dissipative matrix using the formula

$$D_{nn'} \;= \textstyle\sum_{mm'}\gamma_{nm,m'n'}\;\tilde{\varrho}_{mm'} \;-\; 1/2\,\sum_m(\,\gamma_{nm}\,\tilde{\varrho}_{mn'} + \tilde{\varrho}_{nm}\,\gamma_{mn'}\,)\,. \quad \textbf{(8.2.10.2)}$$

Now we need to give a physical meaning to the operators $\hat{a}_j$.

In a two-level scheme, let the operator $\hat{a}_1$ be equal to

$$\hat{a}_1 = \begin{Vmatrix} 0 & a_{12} \\ a_{21} & 0 \end{Vmatrix}. \quad \textbf{(8.2.10.3)}$$

Substitute this operator in formula (8.2.10.2) and write the dissipative matrix $D^{(1)}_{\kappa\kappa'}$:

$$D^{(1)}_{11} = 1/2\,C_1 \textstyle\sum_{mm'}\;(\,2\,a_{1m}\,\tilde{\varrho}_{mm'}\,a^{+}_{m'1} - a^{+}_{1m}\,a_{mm'}\,\tilde{\varrho}_{m'1} - \tilde{\varrho}_{1m}\,a^{+}_{mm'}\,a_{m'1}) =$$

$$= 1/2\,C_1\,(\,2\,a_{12}\,\tilde{\varrho}_{22}\,a^{+}_{21} - a^{+}_{12}\,a_{21}\,\varrho_{11} - \tilde{\varrho}_{11}\,a^{+}_{12}\,a_{21}) = C_1\,(\,a^{2}_{12}\,\tilde{\varrho}_{22} - a^{2}_{21}\,\tilde{\varrho}_{11}),$$

$$D^{(1)}_{12} = 1/2\,C_1 \textstyle\sum_{mm'}\;(2\,a_{1m}\,\tilde{\varrho}_{mm'}\,a^{+}_{m'2} - a^{+}_{1m}\,a_{mm'}\,\tilde{\varrho}_{m'2} - \tilde{\varrho}_{1m}\,a^{+}_{mm'}\,a_{m'2}) =$$

$= 1/2\, C_1\, (\, 2\, a_{12}\, \tilde{\varrho}_{21}\, a_{12}^{+} - a_{12}^{+}\, a_{21}\, \tilde{\varrho}_{12} - \tilde{\varrho}_{12}\, a_{21}^{+}\, a_{12}) = 1/2\, C_1\, [\, 2\, a_{12}\, a_{21}\, \tilde{\varrho}_{21} - (\, a_{12}^{2} + a_{21}^{2}\,)\, \tilde{\varrho}_{12}\,]\,,$

where

$$a_{12}^{2} = p_{12}\,,\; a_{21}^{2} = p_{21}\,.$$

According to formula 8.2. (5.4), we will have

$$p_{12} = p_{12}^{(o)}\, e^{-\beta\,(\varepsilon_1 - \varepsilon_2)/2},\; p_{21} = p_{21}^{(o)}\, e^{-\beta\,(\varepsilon_2 - \varepsilon_1)/2},\; p_{12}^{(o)} = p_{21}^{(o)} = p\,.$$

Let the operator $\hat{a}_2$ have the form

$$\hat{a}_2 = \begin{Vmatrix} 0 & a_1 \\ a_2 & 0 \end{Vmatrix}. \qquad \textbf{(8.2.10.4)}$$

The dissipative matrix $D_{\kappa\kappa'}^{(2)}$ will now be

$$D_{11}^{(2)} = C_2\, (\, a_1^2\, \tilde{\varrho}_{22} - a_2^2\, \tilde{\varrho}_{11})\,, D_{12}^{(2)} = 1/2\, C_2\, [\, 2\, a_1\, a_2\, \tilde{\varrho}_{21} - (\, a_1^2 + a_2^2\,)\, \tilde{\varrho}_{12}]\,. \qquad \textbf{(8.2.10.5)}$$

Denote

$$a_1^2 = \gamma_1\,,\; a_2^2 = \gamma_2\,. \qquad \textbf{(8.2.10.6)}$$

If the operator $\hat{a}_3$ is such that

$$\hat{a}_3 = \begin{Vmatrix} 0 & a \\ a & 0 \end{Vmatrix}. \qquad \textbf{(8.2.10.7)}$$

The third dissipative matrix $D_{\kappa\kappa'}^{(3)}$ according to formula (8.2.10.5) will be

$$D_{11}^{(3)} = C_3\, a^2\, (\, \tilde{\varrho}_{22} - \tilde{\varrho}_{11})\,, D_{12}^{(3)} = C_3\, a^2\, (\, \tilde{\varrho}_{21} - \tilde{\varrho}_{12}\,)\,.$$

If the operator $\hat{a}_4$ is diagonal, *i.e.*, it has the form

$$a_{\alpha\beta} = b_\alpha\, \delta_{\alpha\beta}\,, \qquad \textbf{(8.2.10.8)}$$

then the dissipative matrix $D_{\kappa\kappa'}^{(4)}$ will be

$$D^{(4)}_{\kappa\kappa'} = 1/2\, C_4\, (\, 2\, b_\alpha\, b_\beta - b_\alpha^2 - b_\beta^2\,)\, \tilde{\varrho}_{\kappa\kappa'}\, .$$

In particular

$$D^{(4)}_{\kappa\kappa} = 0\, ,\, D^{(4)}_{12} = \; 1/2\, C_4\, (\, 2\, b_1\, b_2 - b_1^2 - b_2^2\,)\, \tilde{\varrho}_{12}\, ,$$

$$D^{(4)}_{21} = 1/2\, C_4\, (\, 2\, b_2\, b_1 - b_2^2 - b_1^2\,)\, \tilde{\varrho}_{21}$$

You can put

$$D^{(4)}_{12} = -\, \Gamma\, \tilde{\varrho}_{12}\, , \qquad D^{(4)}_{21} = -\, \Gamma\, \tilde{\varrho}_{21}\, .$$

where

$$\Gamma = 1/2\, C_4\, (\, b_1^2 + b_2^2 - 2\, b_1\, b_2\,)\, .$$

Putting all four variants of dissipative matrices $D^{(l)}_{\kappa\kappa'}$ together:

$$D_{\kappa\kappa'} = \textstyle\sum_{l=1}^{4} D^{(l)}_{\kappa\kappa'}\, .$$

Putting $C_1 = 1$, $C_2 = -\, 2$ and $C_3 = 1$, we get

$$D_{11} = (\, p_{12} + \gamma_1 + a^2\,)\, \tilde{\varrho}_{22} - (\, p_{21} + \gamma_2 + a^2\,)\, \tilde{\varrho}_{11}\, , \qquad \textbf{(8.2.10.9)}$$

$$D_{12} = 1/2\, [\, 2\sqrt{p_{12}\, p_{21}}\, \tilde{\varrho}_{21} - (\, p_{12} + p_{21}\,)\, \tilde{\varrho}_{12}] +$$

$$+ (\, \gamma_1 + \gamma_2\,)\, \tilde{\varrho}_{12} - 2\sqrt{\gamma_1\gamma_2}\, \tilde{\varrho}_{21} + a^2\, (\, \tilde{\varrho}_{21} - \tilde{\varrho}_{12}) - \Gamma\, \tilde{\varrho}_{12}\, ,$$

(8.2.10.10)

where the term is $-\, \Gamma\, \tilde{\varrho}_{12}$, responsible for the relaxation of non-diagonal elements of the density matrix. Thus, a dissipative matrix is found that describes the thermodynamic transitions of atoms between levels 1 and 2. Now, we still need to find the physical meaning of the values γ_1, γ_2 and a. To do this, write equation (8.2.2.2) using the probabilities $\widetilde{w}_1$ and $\widetilde{w}_2$. The numbers of atoms N_1 and N_2 are related to these probabilities by relations

$$N_1 = N\widetilde{w}_1\,, \qquad N_2 = N\widetilde{w}_2\,, \qquad \textbf{(8.2.10.11)}$$

Equation (8.2.2.2) will look like this:

$$\partial\widetilde{w}_1/\partial t = -[\,Q + B\,W(\omega)\,]\;\widetilde{w}_1 + [\,A + B\,W(\omega)\,]\,\widetilde{w}_2\,. \qquad \textbf{(8.2.10.12)}$$

Let's substitute a density matrix in this equation instead of probabilities

$$\widetilde{w}_\kappa = \tilde{\varrho}_{\kappa\kappa}\,,$$

we will have

$$\partial\tilde{\varrho}_{11}/\partial t = -[\,Q + B\,W(\omega)\,]\,\tilde{\varrho}_{11} + [\,A + B\,W(\omega)\,]\,\tilde{\varrho}_{22}\,. \qquad \textbf{(8.2.10.13)}$$

Now compare equation (8.2.10.9) with the right side of equation (8.2.10.13). This comparison gives

$$\gamma_1 = A\,, \gamma_2 = Q\,, \qquad a^2 = B\,W(\omega)\,. \qquad \textbf{(8.2.10.14)}$$

8.2.11. Density Matrix Equation of the First Order of Approximation

Equation (8.2.3.1) for the statistical operator $\hat{\varrho}$ was obtained in the first approximation by the Hamiltonian of the interaction between the system and the thermostat [9]. Therefore this equation can be written as follows:

$$\mathrm{i}\,\hbar\,\partial\hat{\varrho}/\partial t = [\,\widehat{H}\;\hat{\varrho}\,] + \lambda\,\mathrm{i}\,\hbar\,\widehat{D}(\hat{\varrho}) + \dots\,, \qquad \textbf{(8.2.11.1)}$$

here λ is an order parameter, $\widehat{D}(\hat{\varrho})$ is a dissipative operator. Let's write the statistical operator $\hat{\varrho}$ as follows

$$\hat{\varrho} = \hat{\varrho}^{(o)} + \lambda\,\hat{\varrho}^{(1)} + \dots\,, \qquad \textbf{(8.2.11.2)}$$

where $\hat{\varrho}^{(o)}$, $\hat{\varrho}^{(1)}$, ... there are approximations of the density matrix.

Substitute operator (8.2.11.2) in equation (8.2.11.1). We will have

$$\mathrm{i}\,\hbar\,(\,\partial\hat{\varrho}^{(o)}/\partial t + \lambda\;\partial\hat{\varrho}^{(1)}/\partial t + \dots\,) = [\,\widehat{H},\,\hat{\varrho}^{(o)} + \lambda\,\hat{\varrho}^{(1)} + \dots\,] + \lambda\,\mathrm{i}\,\hbar\,\widehat{D}(\,\hat{\varrho}^{(o)} + \lambda\,\hat{\varrho}^{(1)} + \dots\,)\,. \qquad \textbf{(8.2.11.3)}$$

If the order parameter $\lambda = 0$, we get an equation for the undisturbed statistical operator $\hat{\varrho}^{(o)}$:

$$\mathrm{i}\,\hbar\,\partial\hat{\varrho}^{(o)}/\partial t = [\,\hat{H}\;\hat{\varrho}^{(o)}]\,. \quad \textbf{(8.2.11.4)}$$

The first parameter of order $\lambda = 1$ gives

$$\mathrm{i}\,\hbar\,\partial\hat{\varrho}^{(1)}/\partial t = [\,\hat{H}\;\hat{\varrho}^{(1)}] + \mathrm{i}\,\hbar\,\hat{D}(\,\hat{\varrho}^{(o)})\,, \quad \textbf{(8.2.11.5)}$$

So, the оператор operator satisfies this system of equations.

8.2.12. Zero-Order Density Matrix in κ-Representation

The zero density matrix will match the equation

$$\mathrm{i}\,\hbar\,\partial\tilde{\varrho}^{(o)}_{\kappa\kappa'}/\partial t = \textstyle\sum_{\eta}\,(\,\widetilde{H}_{\kappa\eta}\,\tilde{\varrho}^{(o)}_{\eta\kappa'} - \tilde{\varrho}^{(o)}_{\kappa\eta}\,\widetilde{H}_{\eta\kappa'}\,)\,. \quad \textbf{(8.2.12.1)}$$

Substituting the Hamiltonian $\widetilde{H}_{\kappa\kappa'}$ in the κ-representation here, we get

$$\mathrm{i}\,\hbar\,\partial\tilde{\varrho}^{(o)}_{\kappa\kappa'}/\partial t = [\,\varepsilon_{\kappa}(t) - \varepsilon_{\kappa'}(t)\,]\,\tilde{\varrho}^{(o)}_{\kappa\kappa'}\,. \quad \textbf{(8.2.12.2)}$$

Or read more

$$\partial\tilde{\varrho}^{(o)}_{11}/\partial t = 0\,,\ \mathrm{i}\,\hbar\,\partial\tilde{\varrho}^{(o)}_{12}/\partial t = -\,[\,\varepsilon_2(t) - \varepsilon_1(t)\,]\,\tilde{\varrho}^{(o)}_{12}\,. \quad \textbf{(8.2.12.3)}$$

The other members will satisfy the relations

$$\tilde{\varrho}^{(o)}_{11} + \tilde{\varrho}^{(o)}_{22} = 1\,,\ \tilde{\varrho}^{(o)}_{21} = \tilde{\varrho}^{(o)\,*}_{12}\,. \quad \textbf{(8.2.12.4)}$$

Formula (8.2.7.9) leads to

$$\varepsilon_2(t) - \varepsilon_1(t) = \hbar\,\widetilde{\omega}(t)\,, \quad \textbf{(8.2.12.5)}$$

where the frequency $\widetilde{\omega}(t)$ is determined by formula (8.2.7.10). Equation (12.4) can be easily integrated:

$$\mathrm{d}\tilde{\varrho}^{(o)}_{12}/\tilde{\varrho}^{(o)}_{12} = \mathrm{i}\,\widetilde{\omega}(t)\,\mathrm{d}t\,.$$

Integrating, we get

$$\ln \tilde{\varrho}_{12}^{(o)} = \mathrm{i} \int \tilde{\omega}(t)\, \mathrm{d}t \, .$$

Using this function (7.10), we will have

$$\ln \tilde{\varrho}_{12}^{(o)} = \mathrm{i}\, \omega_o \int \{\, 1 + 2\, \xi^2(t) \,\}\, \mathrm{d}t \, .$$

Given that according to the formula (8.2.7.8)

$$\xi(t) = \Omega \cos \omega\, t / \omega_o \, ,$$

we have

$$\tilde{\varrho}_{12}^{(o)}(t) = C_{12}\, \mathrm{e}^{\,\mathrm{i}\, \psi(t)} \, , \qquad \textbf{(8.2.12.6)}$$

where

$$\psi(\mathrm{t}) = (\, \omega_o + \Omega^2/\omega_o \,)\, t + \Omega^2/(\, 2\, \omega_o\, \omega \,) \sin 2\, \omega\, t \, . \qquad \textbf{(8.2.12.7)}$$

Other solutions of the system (8.3) will have the form

$$\tilde{\varrho}_{11}^{(o)} = C_{11} \, , \tilde{\varrho}_{21}^{(o)} = C_{12}^{*}\, \mathrm{e}^{\,-\,\mathrm{i}\, \psi(t)} \, , \tilde{\varrho}_{22}^{(o)} = C_{22} \, . \qquad \textbf{(8.2.12.8)}$$

8.2.13. Equation for First-Order Density Matrix in *κ*-Representation

The required equation looks like this:

$$\mathrm{i}\, \hbar\, \partial \tilde{\varrho}_{\kappa\kappa'}^{(1)} / \partial t = \textstyle\sum_{\eta} \, (\, \widetilde{H}_{\kappa\eta}\, \tilde{\varrho}_{\eta\kappa'}^{(1)} - \tilde{\varrho}_{\kappa\eta}^{(1)}\, \widetilde{H}_{\eta\kappa'} \,) + \mathrm{i}\, \hbar\, D_{\kappa\kappa'}(\tilde{\varrho}_{\kappa\kappa'}^{(o)}) \, .$$

(8.2.13.1)

Substituting the formula $\widetilde{H}_{\kappa\eta}$ here, we get

$$\mathrm{i}\, \hbar\, \partial \tilde{\varrho}_{\kappa\kappa'}^{(1)} / \partial t = [\, \varepsilon_{\kappa}(t) - \varepsilon_{\kappa'}(t) \,]\, \tilde{\varrho}_{\kappa\kappa'}^{(1)} + \mathrm{i}\, \hbar\, D_{\kappa\kappa'}(\tilde{\varrho}_{\kappa\kappa'}^{(o)}) \, .$$

(8.2.13.2)

Now let's write down the equations for the density matrices $\tilde{\varrho}_{11}$ and $\tilde{\varrho}_{12}$, substituting here formulas (8.2.10.9), (8.2.10.10) and (8.2.10.14):

$$\partial \tilde{\varrho}_{11} / \partial t = \, - [\, p_{21} + \, Q + B\, W(\omega) \,]\, \tilde{\varrho}_{11} + [\, p_{12} + A + B\, W(\omega) \,]\, \tilde{\varrho}_{22} \, , \qquad \textbf{(8.2.13.3)}$$

$$\partial\tilde{\varrho}_{12}/\partial t = \mathrm{i}\,\tilde{\omega}\,\tilde{\varrho}_{12} + 1/2\,[\,2\sqrt{p_{12}\,p_{21}}\,\tilde{\varrho}_{21} - (\,p_{12} + p_{21}\,)\,\tilde{\varrho}_{12}] +$$

$$+ (\,A + Q\,)\,\tilde{\varrho}_{12} - 2\sqrt{A\,Q}\,\tilde{\varrho}_{21} + B\,W(\omega)\,(\,\tilde{\varrho}_{21} - \tilde{\varrho}_{12}) - \Gamma\,\tilde{\varrho}_{12}\,.$$

(8.2.13.4)

These equations should be supplemented with the following equations

$$\tilde{\varrho}_{11} + \tilde{\varrho}_{22} = 1\,,\ \tilde{\varrho}_{21} = \tilde{\varrho}^{*}_{12}\,. \qquad \textbf{(8.2.13.5)}$$

We proved that phenomenological equation (8.2.2.2) coincides with the equation obtained using the laws of quantum physics.

From equation (8.2.13.3), we find the condition to which the pump parameter Q must follow, which provides an inversion of the population levels:

$$\tilde{\varrho}_{22} > \tilde{\varrho}_{11}\,. \qquad \textbf{(8.2.13.6)}$$

This is the condition

$$Q > A\ +\ p_{12} -\ p_{21}. \qquad \textbf{(8.2.13.7)}$$

Finally, we write the equations for density matrices $\tilde{\varrho}^{(1)}_{\kappa\kappa'}$ of the first order:

$$\partial\tilde{\varrho}^{(1)}_{11}/\partial t = \ -[\ p_{21} +\ Q + B\,W(\omega)\,]\,\tilde{\varrho}^{(0)}_{11} + [\,p_{12} + A + B\,W(\omega)\,]\,\tilde{\varrho}^{(0)}_{22}\,, \qquad \textbf{(8.2.13.8)}$$

$$\partial\tilde{\varrho}^{(1)}_{12}/\partial t = \mathrm{i}\,\tilde{\omega}\,\tilde{\varrho}^{(1)}_{12} + 1/2\,[\,2\sqrt{p_{12}\,p_{21}}\,\tilde{\varrho}^{(0)}_{21} - (\,p_{12} + p_{21}\,)\,\tilde{\varrho}^{(0)}_{12}\,] +$$

$$+ (\,A + Q\,)\,\tilde{\varrho}^{(0)}_{12} - 2\sqrt{A\,Q}\,\tilde{\varrho}^{(0)}_{21} + B\,W(\omega)\,(\tilde{\varrho}^{(0)}_{21} - \tilde{\varrho}^{(0)}_{12}) - \Gamma\,\tilde{\varrho}^{(0)}_{12}\,.$$

(8.2.13.9)

These equations are much easier to solve than equations (8.2.13.3) and (13.4).

When we find the density matrices $\tilde{\varrho}_{11}$ and $\tilde{\varrho}_{12}$ in the κ-representation, we get the equations for the density matrices ϱ_{11} and ϱ_{12} in the α-representation from equation (8.2).

8.2.14. Spectral Density of Radiation Energy

The spectral density of the radiation energy will follow the equation

$$\mu/N_A \, \mathrm{d}W/\mathrm{d}t = (A + B\,W)\,\varrho_{22} - B\,W\,\varrho_{11}\,, \tag{8.2.14.1}$$

where μ is the coefficient responsible for the equality of the dimensions of both parts of the equation, N_A is the number of active atoms that were on the path of the long h of the coherent radiation in the resonator,

$$W(t,\omega) = W^{(\mathrm{Planck})}(\omega) + \Delta(t,\omega) \tag{8.2.14.2}$$

there is $W^{(\mathrm{Planck})}(\omega)$ – the spectral energy density of Planck radiation, $\Delta(t,\omega)$ – the spectral energy density of coherent radiation that occurs due to the operation of the spaser. Planck radiation is thermal radiation that is emitted by heated bodies. This radiation is directed in all directions. The spaser radiation is directed along the rays coming from the center of the metal (Fig. **8.2.2**). The spectral energy density of coherent radiation $\Delta(t,\omega)$ is much higher than the spectral density of thermal radiation $W^{(\mathrm{Planck})}(\omega)$:

$$\Delta(t,\omega) \gg W^{(\mathrm{Planck})}(\omega)\,. \tag{8.2.14.3}$$

In this case, we can write the equation

$$\mu/N_A \, \mathrm{d}\Delta/\mathrm{d}t = (A + B\,\Delta)\,\varrho_{22} - B\,\Delta\,\varrho_{11}\,. \tag{8.2.14.4}$$

In this equation, the value Δ characterizes the coherent radiation that propagates along with one of the rays that start from the center of the spaser. This radiation moves between the walls of the resonator from one wall to the other, from which the radiation flies out. Note that now in equation (8.2.14.4), there are density matrices in the α-representation.

Let's point the x-axis along the beam. The x-axis starts at a distance from the center of the ball. The second wall of the resonator is located at $x = h$. The spectral density of the radiation consists of two terms. One radiation propagates from the center of the ball, and the other, reflected from the wall at $x = h$, propagates in the opposite direction:

$$\Delta = \Delta^{+} + \Delta^{-}. \tag{8.2.14.5}$$

These densities satisfy the equations

$$\mu/N_A \, \mathrm{d}\Delta^+/\mathrm{d}t = (A/2 + B\,\Delta^+)\,\varrho_{22} - B\,\Delta^+\,\varrho_{11}\,, \tag{8.2.14.6}$$

$$\mu/N_A \, \mathrm{d}\Delta^-/\mathrm{d}t = (A/2 + B\,\Delta^-)\,\varrho_{22} - B\,\Delta^-\,\varrho_{11}\,. \tag{8.2.14.7}$$

Let for $x = 0$ the light is reflected completely, and for $x = h$ the light is reflected with the reflection index R, this fact is expressed by equalities:

$$\Delta^-(x = 0) = \Delta^+(x = 0)\,, \tag{8.2.14.8}$$

$$\Delta^-(x = h) = R\,\Delta^+(x = h)\,. \tag{8.2.14.9}$$

The spectral density of the radiation emitted from the spaser will be equal to

$$\Delta^+_{\mathrm{output}}(x = h) = (1 - R)\,\Delta^+(x = h)\,. \tag{8.2.14.10}$$

In equations (8.2.14.6) and (8.2.14.7), we replace time t by the distances that light travels when moving in one direction or another. Light traveling in the direction of increasing the x coordinate passes in time t distance

$$x = c\,t\,, \tag{8.2.14.11}$$

where c is the speed of light. The light that travels in the opposite direction passes the distance at the same time

$$h - x = c\,t\,. \tag{8.2.14.12}$$

Now we will have

$$\mu\,c/N_A \, \mathrm{d}\Delta^+/\mathrm{d}x = (A/2 + B\,\Delta^+)\,\varrho_{22} - B\,\Delta^+\,\varrho_{11}\,, \tag{8.2.14.13}$$

$$-\mu\,c/N_A \, \mathrm{d}\Delta^-/\mathrm{d}x = (A/2 + B\,\Delta^-)\,\varrho_{22} - B\,\Delta^-\,\varrho_{11}\,. \tag{8.2.14.14}$$

Solving these equations, we find the spectral density of the radiation emitted from the spaser

$$\Delta^+_{\mathrm{output}}(\omega) = F(\alpha, R, h)\,N_A\,\Omega^2/\{\mu\,\omega_0^2\,[(\omega - \omega_0)^2 + \Gamma^2]\}\,, \tag{8.2.14.15}$$

where $F(\alpha, R, h)$ is the coefficient,

$$\alpha = \mu\,N_A B\,(\varrho_{22} - \varrho_{11})/(2\,c)\,.$$

Details of these solutions can be found in the book [7].

8.3. QUANTUM THEORY OF GRAPHENE

When a quantum system is in an equilibrium state, the law that characterizes this state can be found using the variational principle. To do this, you need to know the energy of the system and its entropy. We only know one approximate expression for entropy, which is valid only in the case of a diagonal density matrix. Therefore, the energy of the system must also be diagonal. But the energy of the system, as usual, is recorded in the coordinate representation. In this paper, a unitary matrix is found, by which the energy of graphene is transferred from the coordinate representation to the diagonal one. Further, the variational principle allows us to write equations for the probabilities of the graphene state.

8.3.1. Introduction

Under normal conditions, carbon atoms form solid bodies that differ from each other in their crystal lattice; these are **diamond** and **graphite**. Recently, **graphene** was discovered as a two-dimensional modification of carbon, when carbon atoms form a flat hexagonal crystal lattice with a thickness of one atom (see Fig. **8.3.1**). From this plane, you can form a cylinder and a sphere, which are called a **nanotube** and **fullerene**.

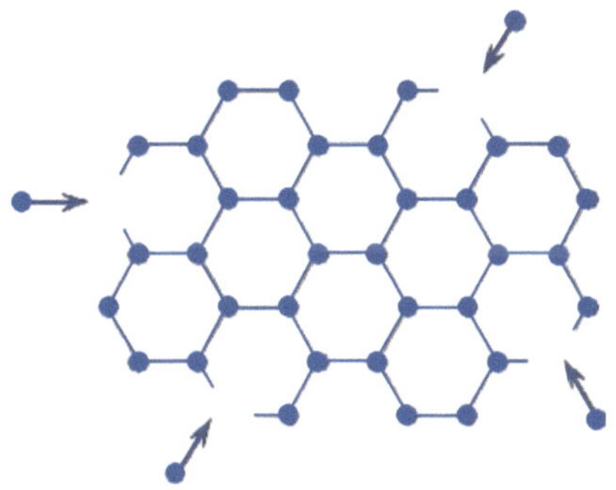

Fig. (8.3.1). Graphene.

Attempts made for chemical delamination of graphite into graphene were unsuccessful. The results of these attempts were fused polyatomic layers. In 2004, K. Novoselov and A. Geim used the method of the mechanical splitting of graphite [12, 13]. The surface of the graphite was tightly lapped to the surface of another crystalline substance. After this, numerous carbon scales of varying thickness remained on the surface of the crystal. With the help of optical and atomic force microscopes, layers that had a monatomic thickness were detected among the scales. To isolate a single monatomic layer, a special substrate was applied to these scales. When heated to high temperatures in an ultra-high vacuum, the substance that is part of the substrate converts from the solid state immediately to the gaseous state, bypassing the melting stage. As a result, a graphene film is formed on the surface of the crystal. For the discovery of the two-dimensional form of carbon, K Novoselov and A. Geim were awarded the 2010 Nobel prize in physics.

Graphene is the first experimentally created two-dimensional material. Thanks to this unique atomic structure, it has amazing properties: huge thermal conductivity, great mechanical strength, flexibility, high electrical conductivity, impermeability to most liquids and gases, transparency. But the most striking property of graphene is its **absolute stability**.

The discovery of graphene so inspired physicists that they received three more two-dimensional materials with similar properties within ten years.

8.3.2. Electronic Structure of Graphene

The carbon atom contains six electrons, which are in States designated by the symbols $1s^2\ 2s^2 2p^2$. Two electrons occupy States with a quantum number $n = 1$. These electrons have the lowest energy $E_1 = -\ Z^2\ R\ \hbar$. All other electrons are in States with $n = 2$ and each of them has energy $E_2 \cong -\ Z^2\ R\ \hbar/4$.

The electrons of two neighboring carbon atoms located in graphene change their position (see Fig. **8.3.2**). These atoms are designated by the letters *A* and *B*. Two electrons with the quantum number $n = 1$ continue to move around the atomic nucleus, only slightly changing their movement. The other four electrons of the atom change their motion significantly. Three electrons from this number cover three neighboring carbon atoms with their movement and form a stable graphene bond with them. There remains the last electron of the atom, which is weaker than the other electrons connected to the atom, having the highest energy and from time to time, passes from one carbon atom to another.

The hexagonal crystal lattice of graphene has *a* period, which is denoted by the letter a. This lattice has a rhombus-shaped unit cell in which two carbon atoms *A* and *B* are included. Therefore, the lattice consists of two Bravais lattices, each of which consists of either *A* or *B* atoms. Two vectors are constructed on the sides of the unit cell

$$\boldsymbol{a}_1 = \sqrt{3}/2\,(\sqrt{3}\,\boldsymbol{i}+\boldsymbol{j})\,a, \qquad \boldsymbol{a}_2 = \sqrt{3}/2\,(\sqrt{3}\,\boldsymbol{i}-\boldsymbol{j})\,a, \tag{8.3.1}$$

where ***i*** and ***j*** are unit orts. Vector

$$\boldsymbol{R} = n_1\,\boldsymbol{a}_1 + n_2\,\boldsymbol{a}_2 \tag{8.3.2}$$

if the length of one of the vectors $\boldsymbol{a}_1$ or $\boldsymbol{a}_2$ is shifted sequentially by a distance, it shows the location of all graphene cells. In formula (2), n_1 and n_2 are integers. The total number of vectors $\boldsymbol{R}$ in graphene is $N/2$, where N is the number of carbon atoms in graphene.

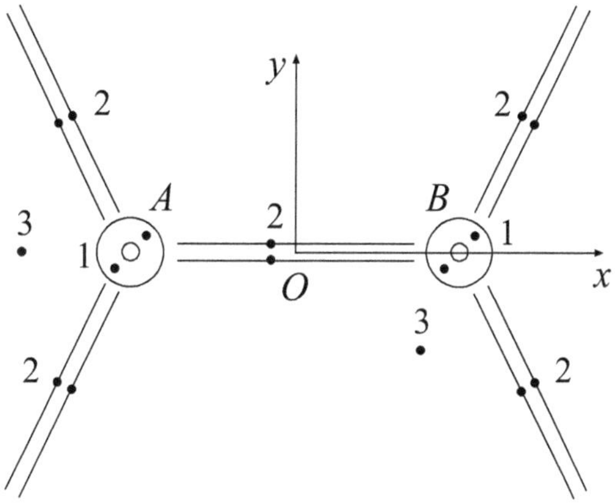

Fig. (8.3.2). Electrons of two neighboring carbon atoms in graphene. These atoms are part of two Bravais lattices *A* and *B*; 1 – two electrons that are rigidly connected to the nucleus of an atom; 2 – two electrons that perform a stable bond of atoms in graphene, each of them a part of an isolated carbon atom; 3 – almost free electrons.

In solids, the energy of atoms when they interact turns into a set of levels that form groups called **energy zones**. Fig. (**3**) shows the energy zones of graphene. The lowest zone with the lowest energies contains $2N$ levels and corresponds to the two electrons in the isolated carbon atom closest to the core. The next zone contains $3N$ levels. Finally, the third zone is partially empty and contains N levels that correspond to the energies of almost free electrons (represented by 3) (see Fig. **8.3.3**). This zone splits into two zones and, therefore, bodies with such zones are called semi-metallic.

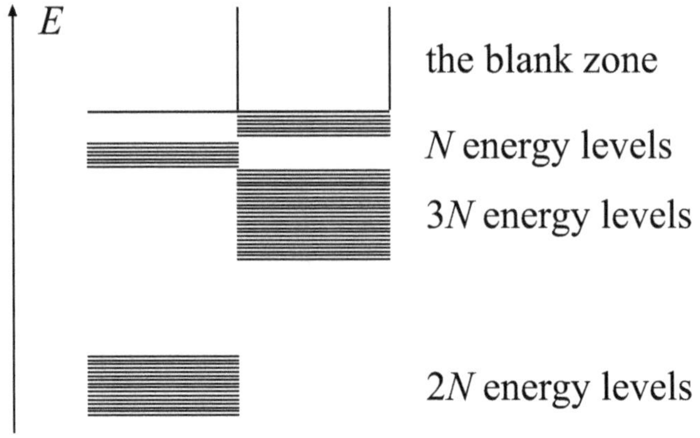

Fig. (8.3.3). Energy zones of graphene (Semi-metal).

8.3.3. Density Matrix and Average Energy of a Simple Crystal

The statistical operator $\hat{\varrho}$ is the most general tool that characterizes any quantum system. Consider a crystal lattice that consists of $q = 1, 2, \ldots$ Bravais lattices. The average energy of such a crystal must be described by a statistical operator that takes into account each of the atoms in the unit cell:

$$E = \mathrm{Tr} \sum_{AB\ldots D} \hat{H}^{AB\ldots D}\, \hat{\varrho}^{AB\ldots D},$$

where the letters A, B, …, D denote each atom in the unit cell, $\hat{H}^{AB\ldots D}$ is the Hamiltonian of the entire crystal, and $\hat{\varrho}^{AB\ldots D}$ is its statistical operator.

If the cell has only one atom with one external electron, the formula for the average energy will look like

$$E = \mathrm{Tr}\,(\,\hat{H}\,\hat{\varrho}\,)\,, \tag{8.3.3}$$

where $\hat{H}$ is the Hamiltonian of all external electrons in the crystal, and $\hat{\varrho}$ is the statistical operator.

The simplest and most commonly accepted form of the Hamiltonian $\hat{H}$ consists of two terms:the operator $\hat{H}^{(1)}$ of one electron and the operator $\hat{H}^{(2)}$ of two electrons that interact with each other.

Based on the statistical operator $\hat{\varrho}$, a hierarchical sequence of operators $\hat{\varrho}^{(1)}$, $\hat{\varrho}^{(2)}$ and so on can be constructed [14]. The operator $\hat{\varrho}^{(1)}$ has a trace equal to the number of external electrons in the system:

$$\mathrm{Tr}\, \hat{\varrho}^{(1)} = N\,. \qquad \textbf{(8.3.4)}$$

The main property of the operator $\hat{\varrho}^{(2)}$ can only be written if it is expressed in terms of the corresponding **density matrix**

$$\varrho(\alpha_1, \alpha_2;\ \alpha_1', \alpha_2') \equiv \varrho_{\alpha_1,\alpha_2;\ \alpha_1',\alpha_2'} \equiv \varrho_{12,1'2'}\,.$$

Here the value α is a quantum variable that characterizes the state of a single particle in some **representation**. For electrons that are **fermions**, this property looks like this

$$\varrho_{12,1'2'} = -\,\varrho_{21,1'2'} = -\,\varrho_{12,2'1'} = \varrho_{21,2'1'}\,. \qquad \textbf{(8.3.5)}$$

If we are interested in only one density matrix $\varrho_{11'}$, then the second density matrix will be:

$$\varrho_{12,1'2'} \cong \varrho_{11'}\,\varrho_{22'} - \varrho_{12'}\,\varrho_{21'}\,. \qquad \textbf{(8.3.6)}$$

This matrix will exactly satisfy the relation (8.3.5).

Now we can write an expression for the average energy of a system of identical particles

$$E = \sum_{\alpha\alpha'} H_{\alpha\alpha'}\,\varrho_{\alpha'\alpha} + 1/2\,\sum_{\{\alpha\}} H_{\alpha_1\alpha_2,\alpha_1'\alpha_2'}\,\varrho_{\alpha_1'\alpha_2',\,\alpha_1\alpha_2}, \qquad \textbf{(8.3.7)}$$

where $\{\alpha\} = \alpha_1, \alpha_2, \alpha_1', \alpha_2'$, $H_{\alpha\alpha'}$ and $H_{\alpha_1\alpha_2,\alpha_1'\alpha_2'}$ are matrix elements of the operators $\hat{H}^{(1)}$ and $\hat{H}^{(2)}$. Matrix elements

$$H_{\alpha_1\alpha_2,\alpha_1'\alpha_2'} \equiv H_{12,1'2'}$$

of the Hamiltonian $\hat{H}^{(2)}_{nm}$; the interactions of two fermions must be antisymmetric:

$$H_{12,1'2'} = -\,H_{21,1'2'} = -\,H_{12,2'1'} = H_{21,2'1'}\,. \qquad \textbf{(8.3.8)}$$

Substituting the approximate expression (8.3.6) in formula (8.3.7) with the antisymmetry of the matrix elements of the Hamiltonian $\hat{H}^{(2)}$, we have

$$E = \sum_{\alpha\alpha'} H_{\alpha\alpha'}\,\varrho_{\alpha'\alpha} + \sum_{\{\alpha\}} H_{\alpha_1\alpha_2,\alpha_1'\alpha_2'}\,\varrho_{\alpha_1'\alpha_1}\,\varrho_{\alpha_2'\alpha_2}\,. \qquad \textbf{(8.3.9)}$$

With matrix elements $H_{\alpha\alpha'}$ and $H_{\alpha_1\alpha_2,\alpha_1'\alpha_2'}$, we find the formula

$$H_{\alpha\alpha'} = \int \varphi^*_{\alpha}(q)\, \hat{H}^{(1)}\, \varphi_{\alpha'}(q)\, \mathrm{d}q\,, \tag{8.3.10}$$

$$H_{\alpha_1\alpha_2,\alpha'_1\alpha'_2} = \int \Phi^*_{\alpha_1\alpha_2}(q_1, q_2)\, \hat{H}^{(2)}\, \Phi_{\alpha'_1\alpha'_2}(q_1, q_2)\, \mathrm{d}q_1\, \mathrm{d}q_2\,. \tag{8.3.11}$$

For electrons, the function $\Phi_{\alpha'_1\alpha'_2}(q_1, q_2)$ must be antisymmetric. This is the function of the Slater.

If the system is in an equilibrium state, then the variational principle should be applied to

solve it. But to do this, you need to find a representation in which the density matrix has a diagonal form:

$$\varrho_{\alpha\alpha'} = w_\alpha\, \delta_{\alpha\alpha'}, \tag{8.3.12}$$

where w_α is the probability that the electron is in the state α, $\delta_{\alpha\alpha'}$ is the Kronecker symbol.

8.3.4. Average Energy of Graphene

In a graphene unit cell, there are two external electrons (represented by 3) (see Fig. **8.3.2**), which are in different States φ^A and φ^B. Now the average energy of graphene will be equal to

$$E = \sum_{ij=1}^{2} \{ \sum_{\alpha\alpha'} H^{(ij)}_{\alpha\alpha'}\, \varrho^{(ji)}_{\alpha'\alpha} + \sum_{\{\alpha\}} H^{(ij)}_{\alpha_1\alpha_2,\alpha'_1\alpha'_2}\, \varrho^{(ji)}_{\alpha'_1\alpha_1}\, \varrho^{(ji)}_{\alpha'_2\alpha_2} \}, \tag{8.3.13}$$

where $i, j = A, B$.

The wave function of electrons in a graphene cell will be equal to the sum of two functions

$$\varphi(\boldsymbol{r}) = c\, \{ \varphi^A(\boldsymbol{r} - \boldsymbol{a}/2) + \varphi^B(\boldsymbol{r} + \boldsymbol{a}/2) \}\,. \tag{8.3.14}$$

We find the constant c from the normalization condition

$$\int \varphi^*(\boldsymbol{r})\, \varphi(\boldsymbol{r})\, \mathrm{d}\boldsymbol{r} = 1\,.$$

It will be approximately equal to $c \cong 1/\sqrt{2}$. All graphene will be described by the function

$$\psi(\boldsymbol{r}) = \sum_{\boldsymbol{R}} \varphi_{\boldsymbol{R}}(\boldsymbol{r}) ,$$

where the function $\varphi_{\boldsymbol{R}}(\boldsymbol{r})$ will be equal to

$$\varphi_{\boldsymbol{R}}(\boldsymbol{r}) = 1/\sqrt{2}\ \{\ \varphi^{A}(\boldsymbol{r} - \boldsymbol{R} - \boldsymbol{a}/2) + \varphi^{B}(\boldsymbol{r} - \boldsymbol{R} + \boldsymbol{a}/2)\ \} = 1/\sqrt{2} \sum_{i=1}^{2} \varphi_{\boldsymbol{R}}^{(i)}(\boldsymbol{r}) ,$$

where

$$\varphi_{\boldsymbol{R}}^{A}(\boldsymbol{r}) = \varphi^{A}(\boldsymbol{r} - \boldsymbol{R} - \boldsymbol{a}/2) , \varphi_{\boldsymbol{R}}^{B}(\boldsymbol{r}) = \varphi^{B}(\boldsymbol{r} - \boldsymbol{R} + \boldsymbol{a}/2) . \quad \textbf{(8.3.15)}$$

Now write the formula (8.3.13) in the coordinate representation

$$E = \sum_{ij=1}^{2} \{ \sum_{\boldsymbol{RR}'} H_{\boldsymbol{RR}'}^{(ij)}\ \varrho_{\boldsymbol{R}'\boldsymbol{R}}^{(ji)} + \sum_{\{\boldsymbol{R}\}} H_{\boldsymbol{R}_1\boldsymbol{R}_2,\boldsymbol{R}_1'\boldsymbol{R}_2'}^{(ij)}\ \varrho_{\boldsymbol{R}_1'\boldsymbol{R}_1}^{(ji)}\ \varrho_{\boldsymbol{R}_2'\boldsymbol{R}_2}^{(ji)} \} ,$$

(8.3.16)

where $\{\boldsymbol{R}\} = \ \boldsymbol{R}_1, \boldsymbol{R}_2, \boldsymbol{R}_1', \boldsymbol{R}_2'$.

8.3.5. The Kinetic Energy of the Graphene

We will first consider the first sum in the formula (8.3.16), which we will call the kinetic energy:

$$E_1 = \sum_{ij=1}^{2} \sum_{\boldsymbol{RR}'} H_{\boldsymbol{RR}'}^{(ij)}\ \varrho_{\boldsymbol{R}'\boldsymbol{R}}^{(ji)} , \quad \textbf{(8.3.17)}$$

where

$$H_{\boldsymbol{RR}'}^{(ij)} = \varepsilon_{\boldsymbol{R}-\boldsymbol{R}'}^{(ij)} = 1/2 \int \varphi_{\boldsymbol{R}}^{(i)}(\boldsymbol{r})\ \widehat{H}^{(1)}\ \varphi_{\boldsymbol{R}'}^{(j)}(\boldsymbol{r})\ \mathrm{d}\boldsymbol{r} , \quad \textbf{(8.3.18)}$$

$$\varepsilon_{\boldsymbol{R}-\boldsymbol{R}'}^{AB} = \varepsilon_{-\boldsymbol{R}+\boldsymbol{R}'}^{BA} = \varepsilon_{\boldsymbol{R}-\boldsymbol{R}'}^{BA} . \quad \textbf{(8.3.19)}$$

Our goal is to make the matrix $H_{\boldsymbol{RR}'}^{(ij)}\ \varrho_{\boldsymbol{R}'\boldsymbol{R}}^{(ji)}$ diagonal. More information formula (8.3.17) will be equal to

$$E_1 = \sum_{\boldsymbol{RR}'} \{ \varepsilon_{\boldsymbol{R}-\boldsymbol{R}'}^{AA}\ \varrho_{\boldsymbol{R}'\boldsymbol{R}}^{AA} + \varepsilon_{\boldsymbol{R}-\boldsymbol{R}'}^{AB}\ \varrho_{\boldsymbol{R}'\boldsymbol{R}}^{AB} + \varepsilon_{\boldsymbol{R}-\boldsymbol{R}'}^{BA}\ \varrho_{\boldsymbol{R}'\boldsymbol{R}}^{BA} + \varepsilon_{\boldsymbol{R}-\boldsymbol{R}'}^{BB}\ \varrho_{\boldsymbol{R}'\boldsymbol{R}}^{BB} \} . \quad \textbf{(8.3.20)}$$

Consider the matrix elements of the Hamiltonian $\widehat{H}^{(1)}$

$$H_{\boldsymbol{RR}'}^{(ij)} = \varepsilon_{R-\boldsymbol{R}'}^{(ij)} = \begin{Vmatrix} \varepsilon_{R-\boldsymbol{R}'}^{AA} & \varepsilon_{R-\boldsymbol{R}'}^{AB} \\ \varepsilon_{R-\boldsymbol{R}'}^{BA} & \varepsilon_{R-\boldsymbol{R}'}^{BB} \end{Vmatrix}. \quad \textbf{(8.3.21)}$$

For brevity, we denote this matrix as

$$\varepsilon^{(ij)} = \begin{Vmatrix} \varepsilon^{AA} & \varepsilon^{AB} \\ \varepsilon^{BA} & \varepsilon^{BB} \end{Vmatrix}. \quad \textbf{(8.3.22)}$$

Using a unitary matrix, we reduce it to a diagonal form [15]. Lets assume that the unitary matrix has the form

$$U_{\kappa i} = \begin{Vmatrix} \cos\vartheta & \sin\vartheta \\ -\sin\vartheta & \cos\vartheta \end{Vmatrix}. \quad \textbf{(8.3.23)}$$

The desired matrix $\tilde{\varepsilon}^{(\kappa\kappa')}$ will be equal to

$$\tilde{\varepsilon}^{(\kappa\kappa')} = \sum_{ij} U_{\kappa i}\, \varepsilon^{(ij)}\, U_{\kappa' j}^{*}. \quad \textbf{(8.3.24)}$$

Or read more

$$\tilde{\varepsilon}^{(\kappa\kappa')} = \begin{Vmatrix} \cos\vartheta & \sin\vartheta \\ -\sin\vartheta & \cos\vartheta \end{Vmatrix} \begin{Vmatrix} \varepsilon^{AA} & \varepsilon^{AB} \\ \varepsilon^{BA} & \varepsilon^{BB} \end{Vmatrix} \begin{Vmatrix} \cos\vartheta & -\sin\vartheta \\ \sin\vartheta & \cos\vartheta \end{Vmatrix}.$$

Multiplying the matrices, we get

$$\tilde{\varepsilon}^{AA} = \varepsilon^{AA}\cos^2\vartheta + (\varepsilon^{AB} + \varepsilon^{BA})\cos\vartheta\sin\vartheta + \varepsilon^{BB}\sin^2\vartheta,$$

$$\tilde{\varepsilon}^{AB} = \tilde{\varepsilon}^{BA} = \varepsilon^{AB}\cos^2\vartheta - (\varepsilon^{AA} - \varepsilon^{BB})\cos\vartheta\sin\vartheta - \varepsilon^{AB}\sin^2\vartheta,$$

$$\tilde{\varepsilon}^{BB} = \varepsilon^{BB}\cos^2\vartheta - (\varepsilon^{AB} + \varepsilon^{BA})\cos\vartheta\sin\vartheta + \varepsilon^{AA}\sin^2\vartheta,$$

(8.3.25)

where it is taken into account that $\varepsilon^{AB} = \varepsilon^{BA}$. Since we are looking for a diagonal matrix $\varepsilon^{(ij)}$, it will be such if

$$\varepsilon^{AB}\cos^2\vartheta - (\varepsilon^{AA} - \varepsilon^{BB})\cos\vartheta\sin\vartheta - \varepsilon^{AB}\sin^2\vartheta = 0. \quad \textbf{(8.3.26)}$$

Lets write this equation as follows

$$b\,\mathrm{tg}^2\vartheta + \mathrm{tg}\,\vartheta - b = 0, \qquad \textbf{(8.3.27)}$$

where

$$b = \varepsilon^{AB}/(\,\varepsilon^{AA} - \,\varepsilon^{BB}\,)\,. \qquad \textbf{(8.3.28)}$$

Equation (8.3.27) has a solution

$$\mathrm{tg} = (\,-1\,\pm\,\sqrt{1+4\,b^2}\,)/(2\,b)\,.$$

Given that the value of b is modulo less than one:

$$|\,b\,| \ll 1\,, \qquad \textbf{(8.3.29)}$$

having fairly approximate equality

$$\sqrt{1+4\,b^2} \cong 1 + 2\,b^2,$$

which gives two solutions:

$$\mathrm{tg}\,\vartheta_1 \cong b\,, \qquad \textbf{(8.3.30)}$$

$$\mathrm{tg}\,\vartheta_2 \cong -\,(\,1 + b^2\,)/b\,.$$

The first solution suits us, but the second one does not.

Since

$$\cos^2\vartheta = 1/(\,1+\vartheta\,)\,,$$

given the inequality (8.3.29), we will have

$$\cos^2\vartheta \cong 1 - b^2.$$

Thus, we get a diagonal matrix

$$\begin{cases} \tilde{\varepsilon}^{AA} = (\,1 - \,b^2)\,(\varepsilon^{AA} \,+ 2\,\varepsilon^{AB}\,b + \varepsilon^{BB}\,b^2)\,, \\ \qquad\quad \tilde{\varepsilon}^{AB} \,= \tilde{\varepsilon}^{BA} \,= \,0\,, \\ \tilde{\varepsilon}^{BB} = (\,1 - \,b^2)(\,\varepsilon^{BB} - 2\,\varepsilon^{AB}\,b + \varepsilon^{AA}\,b^2)\,. \end{cases} \qquad \textbf{(8.3.31)}$$

So, the diagonal matrix is found

$$\tilde{\varepsilon}^{(\kappa\kappa')} = \begin{Vmatrix} \tilde{\varepsilon}^{AA} & 0 \\ 0 & \tilde{\varepsilon}^{BB} \end{Vmatrix}.$$

Find the distance Δ between the zones $\tilde{\varepsilon}^{AA}$ and $\tilde{\varepsilon}^{BB}$. It is equal to

$$\Delta = \tilde{\varepsilon}^{AA} - \tilde{\varepsilon}^{BB} = (\varepsilon^{AA} - \varepsilon^{BB})(1 - b^2)(1 + 3b^2) = 0, \quad \textbf{(8.3.32)}$$

because

$$\varepsilon^{AA} = \varepsilon^{BB},$$

when these zones come into contact (see Fig. **8.3.3**).

Now, using the already found unitary matrix, we will find the density matrix $\varrho^{(\kappa\kappa')}_{\boldsymbol{R}'\boldsymbol{R}} = \tilde{\varrho}^{(\kappa\kappa')}$

in the κ-representation:

$$\tilde{\varrho}^{(\kappa\kappa')} = \begin{Vmatrix} \cos\vartheta & \sin\vartheta \\ -\sin\vartheta & \cos\vartheta \end{Vmatrix} \begin{Vmatrix} \varrho^{AA} & \varrho^{AB} \\ \varrho^{BA} & \varrho^{BB} \end{Vmatrix} \begin{Vmatrix} \cos\vartheta & -\sin\vartheta \\ \sin\vartheta & \cos\vartheta \end{Vmatrix}.$$

in short, put $\varrho^{(ij)}_{\boldsymbol{RR}'} = \varrho^{(ij)}$. Multiplying the matrices, we get

$$\tilde{\varrho}^{AA} = \varrho^{AA}\cos^2\vartheta + 2\,\varrho^{AB}\cos\vartheta\sin\vartheta + \varrho^{BB}\sin^2\vartheta,$$

$$\tilde{\varrho}^{AB} = \tilde{\varrho}^{BA} = \varrho^{AB}\cos^2\vartheta - (\varrho^{AA} - \varrho^{BB})\cos\vartheta\sin\vartheta - \varrho^{AB}\sin^2\vartheta,$$

$$\tilde{\varrho}^{BB} = \varrho^{BB}\cos^2\vartheta - (\varrho^{AB} + \varrho^{BA})\cos\vartheta\sin\vartheta + \varrho^{AA}\sin^2\vartheta.$$

Using the formula (8.3.30), we get

$$\tilde{\varrho}^{AA} = (1 - b^2)(\varrho^{AA} + 2\,\varrho^{AB}b + \varrho^{BB}b^2),$$

$$\tilde{\varrho}^{AB} = \tilde{\varrho}^{BA} = (1 - b^2)(\varrho^{AB} - (\varrho^{AA} - \varrho^{BB})\,b - \varrho^{AB}b^2), \quad \textbf{(8.3.33)}$$

$$\tilde{\varrho}^{BB} = (1 - b^2)(\varrho^{BB} - 2\,\varrho^{AB}b + \varrho^{AA}b^2),$$

where it is taken into account that $\varrho^{AB} = \varrho^{BA}$.

Now that the matrix $\tilde{\varepsilon}^{(\kappa\kappa')}_{R-R'}$ and $\tilde{\varrho}^{(\kappa\kappa')}_{RR'}$ is written in the κ-representation, kinetic energy will look like this

$$E_1 = \sum_{RR'} \{ \tilde{\varepsilon}^{AA}_{R-R'} \tilde{\varrho}^{AA}_{R'R} + \tilde{\varepsilon}^{BB}_{R-R'} \tilde{\varrho}^{BB}_{R'R} \} . \tag{8.3.34}$$

or

$$E_1 = \sum_{\kappa=1}^{2} \sum_{RR'} \tilde{\varepsilon}^{(\kappa\kappa)}_{R-R'} \tilde{\varrho}^{(\kappa\kappa)}_{R'R} .$$

We apply another unitary transformation, whose unitary matrix is equal to

$$U_{kR} = \sqrt{2/N}\, e^{i kR} , \tag{8.3.35}$$

where $\boldsymbol{k}$ is a wave vector, and the vector $\boldsymbol{R}$ belongs to one of the Bravais lattices. The density matrix under the action of this transformation becomes diagonal.

In a simple crystal, this unitary transformation will have the form

$$\varrho_{kk'} = \sum_{RR'} U_{kR}\, \varrho_{RR'}\, U^*_{k'R'} ,$$

where $U_{kR} = 1/\sqrt{N}\, e^{i kR}$. Substituting this matrix gives

$$\varrho_{kk'} = w_k\, \delta_{kk'} , .$$

where w_k is the probability that the electron is in the state $\boldsymbol{k}$. The probability must satisfy the normalization condition

$$\sum_k w_k = N .$$

Applying the matrix (8.3.35) to the graphene density matrices $\tilde{\varrho}^{AA}_{R'R}$ and $\tilde{\varrho}^{BB}_{R'R}$, we have

$$\tilde{\varrho}^{AA}_{k_1k_1'} = w^{AA}_{k_1}\, \delta_{k_1k_1'}, \qquad \tilde{\varrho}^{BB}_{k_2k_2'} = w^{BB}_{k_2}\, \delta_{k_2k_2'} , \tag{8.3.36}$$

where $w^{AA}_{k_1}$ and $w^{BB}_{k_2}$ are the probabilities that the electron is in a state with wave vectors $\boldsymbol{k}_1$ or $\boldsymbol{k}_2$. These probabilities satisfy the conditions

$$\sum_{k_1} w^{AA}_{k_1} = N/2 , \qquad \sum_{k_2} w^{BB}_{k_2} = N/2 . \tag{8.3.37}$$

Repeating the same with the matrix elements $\tilde{\varepsilon}^{AA}_{R\,-\,R'}$ and $\tilde{\varepsilon}^{BB}_{R\,-\,R'}$, we get

$$\tilde{\varepsilon}^{AA}_{k_1} = \sum_{RR'} U_{k_1 R}\, \tilde{\varepsilon}^{AA}_{R\,-\,R'}\, U^{*}_{k_1' R'}\,, \qquad \tilde{\varepsilon}^{BB}_{k_2} = \sum_{RR'} U_{k_2 R}\, \tilde{\varepsilon}^{BB}_{R\,-\,R'}\, U^{*}_{k_2' R'}\,.$$

The first of these quantities is the kinetic energy of the electron in the upper area and the second is the kinetic energy of the electron in the zone, which is below the previous zone at a distance of Δ. Substituting the received elements in the formula (8.3.34), we will have

$$E_1 = \sum_{k_1} \tilde{\varepsilon}^{AA}_{k_1}\, w^{AA}_{k_1} + \sum_{k_2} \tilde{\varepsilon}^{BB}_{k_2}\, w^{BB}_{k_2}\,. \tag{8.3.38}$$

The vectors $\boldsymbol{k}_1$ and $\boldsymbol{k}_2$ belong to the nodes of the inverse lattice, which is built on the basis vectors $\boldsymbol{A}_1$ and $\boldsymbol{A}_2$:

$$\boldsymbol{A}_1 = 2\,\pi/(a\,\sqrt{3})\,(\,\boldsymbol{i}/\sqrt{3} + \boldsymbol{j}\,)\,, \qquad \boldsymbol{A}_2 = 2\,\pi/(a\,\sqrt{3})\,(\,\boldsymbol{i}/\sqrt{3} - \boldsymbol{j}\,)\,.$$

8.3.6. Energy of Interaction of Electrons in Graphene

The next part of the average energy of graphene (16) is related to the matrix elements $H^{(ij)}_{R_1 R_2, R_1' R_2'}$ of the Hamiltonian

$$\hat{H}^{(2)} = e^2/|\,\boldsymbol{r}_1 - \boldsymbol{r}_2\,|$$

repulsions of electrons from each other. Here e is an elementary electric charge, $\boldsymbol{r}_1$and $\boldsymbol{r}_2$ are the radius vectors of two electrons. The interaction energy of the electrons will be equal to

$$E_2 = \sum_{ij\,=\,1}^{2} \sum_{\{R\}} H^{(ij)}_{R_1 R_2, R_1' R_2'}\, \varrho^{(ji)}_{R_1'\, R_1}\, \varrho^{(ji)}_{R_2'\, R_2}\,. \tag{8.3.39}$$

The elements $H^{(ij)}_{R_1 R_2, R_1' R_2'}$ are found by the formula

$$H^{(ij)}_{R_1 R_2, R_1' R_2'} = \int \Phi^{(ij)\,*}_{R_1 R_2}(\boldsymbol{r}_1, \boldsymbol{r}_2)\, \hat{H}^{(2)}\, \Phi^{(ij)}_{R_1' R_2'}(\boldsymbol{r}_1, \boldsymbol{r}_2)\, \mathrm{d}\boldsymbol{r}_1\, \mathrm{d}\boldsymbol{r}_2\,. \tag{8.3.40}$$

For electrons, the function $\Phi^{(ij)}_{R_1 R_2}(\boldsymbol{r}_1, \boldsymbol{r}_2)$ must be antisymmetric. This is

the function of Slater

$$\Phi^{(ij)}_{R_1 R_2}(\boldsymbol{r}_1, \boldsymbol{r}_2) = 1/\sqrt{2}\,\{\,\varphi^{(i)}_{R_1}(\boldsymbol{r}_1)\,\varphi^{(j)}_{R_2}(\boldsymbol{r}_2) - \varphi^{(i)}_{R_1}(\boldsymbol{r}_2)\,\varphi^{(j)}_{R_2}(\boldsymbol{r}_1)\,\}\,. \tag{8.3.41}$$

Now write down the formula (8.3.40)

$$H^{(ij)}_{R_1R_2,R'_1R'_2} = 1/2 \int\{ \varphi^{(i)}_{R_1}(r_1)\, \varphi^{(j)}_{R_2}(r_2) - \varphi^{(i)}_{R_1}(r_2)\, \varphi^{(j)}_{R_2}(r_1) \}\, \hat{H}^{(2)} \cdot \{ \varphi^{(i)}_{R'_1}(r_1)\, \varphi^{(j)}_{R'_2}(r_2) - \varphi^{(i)}_{R'_1}(r_2)\, \varphi^{(j)}_{R'_2}(r_1) \}\, dr_1\, dr_2 = 1/2\, \{ V^{(ij)}_{R_1R_2,R'_1R'_2} - V^{(ji)}_{R_2R_1,R'_1R'_2} - V^{(ji)}_{R_1R_2,R'_2R'_1} + V^{(ij)}_{R_2R_1,R'_2R'_1} \},$$

(8.3.42)

where

$$V^{(ij)}_{R_1R_2,R'_1R'_2} = \int \varphi^{(i)}_{R_1}(r_1)\, \varphi^{(j)}_{R_2}(r_2)\, \hat{H}^{(2)}\, \varphi^{(i)}_{R'_1}(r_1)\, \varphi^{(j)}_{R'_2}(r_2)\, dr_1\, dr_2 . \qquad \textbf{(8.3.43)}$$

Swap r_1 and r_2. Function (8.3.42) is with almost no change. Only equality will be fair

$$H^{(ij)}_{R_1R_2,R'_1R'_2} = H^{(ji)}_{R_2R_1,R'_2R'_1}. \qquad \textbf{(8.3.44)}$$

Using this symmetry, we write

$$H^{(ij)}_{R_1R_2,R'_1R'_2} = V^{(ij)}_{R_1R_2,R'_1R'_2} - V^{(ji)}_{R_2R_1,R'_1R'_2}. \qquad \textbf{(8.3.45)}$$

Now write the matrix

$$V^{(ij)}_{R_1R_2,R'_1R'_2} = \left\| \begin{matrix} V^{AA}_{R_1R_2,R'_1R'_2} & V^{AB}_{R_1R_2,R'_1R'_2} \\ V^{BA}_{R_1R_2,R'_1R'_2} & V^{BB}_{R_1R_2,R'_1R'_2} \end{matrix} \right\| . \qquad \textbf{(8.3.46)}$$

For brevity, we denote this matrix as

$$V^{(ij)} = \left\| \begin{matrix} V^{AA} & V^{AB} \\ V^{BA} & V^{BB} \end{matrix} \right\| \qquad \textbf{(8.3.47)}$$

and present it in a diagonal view

$$\tilde{V}^{(ij)} = \left\| \begin{matrix} \tilde{V}^{AA} & 0 \\ 0 & \tilde{V}^{BB} \end{matrix} \right\| . \qquad \textbf{(8.3.48)}$$

This diagonal matrix is

$$\begin{cases} \tilde{V}^{AA} = (1 - b_1^2)\,(V^{AA} + 2\,V^{AB}\,b_1 + V^{BB}\,b_1^2)\,, \\ \tilde{V}^{AB} = \tilde{V}^{BA} = 0\,, \\ \tilde{V}^{BB} = (1 - b_1^2)(\,V^{BB} - 2\,V^{AB}\,b_1 + V^{AA}\,b_1^2)\,, \end{cases} \quad \textbf{(8.3.49)}$$

where

$$b_1 = V^{AB}/(\,V^{AA} - V^{BB}\,)\,. \quad \textbf{(8.3.50)}$$

Returning to the previous notation, write

$$\tilde{H}^{(ij)}_{R_1R_2,R_1'R_2'} = \tilde{V}^{(ij)}_{R_1R_2,R_1'R_2'} - \tilde{V}^{(ji)}_{R_2R_1,R_1'R_2'}.$$

This matrix will be diagonal:

$$\tilde{H}^{AA}_{R_1R_2,R_1'R_2'} = \tilde{V}^{AA}_{R_1R_2,R_1'R_2'} - \tilde{V}^{AA}_{R_2R_1,R_1'R_2'}\,,\ \tilde{H}^{AB}_{R_1R_2,R_1'R_2'} = \tilde{H}^{BA}_{R_1R_2,R_1'R_2'} = 0\,,$$

$$\tilde{H}^{BB}_{R_1R_2,R_1'R_2'} = \tilde{V}^{BB}_{R_1R_2,R_1'R_2'} - \tilde{V}^{BB}_{R_2R_1,R_1'R_2'}\,. \quad \textbf{(8.3.51)}$$

Substituting matrices (8.3.51) into formula (8.3.39) gives

$$E_2 = \sum_{\{R\}} \{\, \tilde{H}^{AA}_{R_1R_2,R_1'R_2'}\, \tilde{\varrho}^{AA}_{R_1'\,R_1}\, \tilde{\varrho}^{AA}_{R_2'\,R_2} + \tilde{H}^{BB}_{R_1R_2,R_1'R_2'}\, \tilde{\varrho}^{BB}_{R_1'\,R_1}\, \tilde{\varrho}^{BB}_{R_2'\,R_2}\,\}\,. \quad \textbf{(8.3.52)}$$

Again, we apply the unitary matrix (8.3.35). Receive

$$E_2 = 1/2 \sum_{k_1k_2} \{\, \tilde{\varepsilon}^{AA}_{k_1k_2}\, w^{AA}_{k_1}\, w^{AA}_{k_2} + \tilde{\varepsilon}^{BB}_{k_1k_2}\, w^{BB}_{k_1}\, w^{BB}_{k_2}\,\}\,, \quad \textbf{(8.3.53)}$$

where

$$\tilde{\varepsilon}^{AA}_{k_1k_2} = 8/N^2 \sum_{\{R\}} \tilde{H}^{AA}_{R_1R_2,R_1'R_2'}\, \mathrm{e}^{\{\mathrm{i}\,k_1\,(R_1' - R_1) + \mathrm{i}\,k_2\,(R_2' - R_2)\}}\,,$$

$$\tilde{\varepsilon}^{BB}_{k_1k_2} = 8/N^2 \sum_{\{R\}} \tilde{H}^{AA}_{R_1R_2,R_1'R_2'}\, \mathrm{e}^{\{\mathrm{i}\,k_1\,(R_1' - R_1) + \mathrm{i}\,k_2\,(R_2' - R_2)\}}\,. \quad \textbf{(8.3.54)}$$

Now write down the average energy of graphene, which is equal to the sum of the energies (8.3.38) and (8.3.53):

$$E = \sum_k \{ \tilde{\varepsilon}_k^{AA} \, w_k^{AA} + \tilde{\varepsilon}_k^{BB} \, w_k^{BB} \} + 1/2 \sum_{k_1 k_2} \{ \tilde{\varepsilon}_{k_1 k_2}^{AA} \, w_{k_1}^{AA} \, w_{k_2}^{AA} + \tilde{\varepsilon}_{k_1 k_2}^{BB} \, w_{k_1}^{BB} \, w_{k_2}^{BB} \}.$$

(8.3.55)

8.3.7. Variational Principle

Only now we can write an approximate expression for entropy

$$S = - k_{\mathrm{B}} \sum_k \{ (1 - w_k^{AA}) \ln (1 - w_k^{AA}) + w_k^{AA} \ln w_k^{AA} +$$

$$+ (1 - w_k^{BB}) \ln (1 - w_k^{BB}) + w_k^{BB} \ln w_k^{BB} \}.$$

(8.3.56)

The variational principle begins with the thermodynamic potential

$$\Omega = E - S\,T - \mu\, N, \qquad \textbf{(8.3.57)}$$

where T is an absolute temperature and μ is chemical potential. Remember the formula (8.3.37) and

write

$$N = \sum_k (w_k^{AA} + w_k^{BB}) \,.\ (8.3.58)$$

In our case, the thermodynamic potential is a function of w_k^{AA} and w_k^{BB}:

$$\Omega = \Omega(w_k^{AA}, w_k^{BB}) \,.$$

By differentiating this function with probabilities and equating the derivatives to zero,

we get the equations for the desired functions:

$$\partial\Omega / \partial w_k^{AA} = 0 \,, \partial\Omega / \partial w_k^{BB} = 0 \,. \qquad \textbf{(8.3.59)}$$

8.3.8. Equations for Distribution Functions

We get the equations for the probabilities w_k^{AA} and w_k^{BB}:

$$\ln\left[\left(1-w_k^{AA}\right)/w_k^{AA}\right]=\beta\left(\tilde{\varepsilon}_k^{AA}+\sum_{k'}\tilde{\varepsilon}_{kk'}^{AA}w_{k'}^{AA}-\mu\right),$$

$$\ln\left[\left(1-w_k^{BB}\right)/w_k^{BB}\right]=\beta\left(\tilde{\varepsilon}_k^{BB}+\sum_{k'}\tilde{\varepsilon}_{kk'}^{BB}w_{k'}^{BB}-\mu\right).$$

(8.3.60)

Unfortunately, it is not possible to solve these equations, unless you take approximation

formulas for the energies $\tilde{\varepsilon}_{kk'}^{AA}$ and $\tilde{\varepsilon}_{kk'}^{BB}$ of the interaction of two electrons:

$$\tilde{\varepsilon}_{kk'}^{AA}=I^{AA}\,\delta_{k+k'}-J^{AA}\,\delta_{k-k'},\ \tilde{\varepsilon}_{kk'}^{BB}=I^{BB}\,\delta_{k+k'}-J^{BB}\,\delta_{k-k'}\,.$$

(8.3.61)

Substitute formula (8.3.61) in formula (8.3.55). After simple transformations, we get

$$E=\sum_k\{\tilde{\varepsilon}_k^{AA}w_k^{AA}+\tilde{\varepsilon}_k^{BB}w_k^{BB}\}+$$
$$+1/2\sum_k\{(I^{AA}w_{-k}^{AA}-J^{AA}w_k^{AA})\,w_k^{AA}+$$
$$+(I^{BB}w_{-k}^{BB}-J^{BB}w_k^{BB})\,w_k^{BB}\},$$

(8.3.62)

and putting formulas (8.3.61) in equations (8.3.60) gives

$$\ln\left[\left(1-w_k^{AA}\right)/w_k^{AA}\right]=\beta\left(\tilde{\varepsilon}_k^{AA}+I^{AA}w_{-k}^{AA}-J^{AA}w_k^{AA}-\mu\right),$$

$$\ln\left[\left(1-w_k^{BB}\right)/w_k^{BB}\right]=\beta\left(\tilde{\varepsilon}_k^{BB}+I^{BB}w_{-k}^{BB}-J^{BB}w_k^{BB}-\mu\right).$$

(8.3.63)

8.3.9. Conclusion

Equations (8.3.63) can be successfully solved using a computer and they open up new possibilities for designing various experiments with graphene. For example, we can describe the **superconductivity** of graphene.

REFERENCES

[1] B.V. Bondarev, "Equation for the density matrix of the system of identical particles", *Scientific Discussion*, vol. 1, no. 20, pp. 16-20, 2018.
[2] D.J. Bergman, M.I. Stockman, "Surface plasmon amplification by stimulated emission of radiation: Quantum generation of coherent surface plasmons in nanosystems", *Physical Review Letters*, *Lett.*, vol. 90, pp. 027402, 2003.
[3] M.A. Noginov, G. Zhu, A.M. Belgrave, R. Bakker, V.M. Shalaev, E.E. Narimanov, S. Stout, E. Herz, T. Suteewong, U. Wiesner, "Two-dimensional atomic crystals", *Nature*, vol. 460, pp. 1110, 2009.
[4] N.G. Basov, A.M. Prokhorov, "Molecular oscillator and amplifier", *UFN*, 1955, Vol. 57, pp. 485-501.
[5] J.P. Gordon, H.J. Zeiger, C.H. Townes, "The Maser New Type of Microwave Amplifier,Frequency Standard, and Spectrometer", *Phys. Rev.*, vol. 99. pp. 1264-1274, 1955.
[6] B.V. Bondarev, "Vestnik Donetsk National University", *Ser. A: Natural Sciences*, vol. 4, pp. 54-68, 2017.
[7] B.V. Bondarev, My scientific articles, Book 3, M.: Sputnik+, pp. 75, 2018.
[8] G. Lindblad "On the generators of quantum dynamical semigroups", *Commun. Math. Phys.*, vol. 48, pp. 119-130, 1976.
[9] B.V. Bondarev, "Derivation of the quantum kinetic equation from the equation Liouville – von Neumann", *Theor. Math. Phys.*, vol. 100, pp. 33-43, 1994.
[10] B.V. Bondarev "Method of density matrices in the quantum theory of cooperative phenomena". 2nd ed. M.: Sputnik+, pp. 621, 2013.
[11] B.V. Bondarev, "Eight new theories in quantum physics", M.: Sputnik+, pp. 222, 2018.
[12] K.S. Novoselov, A.K. Geim, "Two-dimensional atomic crystals". *Proc. Natl. Acad. Sci.* USA, vol. 102, no. 30, pp. 10451–10453, 2005.
[13] K.S. Novoselov, A.K. Geim "Two-dimensional gas of massless Dirac fermions in graphene". *Nature*, vol. 438, no. 7065, pp. 197–200, 2005.
[14] B.V. Bondarev, "New theory of superconductivity". *Scientific Research Publishing, Inc*, USA, pp. 125, 2017.
[15] B.V. Bondarev "Density matrix method in theory of two-level laser". *Vestnik of Donetsk National University*, vol. 4, pp. 54-68, 2017.

CHAPTER 9

Perspective of Quantum Physics

9.1. THE LOOK INTO FUTURE OF QUANTUM PHYSICS

Quantum mechanics was based on the Schrödinger equation. Soon a statistical operator and a density matrix were invented, for which the Liouville – von Neumann equations were written. But it was impossible to find a statistical operator from this equation. About fifty years passed when the equation for the statistical operator, in which the dissipative operator was present, was phenomenologically written by Lindblad. Two decades later, the author of this article derived the equation for the density matrix. This equation contains a dissipative matrix, knowledge of which makes it possible to find the density matrix. Subsequently the author found the equation for the density matrix of the particle, which is in the system of identical particles [1].

9.1.1. Introduction

The statistical operator and density matrix in quantum physics are the most informative and most accurate tools. They were named after **J. von Neumann** shortly after the quantum theory was constructed [1]. But for a relatively long time, there was no equation that allowed us to find these values. Although many attempts have been made to find this equation, this article is devoted to the history and further development of quantum physics.

9.1.2. Schrödinger Equation

The basis of quantum mechanics is considered to be the **Schrödinger equation**. In this equation, the unknown is the so-called wave function

$$\psi = \psi(q, t)\,, \qquad \textbf{(9.1.2.1)}$$

where t is time and q is a quantum variable that determine the state of the system.

The meaning of the wave function is the product of

$$\psi^*(q, t)\,\psi(q, t) = w(q, t) \qquad \textbf{(9.1.2.2)}$$

Boris V. Bondarev

it is possible to detect the system in the state q at time t. The probability must satisfy the normalization condition:

$$\int \psi^*(q,t)\,\psi(q,t)\,dq = 1\,. \qquad \textbf{(9.1.2.3)}$$

The Schrödinger equation itself can be written as follows

$$i\hbar\,\partial\psi/\partial t = \hat{H}\,\psi\,, \qquad \textbf{(9.1.2.4)}$$

where $\hat{H} = \hat{H}(q, t)$ is the energy operator of the system. This operator explains what to do with the wave function $\psi(q, t)$ so that it gives us the average energy $E(t)$ of the system at time t:

$$E(t) = \int \psi^*(q,t)\,\hat{H}(q,t)\,\psi(q,t)\,dq\,. \qquad \textbf{(9.1.2.5)}$$

9.1.3. Statistical Operator and Density Matrix

But soon **J. von Neumann** came up with the statistical operator

$$\hat{\varrho} = \hat{\varrho}(q,t)\,. \qquad \textbf{(9.1.3.1)}$$

The statistical operator is related to the density matrix $\varrho_{nn'}(t)$ by the formula

$$\varrho_{nn'}(t) = \int \varphi_n^*(q,t)\,\hat{\varrho}(q,t)\,\varphi_{n'}(q,t)\,dq\,. \qquad \textbf{(9.1.3.2)}$$

where functions $\varphi_n(q, t)$ can be found from the Schrödinger equation. The formula (3.2) specifies the density matrix $\varrho_{nn'}(t)$ in the n-representation. The diagonal element ϱ_{nn} of the density matrix is the probability $w_n = w_n(t)$ that the system is in the state n:

$$\varrho_{nn}(t) = w_n(t)\,. \qquad \textbf{(9.1.3.3)}$$

If the statistical operator is

$$\hat{\varrho}(q) = \delta(q - q_0)\,, \qquad \textbf{(9.1.3.4)}$$

where $\delta(q - q_0)$ is Delta function, q_0 – constant, then the state of the system is called **pure**. Formula (9.1.3.2) gives

$$\varrho_{nn'}(t) = \varphi_n^*(q_0, t)\,\varphi_{n'}(q_0, t)\,. \tag{9.1.3.5}$$

Otherwise, the system state is called **mixed**.

The equation for the statistical operator was derived

$$\mathrm{i}\,\hbar\,\partial\hat{\varrho}/\partial t = [\,\hat{H}\,\hat{\varrho}\,]\,, \tag{9.1.3.6}$$

which is called the **Liouville – von Neumann equation**. This equation was derived from the Schrödinger equation. But it turned out that it was impossible **to find a statistical operator from the Liouville – von Neumann equation**. Prove it.

9.1.4. Something is Missing from Liouville – von Neumann Equation

For the density matrix, equation (9.1.3.6) will look like this

$$\mathrm{i}\,\hbar\,\partial\varrho_{nn'}/\partial t = \textstyle\sum_m\,(\,H_{nm}\,\varrho_{mn'} - \varrho_{nm}\,H_{mn'}\,), \tag{9.1.4.1}$$

where $H_{nn'}$ are the matrix elements of the Hamiltonian $\hat{H}$ of the system. By analogy with the formula (9.1.3.2), we write

$$H_{nn'}(t) = \int \varphi_n^*(q,t)\,\hat{H}(q,t)\,\varphi_{n'}(q,t)\,\mathrm{d}q\,. \tag{9.1.4.2}$$

If it turns out that the matrix elements $H_{nn'}$ are diagonal, *i.e.* have the form

$$H_{nn'}(t) = \varepsilon_n(t)\,\delta_{nn'}, \tag{9.1.4.3}$$

where ε_n – energy eigenvalues of the system, $\delta_{nn'}$ – symbols of Kronecker, then the equation (4.1) takes the form

$$\mathrm{i}\,\hbar\,\partial\varrho_{nn'}/\partial t = \{\varepsilon_n(t) - \varepsilon_{n'}(t)\}\,\varrho_{nn'}\,. \tag{9.1.4.4}$$

For $n = n'$ we obtain

$$\partial\varrho_{nn}/\partial t = 0 \quad \text{or} \quad \partial w_n/\partial t = 0\,. \tag{9.1.4.5}$$

According to this equation, the probability of w_n to find a system in the state n is always constant,which was to be proved.

9.1.5. Lindblad Equation for Statistical Operator

It was only in 1976 that Lindblad found an equation that is suitable for all States, both pure and mixed [2]. The equation is given below

$$\mathrm{i}\,\hbar\,\partial\hat{\varrho}/\partial t = [\,\hat{H}\,\hat{\varrho}\,] + \mathrm{i}\,\hbar\,\hat{D}\,, \tag{9.1.5.1}$$

where $\hat{D}$ is the so-called **dissipative operator** describing the dissipative processes occurring in the system. This operator is

$$\hat{D} = \sum_{jk} C_{jk}\{\,[\,\hat{a}_j\,\hat{\varrho}\,,\,\hat{a}_k^{+}] + [\hat{a}_j\,,\,\hat{\varrho}\,\hat{a}_k^{+}]\,\}\,, \tag{9.1.5.2}$$

where C_{jk} is some numbers, $\hat{a}_j$ is an arbitrary operator.

The statistical operator must be normalized at any *t*

$$\mathrm{Tr}\,\hat{\varrho} = 1\,, \tag{9.1.5.3}$$

self-adjoint

$$\hat{\varrho}^{*} = \hat{\varrho} \tag{9.1.5.4}$$

and positively definite. A correct equation describing the evolution of a statistical operator must ensure that these properties are preserved over time which is evident in equation (9.1.5.1).

9.1.6. Equation for Density Matrix

The equation for the density matrix was derived from the Liouville – von Neumann equation in [3]. The nonequilibrium quantum system *S*, which is in thermal contact with the large equilibrium system *R*. the Composite system *R* + *S* obeys the Liouville - von Neumann equation, was considered in the derivation. The density matrix $\varrho_{nn'}$ describes the behavior of the system *S* and satisfies the derived equation

$$\mathrm{i}\,\hbar\,\partial\varrho_{nn'}/\partial t = \sum_m\,(\,H_{nm}\,\varrho_{mn'} - \varrho_{nm}\,H_{mn'}\,) + \mathrm{i}\,\hbar\,D_{nn'}\,, \tag{9.1.6.1}$$

where $D_{nn'}$ is a dissipative matrix which is

$$D_{nn'} = \sum_{mm'} \gamma_{nm,m'n'}\,\varrho_{mm'} - 1/2\sum_m\,(\,\gamma_{nm}\,\varrho_{mn'} + \varrho_{nm}\,\gamma_{mn'}\,). \tag{9.1.6.2}$$

In this formula $\gamma_{nm,m'n'}$ there is some matrix,

$$\gamma_{nn'} = \sum_m \gamma_{mn',nm} \,. \qquad \textbf{(9.1.6.3)}$$

Equation (9.1.6.1) is analogous to Lindblad's equation. Comparing the formula (9.1.5.2) with (6.2), following is established

$$\gamma_{nm,m'n'} = 2 \sum_{jk} C_{jk}\, a_{nm,j}\; a^*_{n'm',k} \,, \qquad \textbf{(9.1.6.4)}$$

where $a_{nm,j}$ is the matrix elements of the $\hat{a}_j$ operator.

The density matrix satisfies the normalization condition

$$\sum_n \varrho_{nn} = 1 \,. \qquad \textbf{(9.1.6.5)}$$

In addition, the density matrix ϱ_{nm} satisfies the condition

$$\varrho^*_{nm} = \varrho_{mn} \,. \qquad \textbf{(9.1.6.6)}$$

The same condition applies to the Hamiltonian:

$$H^*_{nm} = H_{mn} \,.$$

Consider a system where the density matrix ϱ_{nm} at an arbitrary time t was in a diagonal state:

$$\varrho_{nm}(t) = w_n(t)\, \delta_{nm} \,, \qquad \textbf{(9.1.6.7)}$$

Then, from equations (9.1.6.1) and (9.1.6.7) at $n = n'$ we obtain a quantum kinetic equation

$$\partial w_n / \partial t = \sum_m \left(p_{nm}\, w_m - p_{mn}\, w_n \right), \qquad \textbf{(9.1.6.8)}$$

where

$$p_{nm} = \gamma_{nm,mn} = (2\pi/\hbar) \sum_{NM} |\, v_{nN,mM} \,|^2\, W_M\, \delta(\varepsilon_n - \varepsilon_m + E_N - E_M) \qquad \textbf{(9.1.6.9)}$$

And, there is a probability of transition of the system *S* the *R* environment in a unit of time from the state *m* to the state *n*,

$$W_N = \nu \exp (- \beta E_N)$$

there is a probability that the environment *R* is in an equilibrium state with quantum numbers *N*, and E_N is its energy in this state, ν is the normalization factor, $\beta = 1/(k_B T)$ is the inverse temperature; $v_{nN,mM}$ are the matrix elements of the interaction of the system with its environment. Formula (5.9) is **the Golden rule of Fermi**.

This rule together with the Boltzmann principle allows recording the probability of transition in the form

$$p_{nm} = p_{nm}^{(o)} \exp [- \beta (\varepsilon_n - \varepsilon_m)/2] , \qquad \textbf{(9.1.6.10)}$$

where

$$p_{nm}^{(o)} = p_{mn}^{(o)} . \qquad \textbf{(9.1.6.11)}$$

The equilibrium system *S* is characterized by a time-independent probability:

$$\partial w_n / \partial t = 0 . \qquad \textbf{(9.1.6.12)}$$

Substituting the formula (9.1.6.12) in the kinetic equation (9.1.6.8), we obtain

$$\sum_m (p_{nm} w_m - p_{mn} w_n) = 0 .$$

Using the principle of detailed equilibrium, we will have

$$p_{nm} w_m = p_{mn} w_n .$$

Substituting here the formula (9.1.6.10) and (9.1.6.11), we obtain

$$\exp [- \beta (\varepsilon_n - \varepsilon_m)/2] w_m = \exp [- \beta (\varepsilon_m - \varepsilon_n)/2] w_n .$$

Divide the variables:

$$\exp (\beta \varepsilon_n) w_n = \exp (\beta \varepsilon_m) w_m .$$

Since the left and right parts of this equality depend on different arguments, they will be constant and equal to the constant ν. Next, we come to Boltzmann's formula

$$w_n = \nu \exp(-\beta\, \varepsilon_n)\,.$$

Let's use the formula (9.1.6.9):

$$p_{nm} = \gamma_{nm,mn}\,.$$

We express the matrices $\gamma_{nm,m'n'}$ and $\gamma_{nn'}$ through the transition probability p_{nm}.We will have

$$\gamma_{nm,m'n'} = \sqrt{p_{nm}\, p_{n'm'}}\,, \qquad \gamma_{nn'} = \sum_m \sqrt{p_{mn'}\, p_{mn}}\,.$$

Substituting these matrices in the formula (9.1.6.2), we obtain a dissipative matrix in the form

$$D_{nn'} = \sum_{mm'} \{\sqrt{p_{nm}\, p_{n'm'}}\ \varrho_{mm'} - 1/2\ (\sqrt{p_{m'm}\, p_{m'n}}\ \varrho_{mn'} + \varrho_{nm} \sqrt{p_{m'n'}\, p_{m'm}}\ )\}\,.$$

If we put the density matrix $\varrho_{nn'} = w_n\, \delta_{nn'}$ and $n = n'$ in this formula, we obtain a dissipative matrix

$$D_{nn} = \sum_m (p_{nm}\, w_m - p_{mn}\, w_n)\,.$$

If the matrix p_{nm} is such that

$$\pi_{nm} = \gamma_n\, \delta_{nm}\,,$$

then

$$\gamma_{nm,m'n'} = \sqrt{\gamma_n\, \gamma_{n'}}\ \delta_{nm}\, \delta_{n'm'}\,,$$

Herewith

$$\gamma_{nn'} = \gamma_n\, \delta_{nn'},$$

the dissipative matrix will be equal to

$$D^{(r)}_{nn'} = -\,\Gamma_{nn'}\, \varrho_{nn'}\,,$$

where

$$\Gamma_{nn'} = 1/2\,(\,\gamma_n + \gamma_{n'}) - \sqrt{\gamma_n\,\gamma_{n'}} \geq 0\,.$$

value

$$\Gamma_{nn} = 0.$$

The matrix $D^{(r)}_{nn'}$ describes the relaxation of the system *S*, *i.e.* the pursuit of zero by the nondiagonal element of the density matrix.

9.1.7. Equation for Density Matrix is First Order Approximation

In the study [3], it was proved that the equation for the density matrix can be derived from the quantum Liouville – von Neumann equation for a small system *S* interacting with the large equilibrium system *R*. Therefore, this equation can be written as an expansion in powers of order λ. We write the equation for the statistical operator

$$\mathrm{i}\,\hbar\,\partial\hat{\varrho}/\partial t = [\,\hat{H}\,\hat{\varrho}\,] + \lambda\,\mathrm{i}\,\hbar\,\hat{D}(\hat{\varrho}) + \ldots\,, \qquad \textbf{(9.1.7.1)}$$

here λ is the order parameter. The statistical operator $\hat{\varrho}$ is also convenient to write in the form

$$\hat{\varrho} = \hat{\varrho}_0 + \lambda\,\hat{\varrho}_1 + \ldots \qquad \textbf{(9.1.7.2)}$$

Substitute the operator (9.1.7.2) in equation (9.1.7.1), get

$$\mathrm{i}\,\hbar\,(\,\partial\hat{\varrho}_0/\partial t + \lambda\;\partial\hat{\varrho}_1/\partial t + \ldots\,) = [\,\hat{H}\,\hat{\varrho}_0 + \lambda\,\hat{\varrho}_1 + \ldots\,] + \lambda\,\mathrm{i}\,\hbar\,\hat{D}(\,\hat{\varrho}_0 + \lambda\,\hat{\varrho}_1 + \ldots\,)\,. \qquad \textbf{(9.1.7.3)}$$

If the parameter of order is $\lambda = 0$, we obtain the unperturbed statistical operator $\hat{\varrho}_0$:

$$\mathrm{i}\,\hbar\,\partial\hat{\varrho}_0/\partial t = [\,\hat{H}\,\hat{\varrho}_0]\,. \qquad \textbf{(9.1.7.4)}$$

The first value $\lambda = 1$ gives

$$\mathrm{i}\,\hbar\,\partial\hat{\varrho}_1/\partial t = [\,\hat{H}\,\hat{\varrho}_1] + \mathrm{i}\,\hbar\,\hat{D}(\hat{\varrho}_0)\,. \qquad \textbf{(9.1.7.5)}$$

Equilibrium values are subject to equations:

$$[\hat{H}\,\hat{\varrho}_0] = 0\,, \quad \textbf{(9.1.7.6)}$$

$$[\hat{H}\,\hat{\varrho}_1] + \mathrm{i}\,\hbar\,\hat{D}(\hat{\varrho}_0) = 0\,. \quad \textbf{(9.1.7.7)}$$

Equation (9.1.7.7) is equivalent to

$$\hat{D}(\hat{\varrho}_0) = 0\,. \quad \textbf{(9.1.7.8)}$$

Thus, the equilibrium values of the density matrix satisfy the system of the following equations:

$$\begin{cases} [\hat{H}\,\hat{\varrho}_0] = 0\,, \\ \hat{D}(\hat{\varrho}_o) = 0\,. \end{cases} \quad \textbf{(9.1.7.9)}$$

We write the equations (9.1.7.4) and (9.1.7.5) for the density matrix in some *n*-representation:

$$\begin{cases} \mathrm{i}\,\hbar\,\partial\varrho^{(o)}_{nn'}/\partial t = \sum_m (H_{nm}\,\varrho^{(o)}_{mn'} - \varrho^{(0)}_{nm}\,H_{mn'})\,, \\ \mathrm{i}\,\hbar\,\partial\varrho^{(1)}_{nn'}/\partial t = \sum_m (H_{nm}\,\varrho^{(1)}_{mn'} - \varrho^{(1)}_{nm}\,H_{mn'}) + \mathrm{i}\,\hbar\,D_{nn'}(\varrho^{(o)}_{nn'})\,, \end{cases}$$

(9.1.7.10)

where $\varrho^{(o)}_{nn'}$ and $\varrho^{(1)}_{nn'}$ are density matrices in the unperturbed and perturbed states.

We assume that in the *n*-representation we are given the matrix $H_{nn'}$ Hamiltonian. Find the α-representation where the Hamiltonian matrix is diagonal:

$$H_{\alpha\alpha'} = \varepsilon_\alpha\,\delta_{\alpha\alpha'}\,. \quad \textbf{(9.1.7.11)}$$

To do this, you need to find a unitary matrix $U_{n\alpha}$ that asustuse the transition from *n*-views in the α-representation. By definition, the unitary matrix is such that

$$\textstyle\sum_\alpha U_{\alpha n}\,U^*_{\alpha n'} = \delta_{nn'}\,, \qquad \sum_n U^*_{\alpha n}\,U_{\alpha' n} = \delta_{\alpha\alpha'}\,. \quad \textbf{(9.1.7.12)}$$

Now we can write the first equation in the system (9.1.7.10) as follows

$$\mathrm{i}\,\hbar\,\partial\varrho^{(0)}_{\alpha\alpha'}/\partial t = (\varepsilon_\alpha - \varepsilon_{\alpha'})\,\varrho^{(0)}_{\alpha\alpha'}\,. \quad \textbf{(9.1.7.13)}$$

This equation has a simple solution if only their own energy ε_α does not depend on time.

We now write the second equation of the system (9.1.7.10) in the α-representation:

$$\mathrm{i}\,\hbar\,\partial\varrho^{(1)}_{\alpha\alpha'}/\partial t = (\,\varepsilon_\alpha - \varepsilon_{\alpha'}\,)\,\varrho^{(1)}_{\alpha\alpha'} + \mathrm{i}\,\hbar\,D_{\alpha\alpha'}(\varrho^{(0)}_{\alpha\alpha'})\,. \qquad \textbf{(9.1.7.14)}$$

It is required to think about how to solve these equations.

9.1.8. Equation for Density Matrix of System of Identical Particles

In the study [4], the theory of the equation for matrices of density of the particle entering system of identical particles was under construction. Here we will consider the scheme of construction of this theory.

Consider a system consisting of N identical particles. The state of this system is determined by the value q, which represents a set of parameters q_i, where

$$i = 1, 2\,, \ldots, N\,;$$

each of which determines the state of one particle of this system. The statistical operator $\hat{\varrho}$ describing the state of such a system can be written as follows:

$$\hat{\varrho} = \hat{\varrho}(q, t) = \hat{\varrho}(\,q_1, q_2, \ldots, q_N, t\,) \equiv \hat{\varrho}(\,1, 2, \ldots, N, t\,)\,. \qquad \textbf{(9.1.8.1)}$$

Here, the numbers in parentheses indicate the indexes of the variables on which the operator depends. The statistical operator, as well as any other operator describing the state of the system of identical particles, must be symmetric:

$$\hat{\varrho}(\,\ldots, i, \ldots, j, \ldots\,) = \hat{\varrho}(\,\ldots, j, \ldots, i, \ldots\,)\,. \qquad \textbf{(9.1.8.2)}$$

Let us assume the following normalization condition for the statistical operator:

$$\mathrm{Tr}_{12\ldots N}\,\hat{\varrho}(\,1, 2, \ldots, N, t\,) = N!\,. \qquad \textbf{(9.1.8.3)}$$

The Hamilton operator for a system of identical particles can be represented as a sum

$$\hat{H}(\,1, 2, \ldots, N\,) = \sum_{i=1}^{N} \hat{H}(i) + 1/2 \sum_{i=1}^{N} \sum_{j=1,\ j\neq i}^{N} \hat{H}(i,j)\,, \qquad \textbf{(9.1.8.4)}$$

where $\widehat{H}(i)$ is a one-partial Hamiltonian, *i.e.* the energy operator of one particle without taking into account its interaction with other particles; $\widehat{H}(i,j)$ is the interaction operator of two particles.

One-partial Hamiltonian may contain dissipative terms. The summation operation in the expression (9.1.5.2) can be temporarily omitted and written as follows

$$\widehat{D}(\hat{\varrho}) = [\,\hat{a}\,\hat{\varrho}\,,\hat{a}^{+}] + [\,\hat{a}\,,\hat{\varrho}\,\hat{a}^{+}]\,. \qquad \textbf{(9.1.8.5)}$$

Suppose that the dissipative processes occurring in the system are due to the stochastic interaction of particles with the heat reservoir and each particle interacts with it independently of other particles. In this case, the many partial operators $\hat{a}$ in the expression (9.1.3.6) can be represented as the sum of:

$$\hat{a}(\,1,\dots,N\,) = \textstyle\sum_{i=1}^{N}\hat{a}(i)\,, \qquad \textbf{(9.1.8.6)}$$

where operator $\hat{a}(i)$ characterizes the effect of the heat reservoir on one of the particles. Substituting this sum into the expression (9.1.8.5) yields

$$\widehat{D}(\hat{\varrho}) = \textstyle\sum_{i=1}^{N}\sum_{j=1}^{N}\,\{\,[\,\hat{a}(i)\,\hat{\varrho}\,,(j)\,] + [\,\hat{a}(i)\,,\hat{\varrho}\,\hat{a}^{+}(j)\,]\,\}\,. \qquad \textbf{(9.1.8.7)}$$

The two-partial Hamiltonian must be symmetric:

$$\widehat{H}(i,j) = \widehat{H}(j,i)\,. \qquad \textbf{(9.1.8.8)}$$

Under this condition, the Hamiltonian (9.1.3.4) will also be symmetric.

Thus, all parameters included in the Lindblad equation (9.1.5.1) determined

$$\mathrm{i}\,\hbar\,\partial\hat{\varrho}/\partial t = [\,\widehat{H}\,\hat{\varrho}\,] + \mathrm{i}\,\hbar\,\widehat{D}\,,$$

It is very difficult to write down and solve this equation. Therefore, a one-partial operator was introduced

$$\hat{\varrho}(1) = 1/(\,N-1\,)!\,\mathrm{Tr}_{23\dots N}\,\hat{\varrho}(\,1,2,3,\dots,N\,)\,. \qquad \textbf{(9.1.8.9)}$$

From this definition, and condition (9.1.8.3) implies the normalization condition for

$$\mathrm{Tr}_1\, \hat{\varrho}(1) = N\,. \qquad \mathbf{(9.1.8.10)}$$

To obtain the equation for the derivative of the one-partial operator $\hat{\varrho}(1)$, we apply to both parts of the equation (9.1.5.1) the convolution operation $\mathrm{Tr}_{23\ldots N}$. Come to the equation

$$\begin{aligned} \mathrm{i}\,\hbar\, \partial\hat{\varrho}(1)/\partial t = [\, \hat{H}(1)\, \hat{\varrho}(1)] + 1/2\, \mathrm{Tr}_2\, [\, \hat{H}(1,2)\, \hat{\varrho}(1,2)\,] + \\ + \mathrm{i}\,\hbar\, \{\, [\, \hat{a}(1)\, \hat{\varrho}(1)\, ,\, \hat{a}^{+}(1)\,] + [\, \hat{a}(1)\, ,\, \hat{\varrho}(1)\, \hat{a}^{+}(1)\,] + \\ + \mathrm{Tr}_2\, [\hat{a}(2)\, \hat{\varrho}(1,2)\, ,\, \hat{a}^{+}(1)\,] + [\, \hat{a}(1)\, ,\, \mathrm{Tr}_2\, \hat{\varrho}(1,2)\, \hat{a}^{+}(2)\,]\, \}\,. \end{aligned} \qquad \mathbf{(9.1.8.11)}$$

Now we need to get rid of the two-partial statistical operator $\hat{\varrho}(1, 2)$. This can only be done when this operator is written using a two-part density matrix. Therefore, we denote the matrix elements of the operators $\hat{\varrho}(1)$, $\hat{\varrho}(1, 2)$, $\hat{H}(1)$, $\hat{H}(1, 2)$ and $\hat{a}(1)$ in some n-representation as follows

$$\begin{aligned} \varrho_{n_1 n'_1} = \varrho_{11'}\,, \qquad & \varrho_{n_1 n_2, n'_1 n'_2} = \varrho_{12,1'2'}\,, \\ H_{n_1 n'_1} = H_{11'}\,, \qquad & H_{n_1 n_2, n'_1 n'_2} = H_{12,1'2'}\,, \\ a_{n_1 n'_1} = & \; a_{11'}\,, \end{aligned}$$

Two-partial density matrix $\varrho_{12,1'2'}$, describing the state of the boson system should be symmetrical. While the two-partial density matrix of the fermion system should be antisymmetric:

$$\varrho_{12,1'2'} = \pm\, \varrho_{21,1'2'} = \pm\, \varrho_{12,2'1'} = \varrho_{21,2'1'}\,.$$

Approximately this matrix can be expressed as a one-partial density matrix as follows:

$$\varrho_{12,1'2'} \cong \varrho_{11'}\, \varrho_{22'} \pm \varrho_{12'}\, \varrho_{21'}\,, \qquad \mathbf{(9.1.8.12)}$$

where the plus sign corresponds to bosons and the minus sign corresponds to fermions.

Now we write the exact equation (9.1.8.11) in matrix form:

$$\begin{aligned} \mathrm{i}\,\hbar\, \partial\varrho_{11'}/\partial t = \textstyle\sum_{n_2}\, [\, H_{12}\, \varrho_{21'}\,] + 1/2 \textstyle\sum_{n_2 n_3 n_4}\, [\, H_{12,34}\, \varrho_{43\,21'}] + \\ + \mathrm{i}\,\hbar \textstyle\sum_{n_2 n_3}\, (\,[\, a_{12}\, \varrho_{23}\, ,\, a^{+}_{31'}\,] + [\, a_{12}\, ,\, \varrho_{23}\, a^{+}_{31'}\,]\,) + \end{aligned}$$

$$+\mathrm{i}\,\hbar \sum_{n_2 n_3 n_4} ([\, a_{34}\, \varrho_{43,12}\, , a^{+}_{21'}\,] + [\, a_{12}\, , \varrho_{21',34}\, a^{+}_{43}\,])\,. \tag{9.1.8.13}$$

Since the matrix $H_{12,34}$ is symmetric for bosons and antisymmetric for fermions, hence

$$H_{12,1'2'} = \pm\, H_{21,1'2'} = \pm\, H_{12,2'1'} = H_{21,2'1'}\,,$$

the summand in this equation

$$\sum_{n_2 n_3 n_4} [\, H_{12,34}\, \varrho_{43\,21'}]$$

the formula (9.1.8.12) can be used to convert to the form

$$\sum_{n_2 n_3 n_4} [\, H_{12,34}\, \varrho_{43\,21'}] \cong \sum_{n_2 n_3 n_4} [\, H_{12,34}\, (\, \varrho_{42}\, \varrho_{31'} \pm \varrho_{41'}\, \varrho_{32})\,] =$$
$$= 2 \sum_{n_2 n_3 n_4} [\, H_{12,34}\, , \varrho_{42}\, \varrho_{31'}\,]\,. \tag{9.1.8.14}$$

Now substitute in equation (9.1.8.13) instead of $\varrho_{12,1'2'}$ approximate formula (9.1.8.12) and (8.14). We will have

$$\mathrm{i}\,\hbar\, \partial\varrho_{11'}/\partial t = \sum_{n_2} [\, H_{12}\, \varrho_{21'}\,] + \sum_{n_2 n_3 n_4} [\, H_{12,34}\, , \varrho_{42}\, \varrho_{31'}\,] +$$
$$+\mathrm{i}\,\hbar \sum_{n_2 n_3} ([\, a_{12}\, \varrho_{23}\, , a^{+}_{31'}\,] + [\, a_{12}\, , \varrho_{23}\, a^{+}_{31'}\,]) +$$
$$+\mathrm{i}\,\hbar \sum_{n_2 n_3 n_4} \{[\, a_{34}\, (\, \varrho_{41}\, \varrho_{32} \pm \varrho_{42}\, \varrho_{31}\,)\, , a^{+}_{21'}] +$$
$$+ [\, a_{12}\, , \, (\, \varrho_{23}\, \varrho_{1'4} \pm \varrho_{24}\, \varrho_{1'3}\,)\, a^{+}_{43}]\,\}\,. \tag{9.1.8.15}$$

Remember the formula (9.1.6.4). Now this formula will look like this

$$\gamma_{nm,m'n'} = 2\, a_{nm}\, a^{*}_{n'm'}\,. \tag{9.1.8.16}$$

Opening square brackets and substituting this formula, we obtain the desired equation for the density matrix

$$\mathrm{i}\,\hbar\, \partial\varrho_{11'}/\partial t = \sum_{n_2} [\, H_{12}\, \varrho_{21'}\,] + \sum_{n_2 n_3 n_4} [\, H_{12,34}\, , \varrho_{42}\, \varrho_{31'}] +$$
$$+\mathrm{i}\,\hbar/2 \sum_{n_2 n_3} (\, 2\, \gamma_{12,31'}\, \varrho_{23} - \gamma_{23,12}\, \varrho_{31'} - \gamma_{31',23}\, \varrho_{12}\,) +$$

$$+\mathrm{i}\,\hbar \sum_{n_2 n_3 n_4} \{ (\varrho_{14}\, \varrho_{23} \pm \varrho_{24}\, \varrho_{13})\, \gamma_{34,21'} - \gamma_{34,12}\, (\varrho_{24}\, \varrho_{1'3} \pm \varrho_{1'4}\, \varrho_{23}) + + \gamma_{12,43}\, (\varrho_{32}\, \varrho_{41'} \pm \varrho_{42}\, \varrho_{31'}) - (\varrho_{31}\, \varrho_{42} \pm \varrho_{41}\, \varrho_{32})\, \gamma_{21',43} \} \,. \tag{9.1.8.17}$$

Substitute the diagonal matrix (9.1.6.7) into equation (9.1.8.17):

$$\varrho_{12} = w_1\, \delta_{12} \,. \tag{9.1.8.18}$$

Put in the equation $n_2 = n_1$, we have a quantum kinetic equation

$$\partial w_1/\partial t = \sum_{n_2} \{ p_{12}\, w_2\, (1 \pm w_1) - p_{21}\, w_1\, (1 \pm w_2) \} \,. \tag{9.1.8.19}$$

where

$$p_{12} = \gamma_{12,21} \,, \tag{9.1.8.20}$$

there is a probability of transition of a particle per unit of time from the state n_2 to the state n_1.

9.1.9. Energy of System of Identical Particles. Unitary Matrix

We write a known expression for the average energy of the system of identical particles

$$E = \sum_{n_1 n_2} H_{12}\, \varrho_{21} + 1/2 \sum_{\{n\}} H_{12,34}\, \varrho_{34,12} \,, \tag{9.1.9.1}$$

where $\{n\} = n_1,\ n_2,\ n_3,\ n_4$.

Let a κ-representation of the density matrix will be diagonal, *i.e.* have the form

$$\tilde{\varrho}_{\kappa\kappa'} = w_\kappa\, \delta_{\kappa\kappa'}, \tag{9.1.9.2}$$

$$\tilde{\varrho}_{\kappa_1\kappa_2,\kappa_1'\kappa_2'} \equiv \tilde{\varrho}_{12,1'2'} = w_{12}^{(11)}\, (\delta_{11'}\, \delta_{22'} \pm \delta_{12'}\, \delta_{21'}) \,, \tag{9.1.9.3}$$

where the probability w_κ and $w_{12}^{(11)}$ satisfy the conditions

$$\sum_\kappa w_\kappa = N ,$$
$$\sum_{\kappa_1\,\kappa_2 \neq \kappa_1} w_{12}^{(11)} = N (N - 1) . \qquad \textbf{(9.1.9.4)}$$

The probability $w_{12}^{(11)}$ determines that both the states κ_1 and $\kappa_2 \neq \kappa_1$ are occupied by particles.

The transition from the *n*-representation to the *κ*-representation is carried out using the unitary matrix $U_{n\kappa}$, which is determined by the relation

$$\sum_n U_{n\kappa}\, U^*_{n\kappa'} = \delta_{\kappa\kappa'} . \qquad \textbf{(9.1.9.5)}$$

If the unitary matrix is known, then the density matrix in the *κ*-representation is found by the formula

$$\tilde{\varrho}_{\kappa\kappa'} = \sum_{nn'} U_{n\kappa}\, \varrho_{nn'}\, U^*_{n'\kappa'} . \qquad \textbf{(9.1.9.6)}$$

Since the expression (9.1.9.1) in any representation has the same form, substituting expressions (9.1.9.2) and (9.1.9.3) in this formula, we will have

$$E = \sum_\kappa \varepsilon_\kappa\, w_\kappa + 1/2 \sum_{\kappa_1\,\kappa_2 \neq \kappa_1} \varepsilon_{12}\, w_{12}^{(11)} , \qquad \textbf{(9.1.9.7)}$$

where formula (9.1.9.7) is written for the fermion system;

$$\varepsilon_\kappa = \sum_{nn'} U^*_{n\kappa}\, H_{nn'}\, U_{n'\kappa} \qquad \textbf{(9.1.9.8)}$$

is the energy of the particle in the state *κ*,

$$\varepsilon_{12} = 2 \sum_{\{n\}} U^*_{1\kappa_1}\, U^*_{2\kappa_2}\, H_{12,34}\, U_{3\kappa_1} U_{4\kappa_2} \qquad \textbf{(9.1.9.9)}$$

there is the interaction energy of two particles, one of which is in the state κ_1, and the other – in the state $\kappa_2 \neq \kappa_1$.

Let's put the probability $w_{12}^{(11)}$ approximately equal to

$$w_{12}^{(11)} \cong w_1\, w_2\,. \quad \textbf{(9.1.9.10)}$$

Calculate the average energy of the particle by the formula

$$\bar{\varepsilon}_\kappa = \partial E/\partial w_\kappa\,.$$

Differentiating w_κ energy (9.1.9.7) with (9.10), we have

$$\bar{\varepsilon}_\kappa = \varepsilon_\kappa + \sum_{\kappa'} \varepsilon_{\kappa\kappa'}\, w_{\kappa'}\,. \quad \textbf{(9.1.9.11)}$$

9.1.10. Transition Probability

The probability of transition of a particle from one state to another can be found by the formula (9.1.6.10):

$$p_{12} = p_{12}^{(o)} \exp\left[-\beta\,(\,\bar{\varepsilon}_1 - \bar{\varepsilon}_2\,)/2\,\right], \quad \textbf{(9.1.10.1)}$$

where

$$p_{12}^{(o)} = p_{21}^{(o)}.$$

Now the transition probability depends on the average energy $\bar{\varepsilon}_\kappa$ of the particle from other particles in the system.

Let the system of particles come to a state of thermodynamic equilibrium. Since the probability of w_1 is now time-independent, the right side of the equation (9.1.8.19) will be zero. Thus, the principle of detailed equilibrium, in this case, is expressed by equality

$$p_{12}\, w_2\,(\,1 \pm w_1) = p_{21}\, w_1\,(\,1 \pm w_2)\,. \quad \textbf{(9.1.10.2)}$$

Substituting the expression (9.1.10.1) into this equation, after simple transformations we obtain

$$e^{-\beta\,\bar{\varepsilon}_1}\,(\,1 \pm\, w_1)/w_1 = e^{-\beta\,\bar{\varepsilon}_2}\,(\,1 \pm\, w_2)/w_2\,, \quad \textbf{(9.1.10.3)}$$

where the left part depends on κ_1, and the right - κ_2. This is only possible when both parts of equality are equal to the same constant value. Let's denote this value $\exp(\,-\,\beta\,\mu\,)$, where μ is the chemical potential. Come to the equation

$$(1 \pm w_{\kappa})/w_{\kappa} = \exp[\beta(\bar{\varepsilon}_{\kappa} - \mu)], \quad \textbf{(9.1.10.4)}$$

in which the average energy of the particle $\bar{\varepsilon}_{\kappa}$ according to the formula (9.1.9.11) is a functional depending on the distribution function w_{κ}. Only by knowing the dependence of the energy $\bar{\varepsilon}_{\kappa}$ on the probability w_{κ}, it is possible to find the distribution of particles by States.

For equilibrium systems of noninteracting particles, the equation (9.1.10.4) leads to Fermi - Dirac or Bose -- Einstein functions:

$$w = 1/[1 \pm e^{\beta(\varepsilon - \mu)}], \quad \textbf{(9.1.10.5)}$$

where the energy ε is the eigenvalue of a one-partial Hamiltonian.

9.1.11. Variational Principle

If the density matrix does not depend on time, *i.e.* the system is in equilibrium, then the **variational principle** can be applied without resorting to the equation for the density matrix. To carry out this system, known thermodynamic potential is needed

$$\Omega = E - ST - \mu N, \quad \textbf{(9.1.11.1)}$$

where E is the average energy of the system under study, S is the entropy, T is the absolute temperature, μ is the chemical potential, N is the average number of particles in the system.

The energy for the system of fermions in the first approximation has the form

$$E = \sum_{\kappa} \varepsilon_{\kappa} w_{\kappa} + 1/2 \sum_{\kappa_1 \kappa_2 \neq \kappa_1} \varepsilon_{12} w_1 w_2 . \quad \textbf{(9.1.11.2)}$$

Now you need to know the formula for entropy. This formula can be written only in the same approximation. Here is the formula

$$S = -k_B \sum_{\kappa} \{(1 - w_{\kappa}) \ln(1 - w_{\kappa}) + w_{\kappa} \ln w_{\kappa}\} . \quad \textbf{(9.1.11.3)}$$

The number N Express, using the formula (9.1.9.4), using the probability w_{κ}:

$$N = \sum_{\kappa} w_{\kappa} . \quad \textbf{(9.1.11.4)}$$

So, the formula for the thermodynamic potential Ω is obtained. It is expressed only throug9.1.h the probability w_κ:

$$\Omega = \Omega(w_k)\,. \qquad \textbf{(9.1.11.4)}$$

We differentiate this function by probability w_k and equate the derivative to zero, we will have the equation

$$(1 - w_\kappa)/w_\kappa = \exp[\ (\bar{\varepsilon}_\kappa - \mu)\ ]\,.$$

This equation is equivalent to equation (9.1.10.4). Thus, the variational principle is identical to the previously proposed method.

However, the variational principle allows to find not only the particle distribution function w_κ, but also the binary probability $w_{12}^{(11)}$. To do this, the correlation function $\xi_{\kappa\kappa'}$ by means of the relations is introduced

$$w_{\kappa\kappa'}^{(11)} = w_\kappa\, w_{\kappa'}\,(1 - \delta_{\kappa\kappa'}) + \xi_{\kappa\kappa'}\,.$$

9.1.12. Conclusion

This article briefly reviewed the history of the creation and further development of quantum physics. The most striking in this story are the Lindblad equation and the equation for the density matrix, which was obtained by the author of this article. Just as in the Lindblad equation in this equation was the appearance of a dissipative matrix.

REFERENCES

[1] B.V. Bondarev, "The look into the future of quantum physics", *Scientific Discussion*, vol. 1, on. 26, pp. 3. 6-13, 2018.

[2] G. Lindblad, "On the generators of quantum dynamical semigroups", *Commun. Math. Phys.*, vol. 48, pp. 119-130, 1976.

[3] B.V. Bondarev, "Derivation of the quantum kinetic equation from the equation Liouville – von Neumann", *Theor. Math. Phys.*, vol. 100, pp. 33-43, 1994.

[4] B.V. Bondarev, "Equation for the density matrix of the system of identical particles", *Scientific Discussion*, vol. 1, no. 20, pp. 16-20, 2018.

Author List of Scientific Articles and Books

General Relativity

[1] Shirokov M.F., Bondarev B.V. General relativistic equations of motion in topocentric coordinates, Preprint ITF-73-140R, Kiev, 1973, p. 1-14.
[2] Bondarev B.V. Annual variations in the duration of stellar days as the result curvature of the rays of light in the field of Sun, Astronomical journal, 1974, v. 51, № 3, p. 664-665.
[3] Shirokov M.F., Bondarev B.V. New way of checking the effect of the rays in the field Sun, Preprint ITF-74-62R, Kiev, 1974, p. 1-10.
[4] Bondarev B.V. Annual variations in the duration of stellar days as a result curvature of light rays in the field of the Sun, Collection of scientific proceedings of MAI, Research on theoretical and applied physics, M.: News of MAI, 1974, № 290, p. 40-43.
[5] Shirokov M. F., Bondarev B.V. On the stability of finite motions of geodesic the Schwarzschild metric, News of MAI, Series physics, 1974, № 12, p. 52-55.
[6] Bondarev B.V. Some problems of experiment and observations in theory of gravitation of Einstein, The dissertation on competition of scientific degree of the candidate physical and mathematical sciences, Moscow, 1974, p. 1-13.
[7] Bondarev B.V. Shirokov effect in tetrad formulation, Collection of scientific works of MAI, Problems of relativity in terrestrial and cosmic conditions, M.: News of MAI, 1988, p. 7-11.

Capillary Discharge

[8] Bondarev B.V., Muravenko V.G., Shirokov M.F. The one-dimensional theory of capillary discharge with evaporating walls, TVT, 1977, v. 15, № 3, p. 465-470.

Step Kinetics

[9] Bondarev B.V. On the possible mechanism of the kinetics of bimolecular reactions condensed
phase, Collection of scientific proceedings of the MAI.
Physical processes in neutral and ionized gases, M.: Ed. MAI, 1981, p. 41-44.
[10] Bondarev B.V. Correlation theory of kinetics of solid-phase reactions, Kinetics and catalysis, 1982, v. 23, № 2, p. 334-339.

Boris V. Bondarev

[11] Bondarev B.V. Investigation of the kinetics of solid-state reactions involving radicals, Polymer Science, Series A, 1982, v. 24, № 9, p. 1901-1907.
[12] Bondarev B.V. Kinetics of parallel bimolecular reactions in condensed matter phase. Study of defrosting curves, Kinetics and catalysis, 1982, v. 23, № 4, p. 798-804.
[13] Bondarev B.V. Tunnel mechanism of free radical microdiffusion in solids organic matter, Polymer Science, Series A, 1983, v. 25, № 11, p. 2382-2389.
[14] Bondarev B.V. Influence of the tunnel transition of hydrogen atom on the "step" kinetics of solid-phase reactions involving radicals in organic substances, Kinetics and catalysis, 1983, v. 24, № 3, p. 568-571.
[15] Bondarev B.V. Kinetic theory of bimolecular reaction in condensed media, Chem. Phys, 1985, v. 97, № 1, p. 73-86.
[16] Bondarev B.V. Kinetics of the death of stabilized electrons in polyethylene, Polymer Science, Series A, 1985, v. 27, № 12, p. 2589-2593.
[17] Bondarev B.V. A kinetic theory of reaction species accumulation, Chem. Phys, 1986, v. 109, № 2-3, p. 195-206.
[18] Bondarev B.V. A statistical theory of kinetics of bimolecular reactions in condensed media, Chem. Phys, 1987, v. 113, № 3, p. 321-347.
[19] Bondarev, 1988, Statistical theory of systems of created reactive species, Physica A, 1988, v. 148, № 3, p. 456-502.
[20] Bondarev B.V. To the problem of probability of dissociation of molecule in condensed environment, Collection of scientific proceedings of MAI, Experimental and theoretical questions applied physics research, Moscow: News of MAI, 1985, p. 27-29.
[21] Bondarev B.V. Kinetics of diffusion-controlled bimolecular reactions in condensed phase, Kinetics and catalysis, 1985, v. 26, № 2, p. 505-506.
[22] Bondarev B.V. The rate constant of a quantum-diffusion-controled bimolecular reaction, Chem. Phys, 1986, v. 103, № 2-3, p. 183-197.
[23] Bondarev B.V. The probability of the jump-diffusion-controled reaction of a species pair, Physica A, 1987, v. 142, № 1-3, p. 273-308.
[24] Bondarev B.V. The probability of the quantum diffusion-controlled reaction of a species pair, Physica A, 1992, v. 186, p. 388-404.

Binary Alloys

[25] Bondarev B.V. Binary alloys. Statistical theory of nonquilibrium states. Kinetics of ordering, Physica A, 1989, v. 155, № 1, p. 123-166.

Density Matrix

[26] Bondarev B.V. Quantum Markovian master equation for a system of identical particles interacting with the heat reservoir, Physica A, 1991, v. 176, № 2, p. 366-386.
[27] Bondarev B.V. The quantum Markovian master equation and its applications to quasi-particle migration in a stochastic medium, Phys. Lett. A, 1991, v. 153, № 6-7, p. 326-329.
[28] Bondarev B.V. Quantum lattice gas. Method of density matrix, Physica A, 1992, v. 184, p. 205-230.
[29] Bondarev B.V. Quantum Markovian master equation theory of particle migration in a stochastic medium, Physica A, 1992, v. 183, № 1, p. 159-174.
[30] Bondarev B.V. The long-range ordering in a quantum lattice gas, Physica A, 1994, v. 209, p. 477-485.
[31] Bondarev B.V. Derivation of the quantum kinetic equation from the equation Liouville – von Neumann, Theor. Math. Phys. 1994, v. 100, № 1, p. 33-43.
[32] Bondarev B.V. Quantum diffusion of a particle in a stochastic medium, News of MAI, 1999, v. 6, № 2, p. 42-44.
[33] Bondarev B. V. Quantum Markov kinetic equation for the system of identical particles, Vestnik MAI, 2001, v. 8, № 2, p. 61-64.
[34] Bondarev B.V. Density matrix equation for harmonic oscillator, Arxiv: 1302.0303, 1 Feb. 2013, p. 1-8.

Bose Condensation

[35] Bondarev B.V. Density matrix method in the theory of equilibrium bozon states, Vestnik MAI, 1998, v. 5, № 1, p. 34-40.
[36] Bondarev B.V. Application of the variational density matrix method for the description of thermodynamic properties of quantum Bose gas, Vestnik MAI, 1998, v. 5, № 2, p. 53-60.
[37] Bondarev B.V. Theory of superfluidity. Method of equilibrium density matrix, Arxiv: 1412.6004, 25 Dec. 2013, p. 1-12.

Superconductivity

[38] Bondarev B.V. On some features of the electron distribution function the Bloch states. Vestnik MAI, 1996, v. 3, № 2, p. 56-65.
[39] Bondarev B.V. Anisotropy and superconductivity. Arxiv: 1302.5066, 12 Feb. 2013, p. 1-14.
[40] Bondarev B.V. The Fermy – Dirac function and the anisotropic distribution

of the interaction electrons, Arxiv: 1301.4711, 21 Jan. 2013, p. 1-11.
[41] Bondarev B.V. Master equation for a system of identical particles and the anisotropic distribution of the interacting electrons, Arxiv: 1301.4712, 21 Jan. 2013, p. 1-12.
[42] Bondarev B.V. New theory of superconductivity. Method of equilibrium density matrix. Arxiv: 1412.6008, 22 Sep. 2013, p. 1-14.
[43] Bondarev B.V. Fermi – Dirac function and energy gap. Arxiv: 1412.6009 22 Sep. 2013, p. 1-5.
[44] Bondarev B.V. Gapless superconductivity, Int. J. Phys. 2015, v. 3, № 2, p. 88-95.
[45] Bondarev B.V. Method of equilibrium density matrix. Energy of interacting valence electrons in metal, Int. J. Phys. 2015, v. 3, № 3, p. 108-112.
[46] Bondarev B.V. New theory of superconductivity. Method of equilibrium density matrix. Magnetic field in superconductor, Open Access Library Journal, 2015, v. 2: Oalib 1102149, p. 1-20.
[47] Bondarev B.V. Method of eguilibrium density matrix. Anisotropy and superconductivity, Energy gap. Advanced Materials: Manufacturing, Physics, Mechanics and Applications; edt. Ivan A. Parinov. New York, London, Springer Proceedings in Physics, 2016, v. 175, p. 157-178.
[48] Bondarev B.V. New Theory of Superconductivity. Magnetic Field in Superconductor. Effect of Meissner and Ochsenfeld. Open Access Library Journal, 2016, v. 3: Oalib 1102418, p. 1-27.
[49] Bondarev B.V. New Theory of Superconductivity. Does the London Equation Have the Proper Solution? No, It Does Not. Self-Generated and External Magnetic Fields in Superconductors, Open Access Library Journal, 2016, v. 3: Oalib 1103148, p. 1-20.

Theory of Lasers

[50] Bondarev B.V. Investigation of open resonators of optical gas generators. Moscow Institute of physics and technology, Student thesis,1965, p. 46.
[51] Bondarev B.V. New method of density matrices in the theory of a two-level laser. Bulletin Science, Prospects of development of science in the modern world, Ufa: Ed. Dendra, December 2017, Part 1, p. 7-11.
[52] Bondarev B.V. New method of density matrices. Spectral energy density two-level laser radiation. Scientific discussion, 2017, v. 1, № 14, p. 16-17.
[53] Bondarev B.V. New method of density matrices. Spectral energy density two-level laser radiation. Bulletin of Science, Topical issues in science and practice, Ufa: Ed. Dendra, February 2018, Part 1, p. 203-207.
[54] Bondarev B.V. Density matrix method in two level laser theory. European

multi science journal, 2018, №12, p. 18-26.
[55] Bondarev B.V. The density matrix Method in the new laser theory. Spectral density energy of laser radiation. European multi science journal, 2018, № 14, p. 44-53.
[56] Bondarev B.V. The density matrix. The theory of three-level laser. Scientific discussion, 2018, v. 1, № 16, p. 28-34.
[57] Bondarev B.V. The density matrix method in the theory of a two-level laser. Vestnik of Donetsk Nationa Vestnik l University, 2017, № 4, p. 54-68.
[58] Bondarev B.V. New theory of laser. Method of density matrix. Advanced Materials: Manufacturing, Physics, Mechanics and Applications; edt. Ivan A. Parinov. New York, London, Springer Proceedings in Physics 2019, v. 224, p. 163-186.

Distribution of Electrons in Arbitrary Atom

[59] Bondarev B.V. Density Matrix Method in Theory of Atom. Advanced Materials: Manufacturing, Physics, Mechanics and Applications; edt. Ivan A. Parinov. New York, London, Springer Proceedings in Physics, 2018, v. 207, p. 145-159.
[60] Bondarev B.V. Density matrix method. New calculation of electron energy levels in atom, Scientific discussion, 2018, v. 1, № 18, p. 32-36.

Theory of Quantum Oscillator. Dissipative Matrix

[61] Bondarev B.V. Lindblad Equation for the Harmonic Oscillator. Uncertainty Relation Depending on Temperature, Applied Mathematics, 2017, v. 8, p. 1529-1538.
[62] Bondarev B.V. Statistical operator in the theory of quantum oscillator. Dissipative decay operator, Scientific discussion, 2018, v. 1, № 17, p. 49-51.
[63] Bondarev B.V. The density matrix method in quantum oscillator theory. Dissipative matrix. Scientific discussion, 2018, v. 1, № 19, p. 52-55.

Theory of Nanophysics

[64] Bondarev B.V. Equation for the density matrix of the system of identical particles. Scientific discussion, 2018, v. 1, № 20, p. 16-20.
[65] Bondarev B.V. The look into the future of quantum physics. Scientific discussion. 2018, v. 1, № 26, p. 3. 6-13.

[66] Bondarev B.V. Equation for the density matrix of the system of identical particles. Advanced Materials: Manufacturing, Physics, Mechanics and Applications; edt. Ivan A. Parinov. New York, London, Springer Proceedings in Physics, 2020, p. 141.

Theory of Light Emitting Diode

[67] Bondarev B.V. Density matrix method in quantum theory of light emitting diode (LED). Scientific discussion, 2019, v. 1, № 31, p. 43-50.

Theory of Ball Lightning

[68] Bondarev B.V. Ball lightning. Density matrix method. Scientific discussion, 2019, v. 1, № 38, p. 37-44.
[69] Bondarev B.V. Equation for statistical operator in theory of ball lightning. Scientific
discussion. Scientific discussion, 2020, v. 1, № 39, p. 10-16.
[70] Bondarev B.V. Equation for density matrix of atomic nuclei in quantum theory of ball lightning. Scientific discussion, 2020, v. 1, № 40, p. 19-26.

Theory of Spaser

[71] Bondarev B.V. Spaser's quantum theory. Method of density matrix. Scientific discussion, 2020, v. 1, № 41. p. 21-27.

Theory of Graphene

[72] Bondarev B.V. Method of density matrix. Quantum theory of graphene. Superconductivity. Scientific discussion, 2020, v. 1, № 42, p. 6-12.

Last Article

[73] Bondarev B.V. The equation for statistical operator and state of nucleons in atomic nucleus. Scientific discussion, 2020, v. 1, № 46, p. 34-41.
[74] Bondarev B.V. The statistical operator and density matrix of free particle. Scientific discussion, 2020, v. 1, № 47, in print

List of Scientific Books

[1] Bondarev B.V. Method of density matrices in the quantum theory of

cooperative phenomena. 2 nd ed. M.: Sputnik+, 2013, p. 621.
[2] Bondarev B.V. My scientific articles. Moscow: Sputnik+, 2013, p. 293.
[3] Bondarev B.V. General theory of relativity. M.: Sputnik+, 2014, p. 74.
[4] Bondarev B.V. The density matrix method in quantum theory of superfluidity. M.: Sputnik+, 2014, p. 91.
[5] Bondarev, The density matrix Method in quantum superconductivity theory. 2nd ed, M.: Sputnik+, 2016, p. 112.
[6] Bondarev B.V. What is superconductivity? M.: Sputnik+, 2016, p. 34.
[7] Bondarev B.V. New theory of superconductivity. Scientific Research Publishing, Inc, USA, 2017, p. 125.
[8] Bondarev B.V. My scientific articles, Book 2, M. Sputnik+, 2017, p. 65.
[9] Bondarev B.V. My scientific articles, Book 3, M. Sputnik+, 2018, p. 75.
[10] Bondarev B.V. Density matrix method. Seven new theories in quantum physics. M. Sputnik+, 2019, p. 104.
[11] Bondarev B.V. Step kinetics of reactions in solid media. Correlation theory. M. Sputnik+, 2019, p. 40.
[12] Bondarev B.V. Eight new theories in quantum physics, M.: Sputnik+, 2018, p. 222.

List of Training Books

[1] Bondarev B.V., Nikolaev F.A. Electrical phenomena and electrical properties aviation materials: Studies. Benefit. M.: MAI, 1984, p. 92.
[2] Bondarev B.V., Nikolaev F.A. Magnetic phenomena and magnetic properties of aircraft materials: Studies. benefit. M.: MAI, 1985. p 71.
[3] Bondarev B.V., Kalashnikov N. P., Spirin G. G. Course of General physics in 3 books, M.: High school, 2003.
[4] Bondarev B.V. Elementary physics, Moscow, Higher school, 2004, p. 230.
[5] Bondarev B.V., Spirin G. G. Course of General physics. M.: Higher school, 2005, p. 560.
[6] Bondarev B.V. General physics Course. M.: Sputnik+, 2015, p. 420.

SUBJECT INDEX

Boris V. Bondarev

D

E

I

K

L

M

R

S

T

V

W

www.ingramcontent.com/pod-product-compliance
Lightning Source LLC
Chambersburg PA
CBHW050801220326
41598CB00006B/83

* 9 7 8 9 8 1 1 4 7 5 4 0 5 *